Mathematical Techniques of Applied Probability

Volume 2
Discrete Time Models: Techniques and Applications

OPERATIONS RESEARCH
AND INDUSTRIAL ENGINEERING

Consulting Editor: J. William Schmidt

**Virginia Polytechnic Institute and State University
Blacksburg, Virginia**

Mathematical Techniques of Applied Probability

Volume 2
Discrete Time Models: Techniques and Applications

Jeffrey J. Hunter

Department of Mathematics and Statistics
The University of Auckland
Auckland, New Zealand

ACADEMIC PRESS   1983
A Subsidiary of Harcourt Brace Jovanovich, Publishers

New York   London
Paris San Diego San Francisco São Paulo Sydney Tokyo Toronto

ACADEMIC PRESS, INC.
111 Fifth Avenue, New York, New York 10003

United Kingdom Edition published by
ACADEMIC PRESS, INC. (LONDON) LTD.
24/28 Oval Road, London NW1   7DX

Library of Congress Cataloging in Publication Data

Hunter, Jeffrey J.
  Mathematical techniques of applied probability.

  (Operations research and industrial engineering)
  Includes bibliographical references and index.
  Contents: v. 1. Discrete time models: basic theory --
v. 2. Discrete time models: techniques and applications
  1. Probabilities. I. Title. II. Series.
QA273.H87   1983   519.2   82-22642
ISBN 0-12-361802-9 (v. 2)

PRINTED IN THE UNITED STATES OF AMERICA

83 84 85 86   9 8 7 6 5 4 3 2 1

Contents

Chapter 9   **Discrete Time Queueing Models**

Preface

During the first few years of teaching a course on applied probability, to advanced undergraduate and first-year graduate students at the University of Auckland, I experienced problems in putting together a suitable program. The students I had to work with had all taken a year-long introductory course in probability and statistics and had passed mathematics courses including linear algebra, some real analysis, and elementary differential equations. Although the students were in the main majoring in mathematics, many had interests in operation research, some of whom were enrolled in engineering science programs. Initially, I presented a course which exposed the students to many varied topics in stochastic processes, both in discrete and continuous time. It soon became obvious, however, that such a form of presentation raised more problems than it solved. The students gained a superficial knowledge of a wide variety of stochastic models but often had considerable difficulty in being able to use the material at their disposal. On many occasions in my teaching I had to resort to the familiar handwaving expression "it can be shown that." Examinations tended to become regurgitation of material memorized rather than understood. From a pedagogical point of view I found this most undesirable. Traditional textbooks did not seem to offer me much help. The well-known classic work by William Feller contains a wealth of material, but much of the first half of the book deals with material covered in the earlier prerequisite probability course and the latter half requires very careful sifting.

As a consequence of these observations, I set about to present a course that would lead students through a variety of stochastic models, increasing in complexity, but with rigor and thoroughness so that not only could they appreciate the systematic development but also gain sufficient insight and knowledge that they themselves would be armed with tools and techniques to set about solving problems of a related nature.

These books are an outgrowth of the lecture notes that evolved from my presentation. The basic prerequisites assumed are an introductory probability course and some acquaintance with real analysis and linear algebra.

The first two volumes focus attention on discrete time models. Although it was the original intention of the author to have the material of these initial two volumes published as a single work the author and the publisher agreed to a split of the material with the first volume devoted to a presentation of the basic theory of discrete time models concentrating on a thorough introduction to Markov chains preceded by an in-depth examination of the recurrent event model. The tools of generating functions and matrix theory are also introduced to facilitate a detailed study of such models. While Volume 1, consisting of the first five chapters, forms a natural unit, Volume 2 is heavily dependent upon the theory and tools introduced in the initial work. This continuation gives a systematic presentation of techniques for determining the key properties of Markov chains and ties this in with two main application areas, branching chains and discrete time queueing models. A sequel is planned whereby the more general continuous time analogs will be considered. Renewal processes, Markov renewal processes including the special cases of Markov chains in continuous time and birth–death processes and their application to queueing models will be covered in this subsequent treatise.

Volume 1 is ideally suited for a semester (or half year) course leading to a thorough understanding of the basic ideas of Markov chains while both of the first two volumes are designed as a text for a year-long course on discrete time stochastic modeling. The follow-up volume, used in conjunction with these initial two parts, will give the instructor ample flexibility to provide a variety of courses in applied probability.

The style of the work is deliberately formal. I have used the "definition–theorem–proof" format intentionally. To do otherwise would have meant an already large text would have been excessive in size. I have attempted to remove "woolly" arguments that some authors use and made sure that a sound logical presentation is given. When a proof is omitted, a reference is usually provided to aid the reader in his probing. Sometimes I have pursued avenues in a little more depth than is necessary for a teaching textbook. To indicate such material I have starred certain sections, theorems, and proofs. The occasional exercise has also been given the same designation. A secondary use for such starring is to denote that material which can safely be omitted when teaching from the book without destroying any continuity of ideas and development.

There are a few features that readers of the text will find new. In Volume 1, the definition of a recurrent event (Definition 3.1.1) gives a precise formulation upon which our presentation is based. Many texts use a descriptive argument and consequently "waffle" their way through this topic without a formal statement. The beauty of our approach is evident when we examine the embedded recurrent events present in Markov chains (Theorem 5.2.1). The use of generalized inverses in the study of Markov chains has recently been exploited by the author and others and in Volume 2 (Chapter 7) this technique is used to derive stationary distributions and other properties of finite Markov chains. The last chapter of Volume 2 presents a systematic modeling of discrete time queueing systems. This survey contains some new ideas and explains in detail how care must be taken in examining embedded processes. It also shows that we may examine the structure of well-known Markov chains found in such queues without having to present Laplace or Laplace–Stieltjes transforms and the associated complex variable procedures that often tend to disguise the mathematics of the modeling.

Instructors intending to use this material as a textbook may wish to take into consideration the following observations. Chapter 1 is intended solely as a brief review of the elementary theory of probability for discrete random variables, and entering students should be able to move rather quickly through the material contained therein. Because of the importance of generating functions and their use right through both volumes, considerable stress should be put on grasping the concepts presented in Chapter 2. I have included Chapter 4 within the main body of the text rather than relegating it to an appendix. It is inserted at that position because it offers students an opportunity to survey and perhaps extend their basic linear algebra prerequisite material before venturing into Markov chains where such ideas are used. I suggest that no formal instruction be given on this chapter but rather that the techniques be referred to as they arise. In using Volumes 1 and 2 for a year-long course some teachers may feel that some continuous time models should be introduced. In such a case, the chapter on discrete time queueing models could be replaced by some material on Poisson processes and birth and death processes. I had to make a conscious decision to delay the treatment of such topics to a subsequent volume. This will enable me to provide the basic review material on differential equations and transforms very much in an analogous fashion to that carried out for difference equations and generating functions in the first volume. It is my intention that Volume 3 (to appear) may also be paired with Volume 1 to give an alternative year-long course covering a balance between discrete and continuous time processes.

Acknowledgments

Although the compilation of the material in this work is my own there are those who have had considerable influence on my approach. During my days as a graduate student at the University of North Carolina I learned much about stochastic processes from Professors Walter L. Smith and M. Ross Leadbetter. In fact, the idea of a formal definition for a recurrent event process arose in courses taught by the latter. I am much indebted to both of these former advisors and wish to express my appreciation for all that they gave me in time and assistance. The final polishing of the manuscript was the result of teaching this material for some five or six years to students at the University of Auckland. To my colleagues, Alastair Scott and Christopher Wild, who offered to use the material in teaching segments of the course, I wish to express my thanks for their advice and help, which also included the detection of errors and misprints in the draft manuscript. Typing assistance has been provided for cheerfully by many within the secretarial section of the Department of Mathematics, in particular, Helen Bedford, Joanne Hunter, and, more recently, Eve Malbon. No task such as this could have been undertaken without the tremendous support and encouragement given to this project by my wife, Hazel, and our children, Mark and Michelle. I have tried not to neglect their many activities, but at the same time they did not pressure me to neglect mine. All in all, it has been a most satisfying project. My heartfelt thanks to all who saw it finally brought to fruition.

Since the book has developed from my early lecture notes there will be examples, exercises, and naturally ideas that I have culled from many books and papers. The pioneering text of William Feller, mentioned earlier, gave many of us our first glimpse into this field and naturally his text has colored our views. Some of the exercises that are acknowledged are not quoted verbatim from their source and many appear with notational differences. However, where possible I have indicated the source of the material. There may be instances where I have failed to give credit where it is due. Such an instance is not intentional and I offer my apologies to any who have been slighted. I can assure them that no oversight is intended.

In particular I would like to thank the following for permission to reproduce the material as specified below:

W. Feller, "An Introduction to Probability Theory and its Applications," Vol. 1, 3rd ed. Copyright © 1968 by John Wiley & Sons Inc. (Exercises 6.1.5, 6.3.10). Reprinted by permission of John Wiley & Sons Inc., New York.

N. T. J. Bailey, "The Elements of Stochastic Processes with Applications to the Natural Sciences." Copyright © 1964 by John Wiley & Sons Inc. (Exercise 6.3.4). Reprinted by permission of John Wiley & Sons Inc., New York.

S. Karlin, "A First Course in Stochastic Processes." Copyright © 1966, 1968 by Academic Press, Inc. (Exercise 6.3.5). Reprinted by permission of Academic Press, Inc., New York.

James R. Gray, "Probability," Oliver and Boyd, Ltd., Edinburgh. Copyright © 1967 by J. R. Gray (Exercises 6.1.2, 6.1.10, 6.1.11). Reprinted by permission of J. R. Gray.

Contents of Volume 1

Chapter 6

Markov Chains in Discrete Time— General Techniques

6.1 Determination of the n-Step Transition Probabilities $p_{ij}^{(n)}$

The evaluation of the n-step transition probabilities $p_{ij}^{(n)}$ or, equivalently, the n-step transition matrix $P^{(n)} = [p_{ij}^{(n)}]$ is of considerable importance to our study of Markov chains (MC's). In this section we present a variety of techniques to effect their determination, given the transition matrix $P = [p_{ij}]$ of the MC.

Theorems 5.1.2 and 5.1.5 outlined the main properties of $P^{(n)}$, the central result being that

$$P^{(n)} = P^n \qquad (n \geq 1). \tag{6.1.1}$$

From Eq. (6.1.1), with $P^{(0)} \equiv I$, we deduce the following recursive results:

$$P^{(n)} = \begin{cases} PP^{(n-1)} & (n \geq 1), \\ P^{(n-1)}P & (n \geq 1). \end{cases} \tag{6.1.2} \tag{6.1.3}$$

The elementwise definition of matrix multiplication then implies that

$$p_{ij}^{(n)} = \begin{cases} \displaystyle\sum_{k \in S} p_{ik} p_{kj}^{(n-1)} & (n \geq 1), \\ \displaystyle\sum_{k \in S} p_{ik}^{(n-1)} p_{kj} & (n \geq 1). \end{cases} \tag{6.1.4} \tag{6.1.5}$$

1

Equations (6.1.2) and (6.1.3) or, equivalently, (6.1.4) and (6.1.5) form the basis of inductive or recursive methods. Equation (6.1.1) means that we can use some of the matrix techniques presented in Chapter 4. Alternatively we can utilize generating functions or, more generally, matrix generating functions.

Before examining these techniques it is worth noting that in special circumstances the multistep transition probabilities can sometimes be deduced by elementary probability considerations.

EXAMPLE 6.1.1:   *Unrestricted Random Walk.*   For this model the state space is $S = \{0, \pm 1, \pm 2, \ldots\}$ and the transition probabilities are, for all $i \in S$, $p_{i,i+1} = p, p_{i,i-1} = q, p_{i,j} = 0$ for $j \neq i - 1, i + 1$ $(0 < p < 1, q = 1 - p)$. Then

$$p_{ij}^{(n)} = \binom{n}{\dfrac{n + j - i}{2}} p^{(n+j-i)/2} q^{(n-j+i)/2}.$$

A simple justification of this result is that after n successive steps the process has moved $j - i$ more steps to the right than it has to the left so that it must have taken $(n + j - i)/2$ steps to the right and $(n - j + i)/2$ steps to the left. At each trial the process moves one step to the right with probability p or one step to the left with probability q, independent of the position at the commencement of the trial, so that we can regard the process as being generated by n Bernoulli trials and hence the answer above.

Observe that all the states communicate and thus the MC is irreducible (Definition 5.3.3). All the states are either transient, persistent nonnull, or are persistent null and have the same period (Theorem 5.3.3). Consider a typical state, say state i.

$$p_{ii}^{(n)} = \begin{cases} \dbinom{n}{n/2} (pq)^{n/2}, & n \text{ even.} \\ 0, & n \text{ odd.} \end{cases}$$

Obviously, state i is periodic, period 2. Furthermore, it is easily seen (cf. Example 3.3.2 concerning return to equilibrium in Bernoulli trials) that

$$\sum_{n=0}^{\infty} p_{ii}^{(n)} = \begin{cases} |p - q|^{-1} < \infty, & p \neq \tfrac{1}{2}, \\ \infty, & p = \tfrac{1}{2}, \end{cases}$$

and hence the states are transient if $p \neq \tfrac{1}{2}$ and persistent (null) if $p = \tfrac{1}{2}$.
□

Let us consider recursive or inductive techniques. We have already seen an application of this method in the proof of Theorem 5.1.7 where the multistep transition probabilities were obtained for the general two-state MC by utilizing Eqs. (6.1.5).

Another important application involves obtaining a general form for $P^{(n)}$ in the case of an absorbing MC.

THEOREM 6.1.1:   For an absorbing MC with transition matrix

$$P = \begin{bmatrix} I & 0 \\ R & Q \end{bmatrix}, \qquad P^{(n)} = \begin{bmatrix} I & 0 \\ \left(\sum_{k=0}^{n-1} Q^k\right)R & Q^n \end{bmatrix} \qquad (n \geq 1).$$

Proof: This can be established using Eq. (6.1.2). By letting

$$P^{(n)} = \begin{bmatrix} S_n & T_n \\ R_n & Q_n \end{bmatrix},$$

since $P^{(n)} = PP^{(n-1)}$,

$$\begin{bmatrix} S_n & T_n \\ R_n & Q_n \end{bmatrix} = \begin{bmatrix} I & 0 \\ R & Q \end{bmatrix}\begin{bmatrix} S_{n-1} & T_{n-1} \\ R_{n-1} & Q_{n-1} \end{bmatrix}.$$

By block multiplication we see immediately that

$$S_n = IS_{n-1} = S_{n-1} = S_{n-2} = \cdots = S_1 = I,$$

$$T_n = IT_{n-1} = T_{n-1} = T_{n-2} = \cdots = T_1 = 0,$$

$$Q_n = RT_{n-1} + QQ_{n-1} = QQ_{n-1}$$
$$= Q^2 Q_{n-2} = Q^3 Q_{n-3} = \cdots = Q^{n-1}Q_1 = Q^n,$$

$$R_n = RS_{n-1} + QR_{n-1} = R + QR_{n-1}$$
$$= R + Q(R + QR_{n-2}) = R + QR + Q^2(R + QR_{n-3}) = \cdots$$

$$= (I + Q + \cdots + Q^{n-2})R + Q^{n-1}R_1 = \left(\sum_{k=0}^{n-1} Q^k\right)R. \quad \square$$

COROLLARY 6.1.1A: Let

$$P = \begin{bmatrix} I & 0 \\ R & Q \end{bmatrix}$$

be the transition matrix of an absorbing MC. Then

(a) $\lim_{n \to \infty} Q^n = 0$,

(b) $I - Q$ is nonsingular and hence $N \equiv (I - Q)^{-1}$ exists,

(c)
$$P^{(n)} = \begin{bmatrix} I & 0 \\ N(I - Q^n)R & Q^n \end{bmatrix}.$$

Proof: (a) From Theorem 6.1.1, $Q^n = [p_{ij}^{(n)}]$ where $i, j \in T$, the set of transient states. Now from Theorem 5.2.8(a) $\lim_{n \to \infty} p_{ij}^{(n)} = 0$, for $i, j \in T$ and the result follows.

(b) This follows from Theorem 4.5.4 and (a) above.

(c) Let $R_n = (\sum_{k=0}^{n-1} Q^k)R$. Then

$$(I - Q)R_n = \sum_{k=0}^{n-1} (Q^k - Q^{k+1})R = (I - Q^n)R$$

and hence

$$R_n = (I - Q)^{-1}(I - Q^n)R = N(I - Q^n)R.$$

The result follows from Theorem 6.1.1. □

Some of the more general techniques used for obtaining expressions of P^n as outlined later in this section (e.g., matrix diagonalization) can also be applied to Q^n. Note, however, that Q is not a stochastic matrix. (See Example 6.1.4.)

In passing one should note that for all transition matrices P, $I - P$ is a singular matrix since $\det(I - P) = 0$ by virtue of the fact that the row sums are all identically zero. This is in contrast to the above corollary, which shows that $I - Q$ is a nonsingular matrix.

An extension of Theorem 6.1.1, to obtain a general form for $P^{(n)}$ when the transition matrix P is expressed in its general canonical form (as given by Theorem 5.4.1) is as follows.

THEOREM 6.1.2: If

$$P = \begin{bmatrix} P_1 & 0 & \cdots & 0 & 0 \\ 0 & P_2 & \cdots & 0 & 0 \\ \vdots & \vdots & & \vdots & \vdots \\ 0 & 0 & \cdots & P_k & 0 \\ R_1 & R_2 & \cdots & R_k & Q \end{bmatrix}$$

then

$$P^{(n)} = \begin{bmatrix} P_1^n & 0 & \cdots & 0 & 0 \\ 0 & P_2^n & \cdots & 0 & 0 \\ \vdots & \vdots & & \vdots & \vdots \\ 0 & 0 & \cdots & P_k^n & 0 \\ R_1^{(n)} & R_2^{(n)} & \cdots & R_k^{(n)} & Q^n \end{bmatrix}.$$

Proof: We give an inductive proof. Observe that the result is true for $n = 1$. Assume that the general form is valid for $n = m$. Then by block multiplication it is easy to verify that

$$P^{(m+1)} = PP^{(m)}$$

$$= \begin{bmatrix} P_1^{m+1} & 0 & \cdots & 0 & 0 \\ 0 & P_2^{m+1} & \cdots & 0 & 0 \\ \vdots & \vdots & & \vdots & \vdots \\ 0 & 0 & \cdots & P_k^{m+1} & 0 \\ R_1^{(m+1)} & R_2^{(m+1)} & \cdots & R_k^{(m+1)} & Q^{m+1} \end{bmatrix},$$

where, in fact, for $i = 1, \ldots, k$, $R_i^{(m+1)} = R_i P_i^m + Q R_i^{(m)}$. (It is an easy exercise to verify that $R_i^{(n)} = \sum_{j=0}^{n-1} Q^j R_i P_i^{n-1-j}$.) □

The following example gives an application of Eq. (6.1.5) in solving for the transition probabilities. It should be remarked that only in special cases, usually when there is some symmetry of transitions present, is it possible to solve directly for the $p_{ij}^{(n)}$ by solving equations of the type (6.1.4) or (6.1.5).

EXAMPLE 6.1.2: A Markov chain consisting of three states is governed by a random mechanism so that at each trial one of the states not currently occupied is selected with equal chance. Find the probability that after n steps the chain is in its original state. Find also $p_{ij}^{(n)}$ for all $i \neq j$ where $i, j \in S = \{1, 2, 3\}$.

The transition graph is given in Fig. 6.1.1 and thus the transition matrix is

$$P = \begin{bmatrix} 0 & \frac{1}{2} & \frac{1}{2} \\ \frac{1}{2} & 0 & \frac{1}{2} \\ \frac{1}{2} & \frac{1}{2} & 0 \end{bmatrix}.$$

Observe by symmetry that $p_{11}^{(n)} = p_{22}^{(n)} = p_{33}^{(n)}$. Thus we consider finding $p_{11}^{(n)}$. Since

$$p_{11}^{(n)} = \sum_{k=1}^{3} p_{1k}^{(n-1)} p_{k1} = \tfrac{1}{2} p_{12}^{(n-1)} + \tfrac{1}{2} p_{13}^{(n-1)},$$

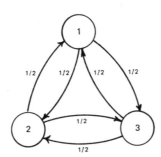

FIGURE 6.1.1.

i.e., $p_{11}^{(n)} = \frac{1}{2}[1 - p_{11}^{(n-1)}]$, since $P^{(n-1)}$ is a stochastic matrix. Using the usual techniques for solving difference equations (Section 2.9) it is easy to show

$$p_{11}^{(n)} = A\lambda^n + B,$$

where $\lambda = -\frac{1}{2}$ arises as the complementary solution and $B = \frac{1}{3}$ as the particular solution.

When $n = 1$, $p_{11}^{(1)} = 0 = A(-\frac{1}{2}) + \frac{1}{3}$ implies $A = \frac{2}{3}$, giving

$$p_{11}^{(n)} = \frac{1}{3}[1 - (-\frac{1}{2})^{n-1}], \qquad n \geq 1.$$

Consider now finding $p_{12}^{(n)}$. As earlier

$$p_{12}^{(n)} = \frac{1}{2}p_{11}^{(n-1)} + \frac{1}{2}p_{13}^{(n-1)} = \frac{1}{2}[1 - p_{12}^{(n-1)}],$$

implying $p_{12}^{(n)} = A(-\frac{1}{2})^n + \frac{1}{3}$, as above. In this case $p_{12}^{(1)} = \frac{1}{2} = A(-\frac{1}{2}) + \frac{1}{3}$ requires $A = -\frac{1}{3}$, giving

$$p_{12}^{(n)} = \frac{1}{3}[1 - (-\frac{1}{2})^n] \qquad (n \geq 1).$$

It is easily verified that $p_{13}^{(n)} = p_{12}^{(n)}$ using the fact that $p_{11}^{(n)} + p_{12}^{(n)} + p_{13}^{(n)} = 1$. However, by symmetry, it is obvious that $p_{ij}^{(n)}$ is the same for all $i \neq j$ and hence the same as $p_{12}^{(n)}$. $\square$

When the transition matrix has a rather general structure we need suitable techniques to find the multistep transition probabilities. In many instances the procedures elaborated in Section 4.4 concerning the diagonalization of matrices can be usefully applied to transition matrices governing finite Markov chains. A restatement of Theorem 4.4.8 to this situation is as follows.

THEOREM 6.1.3:   Let P be the transition matrix of a finite Markov chain whose state space is represented as $\{1, 2, \ldots, m\}$. If P is diagonalizable, then for $k = 1, 2, \ldots$

$$P^k = \sum_{l=1}^{m} \lambda_l^k \mathbf{x}_l \mathbf{y}_l', \qquad (6.1.6)$$

where $\mathbf{x}_1, \ldots, \mathbf{x}_m$ are linearly independent right eigenvectors corresponding to the eigenvalues $\lambda_1, \ldots, \lambda_m$ of P, and $\mathbf{y}_1', \ldots, \mathbf{y}_m'$ are the corresponding linearly independent left eigenvectors chosen so that $\mathbf{y}_i' \mathbf{x}_j = \delta_{ij}$ $(i, j = 1, \ldots, m)$. $\square$

Theorem 4.4.7 gives conditions for the diagonalizability of matrices. The case when the eigenvalues are all different, which occurs frequently in many applications of interest, is worthy of special consideration. In Corollary 4.4.8B we saw that for this situation there is no need to be specific in our choice of the eigenvectors. In fact, any eigenvectors can be used. We now restate Corollary 4.4.8B as a procedure that we can use to derive either P^k or the $p_{ij}^{(k)}$ given the transition matrix P.

COROLLARY 6.1.3A: Let P be the transition matrix of an m-state Markov chain with eigenvalues $\lambda_1, \ldots, \lambda_m$ that are all distinct. Let $\mathbf{x}_l$ and $\mathbf{y}_l'$ be *any* nonzero right and left eigenvectors belonging to the eigenvalue λ_l.
   Let

$$\mathbf{x}_l' = (x_{1l}, x_{2l}, \ldots, x_{ml}), \qquad \mathbf{y}_l' = (y_{l1}, y_{l2}, \ldots, y_{lm}),$$

and

$$c_l = \mathbf{y}_l' \mathbf{x}_l = \sum_{r=1}^{m} y_{lr} x_{rl} \ (\neq 0).$$

   Let

$$T = [x_{ij}] = [\mathbf{x}_1, \mathbf{x}_2, \ldots, \mathbf{x}_m],$$

$$S = [y_{ij}] = \begin{bmatrix} \mathbf{y}_1' \\ \mathbf{y}_2' \\ \vdots \\ \mathbf{y}_m' \end{bmatrix},$$

$$\Lambda = \text{diag}(\lambda_1, \lambda_2, \ldots, \lambda_m),$$

$$C = \text{diag}(c_1, c_2, \ldots, c_m) = ST.$$

Then

$$P^k = T\Lambda^k T^{-1} = T\Lambda^k C^{-1} S \tag{6.1.7}$$

$$= \sum_{l=1}^{m} c_l^{-1} \lambda_l^k \mathbf{x}_l \mathbf{y}_l'. \tag{6.1.8}$$

In particular, the k-step transition probabilities are given as

$$p_{ij}^{(k)} = \sum_{l=1}^{m} c_l^{-1} \lambda_l^k x_{il} y_{lj} \qquad (i, j = 1, \ldots, m). \quad \square \tag{6.1.9}$$

When P is not diagonalizable an expression for P^k can be found by using the Jordan canonical form as given in Theorem 4.4.10.

As an illustration of the procedure outlined by Theorem 6.1.3 and its corollary we examine the derivation of the multistep transition probabilities for a two-state Markov chain.

EXAMPLE 6.1.3:   *Two-State Markov Chain.*   Let

$$P = \begin{bmatrix} 1-a & a \\ b & 1-b \end{bmatrix} \qquad (0 \leq a \leq 1, 0 \leq b \leq 1)$$

be the transition matrix of a two-state MC and let $d = 1 - a - b$. Obtain the results of Theorem 5.1.7 under the assumptions $a \neq 0$ and $b \neq 0$, and thus $d \neq 1$. (If either of a or b are zero, there is an absorbing state and the derivation of P^k is trivial.)

(i) To find the eigenvalues of P we solve the characteristic equation

$$\det(P - \lambda I) = \begin{vmatrix} 1 - a - \lambda & a \\ b & 1 - b - \lambda \end{vmatrix} = (1 - a - \lambda)(1 - b - \lambda) - ab$$

$$= \lambda^2 - \lambda(2 - a - b) + (1 - a - b) = (\lambda - 1)(\lambda - d) = 0,$$

giving $\lambda_1 = 1$ and $\lambda_2 = d$, which are different under the stated assumptions.

(ii) The next step is to obtain a set of right eigenvectors $\mathbf{x}_1$ and $\mathbf{x}_2$ for P, such that $P\mathbf{x}_j = \lambda_j \mathbf{x}_j$ for $j = 1, 2$. Let

$$\mathbf{x}_j = \begin{pmatrix} x_{1j} \\ x_{2j} \end{pmatrix}.$$

Then

$$\begin{bmatrix} 1 - a & a \\ b & 1 - b \end{bmatrix} \begin{bmatrix} x_{1j} \\ x_{2j} \end{bmatrix} = \lambda_j \begin{bmatrix} x_{1j} \\ x_{2j} \end{bmatrix}$$

implies

$$(1 - a)x_{1j} + ax_{2j} = \lambda_j x_{1j},$$

and

$$bx_{1j} + (1 - b)x_{2j} = \lambda_j x_{2j}.$$

We require only one of these equations since one equation will always be redundant (having been used in effect to obtain the eigenvalue λ_j). When $j = 1$ the first equation becomes $(1 - a)x_{11} + ax_{21} = x_{11}$, implying, since $a \neq 0$, that $x_{11} = x_{21}$. We do not need the full generality of x_{11} and may take any value for x_{11} ($\neq 0$), e.g., $x_{11} = 1$, giving $\mathbf{x}_1 = \binom{1}{1}$. When $j = 2$ the first (or second) equation implies that $bx_{12} + ax_{22} = 0$. For a specific choice take, say, $x_{12} = -a$. Then $x_{22} = b$ and $\mathbf{x}_2 = \binom{-a}{b}$, giving

$$T = [\mathbf{x}_1, \mathbf{x}_2] = \begin{bmatrix} 1 & -a \\ 1 & b \end{bmatrix}.$$

(iii) The third step requires the derivation of the left eigenvectors $\mathbf{y}'_j$. Let $\mathbf{y}'_j = (y_{j1}, y_{j2})$. Then

$$(y_{j1}, y_{j2}) \begin{bmatrix} 1 - a & a \\ b & 1 - b \end{bmatrix} = \lambda_j(y_{j1}, y_{j2}),$$

giving $y_{j1}(1 - a) + y_{j2}b = \lambda_j y_{j1}$ (together with the redundant equation $y_{j1}a + y_{j2}(1 - b) = \lambda_j y_{j2}$).

For $j = 1$ we require $ay_{11} - by_{12} = 0$, and a suitable choice for $\mathbf{y}'_1$ is $\mathbf{y}'_1 = (b, a)$.

For $j = 2$ we require $y_{21} + y_{22} = 0$ (since $b \neq 0$) and we can take $\mathbf{y}'_2 = (1, -1)$, giving

$$S = \begin{bmatrix} \mathbf{y}'_1 \\ \mathbf{y}'_2 \end{bmatrix} = \begin{bmatrix} b & a \\ 1 & -1 \end{bmatrix}.$$

(iv)   The only terms not evaluated yet are the c_1 and c_2 or equivalently the diagonal matrix

$$C = ST = \begin{bmatrix} b & a \\ 1 & -1 \end{bmatrix} \begin{bmatrix} 1 & -a \\ 1 & b \end{bmatrix} = \begin{bmatrix} 1 - d & 0 \\ 0 & -(1 - d) \end{bmatrix}.$$

Alternatively, $c_1 = \mathbf{y}'_1 \mathbf{x}_1 = 1 - d$ and $c_2 = \mathbf{y}'_2 \mathbf{x}_2 = -(1 - d)$.

(v)   Using either Eq. (6.1.7) or Eq. (6.1.8), we obtain

$$P^k = \frac{1}{1 - d} \begin{bmatrix} b + ad^k & a - ad^k \\ b - bd^k & a + bd^k \end{bmatrix},$$

as obtained in Theorem 5.1.7.

Note that we could have used the expression $P^k = T\Lambda^k T^{-1}$ by showing that

$$T^{-1} = \frac{1}{1 - d} \begin{bmatrix} b & a \\ -1 & 1 \end{bmatrix}.$$

The technique used above avoided the computation of a matrix inverse.   □

Some useful techniques for finding eigenvalues and eigenvectors were presented in Section 4.4 (just after Theorem 4.4.5) and are worth utilizing.

For example, the selection of any nonzero column (row) or a multiple thereof of $\mathrm{adj}(P - \lambda_j I)$ as a right (left) eigenvector $\mathbf{x}_j$ ($\mathbf{y}'_j$) associated with the eigenvalue λ_j obviates the need of solving systems of equations.

In Example 6.1.3,

$$\mathrm{adj}(P - \lambda_j I) = \begin{bmatrix} 1 - b - \lambda_j & -a \\ -b & 1 - a - \lambda_j \end{bmatrix}$$

and with $\lambda_1 = 1$ and $\lambda_2 = 1 - a - b$ it is easy to see that the $\mathbf{x}_1$ chosen was $-1/b$ times column 1, or $-1/a$ times column 2 of $\mathrm{adj}(P - \lambda_1 I)$ and $\mathbf{x}_2$ as $-$column 1 or column 2 of $\mathrm{adj}(P - \lambda_2 I)$. Similarly $\mathbf{y}'_1$ was chosen as $-$row 1 of $\mathrm{adj}(P - \lambda_1 I)$ and $\mathbf{y}'_2$ as $1/a$ times row 1 of $\mathrm{adj}(P - \lambda_2 I)$.

EXAMPLE 6.1.4:   *Classical Gambler's Ruin Model.*   In Example 5.1.2 we described the classical gambler's ruin model as a special case of a simple random walk on the nonnegative integers with two absorbing barriers. With the state space taken as $\{0, 1, \ldots, a-1, a\}$, the Markov chain has transition matrix given by

$$
\begin{bmatrix}
1 & 0 & 0 & \cdots & 0 & 0 & 0 \\
q & 0 & p & \cdots & 0 & 0 & 0 \\
0 & q & 0 & \cdots & 0 & 0 & 0 \\
\vdots & \vdots & \vdots & & \vdots & \vdots & \vdots \\
0 & 0 & 0 & \cdots & 0 & p & 0 \\
0 & 0 & 0 & \cdots & q & 0 & p \\
0 & 0 & 0 & \cdots & 0 & 0 & 1
\end{bmatrix}.
$$

Let us reorder the states and take $S = \{0, a, 1, 2, \ldots, a-1\}$. Consequently we express the transition matrix in canonical form as

$$
P = \begin{bmatrix}
1 & 0 & | & 0 & 0 & 0 & \cdots & 0 & 0 & 0 \\
0 & 1 & | & 0 & 0 & 0 & \cdots & 0 & 0 & 0 \\
\hline
q & 0 & | & 0 & p & 0 & \cdots & 0 & 0 & 0 \\
0 & 0 & | & q & 0 & p & \cdots & 0 & 0 & 0 \\
\vdots & \vdots & | & \vdots & \vdots & \vdots & & \vdots & \vdots & \vdots \\
0 & 0 & | & 0 & 0 & 0 & \cdots & q & 0 & p \\
0 & p & | & 0 & 0 & 0 & \cdots & 0 & q & 0
\end{bmatrix}
= \begin{bmatrix} I & 0 \\ R & Q \end{bmatrix},
\qquad (6.1.10)
$$

where R is an $(a-1) \times 2$ matrix and Q an $(a-1) \times (a-1)$ matrix, as displayed.

We seek expressions for $p_{rs}^{(n)}$ for all $r, s = 0, 1, \ldots, a$. From Theorem 6.1.1 we can express the n-step transition matrix as

$$
P^{(n)} = \begin{bmatrix}
I & 0 \\
\left(\sum_{k=0}^{n-1} Q^k \right) R & Q^n
\end{bmatrix}
\qquad (n \geq 1)
$$

so that the $p_{rs}^{(n)}$ are all determined once $Q^n \; (= [p_{ij}^{(n)}]$ for $i, j = 1, 2, \ldots, a-1)$ is found.

Although Q is not a stochastic matrix, we can use the method of Corollary 6.1.3A, provided the eigenvalues of Q are all distinct.

The equation $\mathbf{y}'Q = \lambda \mathbf{y}'$ has a nonzero solution $\mathbf{y}'$ if and only if λ is an eigenvalue of Q (and hence $\mathbf{y}'$ is the left eigenvector associated with λ). If we let $\mathbf{y}' = (y_1, y_2, \ldots, y_{a-1})$, then we require that

$$
py_{k-1} + qy_{k+1} = \lambda y_k \qquad (k = 1, 2, \ldots, a-1),
$$

where we have defined $y_0 \equiv 0$ and $y_a \equiv 0$.

The solution to this homogeneous second-order difference equation is

$$y_k = C\omega_1^k + D\omega_2^k,$$

where ω_1 and ω_2 are roots of the equation $q\omega^2 - \lambda\omega + p = 0$.

Since $y_0 = y_a = 0$, we require that $C + D = 0$ and $C\omega_1^a + D\omega_2^a = 0$. This implies that $\omega_1^a = \omega_2^a$ and hence that $\omega_1 = \omega_2 e^{2\pi i l/a}$ $(i = \sqrt{-1})$ for some integer l, which, without loss of generality, assumes one of the values $0, 1, 2, \ldots, a - 1$. Since $\omega_1 \omega_2 = p/q$,

$$\omega_1 = \left(\frac{p}{q}\right)^{1/2} e^{\pi i l/a}, \qquad \omega_2 = \left(\frac{p}{q}\right)^{1/2} e^{-\pi i l/a}.$$

Furthermore, $\omega_1 + \omega_2 = \lambda/q$, giving

$$\lambda = (pq)^{1/2}(e^{\pi i l/a} + e^{-\pi i l/a}) = 2(pq)^{1/2} \cos(\pi l/a).$$

The solution for y_k can now be expressed as

$$y_k = C\left(\frac{p}{q}\right)^{k/2} \{e^{\pi i l k/a} - e^{-\pi i l k/a}\}$$

$$= A\left(\frac{p}{q}\right)^{k/2} \sin(\pi l k/a), \qquad k = 1, 2, \ldots, a - 1.$$

Since Q is an $(a - 1)$th-order matrix it must have exactly $(a - 1)$ eigenvalues. These are given by λ above taking the values $1, 2, \ldots, a - 1$ for the index l. We cannot take $l = 0$ since in this case $y_k = 0$ for all k and $\mathbf{y}'$ is the zero vector and hence not an eigenvector.

Thus the eigenvalues of Q are given by

$$\lambda_l = 2(pq)^{1/2} \cos(\pi l/a), \qquad l = 1, 2, \ldots, a - 1,$$

with associated left eigenvectors $\mathbf{y}_l' = (y_{l1}, y_{l2}, \ldots, y_{l,a-1})$, where

$$y_{lk} = A_l\left(\frac{p}{q}\right)^{k/2} \sin(\pi l k/a), \qquad k = 1, \ldots, a - 1.$$

It is easy to verify that the difference equations resulting from $Q\mathbf{x} = \lambda\mathbf{x}$ are of the same form as those solved above but with p and q interchanged. Consequently the solution to $Q\mathbf{x}_l = \lambda_l \mathbf{x}_l$ $(l = 1, 2, \ldots, a - 1)$ is given by

$$\mathbf{x}_l' = (x_{1l}, x_{2l}, \ldots, x_{a-1,l})$$

where

$$x_{kl} = B_l\left(\frac{q}{p}\right)^{k/2} \sin(\pi k l/a), \qquad k = 1, \ldots, a - 1.$$

Corollary 6.1.3A gives for $r, s = 1, 2, \ldots, a - 1$

$$p_{rs}^{(n)} = \sum_{l=1}^{a-1} c_l^{-1} \lambda_l^n x_{rl} y_{ls},$$

where

$$c_l = \sum_{k=1}^{a-1} y_{lk} x_{kl}$$

$$= A_l B_l \sum_{k=1}^{a-1} \sin^2(\pi l k / a)$$

$$= A_l B_l a / 2$$

(by Exercise 6.1.7).

Substitution and simplification gives for $r, s = 1, 2, \ldots, a - 1$,

$$p_{rs}^{(n)} = \frac{2^{n+1}}{a} p^{(n+s-r)/2} q^{(n-s+r)/2} \sum_{l=1}^{a-1} \cos^n(\pi l / a) \sin(\pi r l / a) \sin(\pi s l / a). \quad (6.1.11)$$

The only multistep transition probabilities yet to be determined are $p_{r0}^{(n)}$ and $p_{ra}^{(n)}$ ($r = 1, \ldots, a - 1$). Equating the appropriate blocks of $P^{(n)}$ we see that

$$\begin{bmatrix} p_{10}^{(n)} & p_{1a}^{(n)} \\ \vdots & \vdots \\ p_{r0}^{(n)} & p_{ra}^{(n)} \\ \vdots & \vdots \\ p_{a-1,0}^{(n)} & p_{a-1,a}^{(n)} \end{bmatrix} = \sum_{k=0}^{n-1} \begin{bmatrix} p_{11}^{(k)} & \cdots & p_{1,a-1}^{(k)} \\ \vdots & & \vdots \\ p_{r1}^{(k)} & \cdots & p_{r,a-1}^{(k)} \\ \vdots & & \vdots \\ p_{a-1,1}^{(k)} & \cdots & p_{a-1,a-1}^{(k)} \end{bmatrix} \begin{bmatrix} q & 0 \\ 0 & 0 \\ \vdots & \vdots \\ 0 & 0 \\ 0 & p \end{bmatrix}.$$

Thus

$$p_{r0}^{(n)} = \sum_{k=0}^{n-1} p_{r1}^{(k)} q, \qquad p_{ra}^{(n)} = \sum_{k=0}^{n-1} p_{r,a-1}^{(k)} p. \quad (6.1.12)$$

Substituting the expressions for $p_{r1}^{(k)}$ and $p_{r,a-1}^{(k)}$ as derived above (and which hold for $k = 0$ by making use of Exercise 6.1.7), we obtain, for $r = 1, \ldots, a - 1$,

$$p_{r0}^{(n)} = \frac{2}{a} p^{(1-r)/2} q^{(1+r)/2} \sum_{l=1}^{a-1} \frac{\sin(\pi r l / a) \sin(\pi l / a)}{1 - 2(pq)^{1/2} \cos(\pi l / a)} [1 - \{2(pq)^{1/2} \cos(\pi l / a)\}^n],$$

$$p_{ra}^{(n)} = \frac{2}{a} p^{(a+1-r)/2} q^{(-a+1+r)/2} \sum_{l=1}^{a-1} \frac{\sin(\pi r l / a) \sin(\overline{\pi a - 1} l / a)}{1 - 2(pq)^{1/2} \cos(\pi l / a)}$$
$$\times [1 - \{2(pq)^{1/2} \cos(\pi l / a)\}^n]. \quad \square$$

The behavior of a MC is, of course, intimately connected with the nature of its transition matrix. It is, however, possible to deduce the main properties concerning the decomposition of the state space directly from a knowledge of the location of the eigenvalues of the transition matrix. The following theorems detail some of the relevant results.

THEOREM 6.1.4:   For any stochastic matrix P, the eigenvalues λ satisfy $|\lambda| \leq 1$.

Moreover, $\lambda = 1$ is an eigenvalue with corresponding right eigenvector $\mathbf{e}$ where $\mathbf{e}' = (1, 1, \ldots, 1)$.

Proof: Suppose the transition matrix is of order m. Then if λ is an eigenvalue there exists an $\mathbf{x} \neq \mathbf{0}$ such that $P\mathbf{x} = \lambda \mathbf{x}$. If $\mathbf{x}' = (x_1, x_2, \ldots, x_m)$, then

$$\sum_{j=1}^{m} p_{ij} x_j = \lambda x_i \qquad \text{for} \quad i = 1, 2, \ldots, m.$$

Now choose i such that $|x_i| = \max_{1 \leq j \leq m} |x_j|$. Then

$$|\lambda| = \left| \sum_{j=1}^{m} p_{ij} \frac{x_j}{x_i} \right| \leq \sum_{j=1}^{m} p_{ij} \left| \frac{x_j}{x_i} \right| \leq \sum_{j=1}^{m} p_{ij} = 1.$$

The last part of the theorem is immediate. In fact

$$P\mathbf{e} = \begin{bmatrix} \sum_j p_{ij} \\ \vdots \\ \sum_j p_{mj} \end{bmatrix} = \begin{bmatrix} 1 \\ \vdots \\ 1 \end{bmatrix} = \mathbf{e},$$

and hence the equation $P\mathbf{x} = \mathbf{x}$ has a solution $\mathbf{x} = \mathbf{e}$, implying that $\lambda = 1$ is an eigenvalue (Theorem 4.4.4) with associated eigenvector $\mathbf{e}$.   □

THEOREM 6.1.5:   If P is the transition matrix of a regular Markov chain and if λ is an eigenvalue with $|\lambda| = 1$ and $\mathbf{x}$ its corresponding right eigenvector, then $\mathbf{x} = c\mathbf{e}$ where $\mathbf{e}' = (1, 1, \ldots, 1)$ and λ must be 1.

*Proof: Observe that for all $n = 1, 2, \ldots, P^n \mathbf{x} = \lambda^n \mathbf{x}$. (This is established by induction since it is true for $n = 1$, and if it is true for $n = k$, then $P^{k+1} \mathbf{x} = P P^k \mathbf{x} = P \lambda^k \mathbf{x} = \lambda^{k+1} \mathbf{x}$, i.e., it is true for $n = k + 1$ and hence true generally.)

Since the chain is regular (by Theorem 5.3.12), we may choose n so large that all the elements of P^n are positive. For such an n, with $\mathbf{x}' = (x_1, x_2, \ldots, x_n)$,

$$\sum_{j=1}^{m} p_{ij}^{(n)} x_j = \lambda^n x_i \qquad \text{for} \quad i = 1, 2, \ldots, m.$$

As in Theorem 6.1.4, choose i such that $|x_i| = \max_{1 \le j \le m} |x_j|$. Then

$$1 = |\lambda^n| = \left| \sum_{j=1}^{m} p_{ij}^{(n)} \frac{x_j}{x_i} \right| \le \sum_{j=1}^{m} p_{ij}^{(n)} \left| \frac{x_j}{x_i} \right| \le \sum_{j=1}^{m} p_{ij}^{(n)} = 1.$$

Thus we have equality throughout the above inequalities and so

$$\sum_{j=1}^{m} p_{ij}^{(n)} |x_j| = \sum_{j=1}^{m} p_{ij}^{(n)} |x_i| \qquad (6.1.13)$$

and

$$\left| \sum_{j=1}^{m} p_{ij}^{(n)} x_j \right| = \sum_{j=1}^{m} p_{ij}^{(n)} |x_j|. \qquad (6.1.14)$$

Since $p_{ij}^{(n)} > 0$ for all i, j and $|x_i| - |x_j| \ge 0$ for all j from Eq. (6.1.13), i.e., $\sum_{j=1}^{m} p_{ij}^{(n)}(|x_i| - |x_j|) = 0$, we deduce that $|x_i| = |x_j|$ for all j, i.e.,

$$|x_1| = |x_2| = \cdots = |x_m| = M,$$

say.

From Eq. (6.1.14) and Exercise 6.1.8 we also have that for $j = 1, 2, \ldots, m$

$$x_j = |x_j| e^{i\theta} \qquad (i = \sqrt{-1}).$$

These two above results thus imply that $x_j = M e^{i\theta} = c$, say, and consequently $\mathbf{x} = c\mathbf{e}$.

That λ must indeed be 1 now follows since

$$\lambda c \mathbf{e} = \lambda \mathbf{x} = P\mathbf{x} = Pc\mathbf{e} = cP\mathbf{e} = c\mathbf{e}$$

and $c\mathbf{e} = \mathbf{x}$ is not a zero vector. □

THEOREM 6.1.6: If P is the transition matrix of a finite irreducible periodic Markov chain with period d, then P has eigenvalues the dth roots of unity, each with multiplicity one, and there are no other eigenvalues of modulus one. If λ is a dth root of unity, then its corresponding eigenvector is given by any constant multiple of the vector $\mathbf{x}$ where $\mathbf{x}' = (\mathbf{e}', \lambda \mathbf{e}', \ldots, \lambda^{d-1} \mathbf{e}')$, where the vector is partitioned according to the subclasses of the chain. In particular, the eigenvalue $\lambda = 1$ is simple with a corresponding right eigenvector $\mathbf{e}$.

Proof: Let $G_0, G_1, \ldots, G_{d-1}$ be the subclasses of states of the process as established in Theorem 5.3.14, i.e., $i \in G_r$ implies $p_{ij} = 0$ for every $j \notin G_{r+1}$. It is no loss of generality to assume that $G_0 = \{1, \ldots, m_1\}$, $G_1 = \{m_1 + 1, \ldots, m_1 + m_2\}, \ldots, G_{d-1} = \{m - m_d + 1, \ldots, m\}$ $(m = m_1 + \cdots + m_d)$.

From Theorem 5.4.4

$$P = \begin{bmatrix} 0 & P_1 & 0 & \cdots & 0 \\ 0 & 0 & P_2 & \cdots & 0 \\ \vdots & \vdots & \vdots & & \vdots \\ 0 & 0 & 0 & \cdots & P_{d-1} \\ P_d & 0 & 0 & \cdots & 0 \end{bmatrix},$$

giving

$$P^d = \begin{bmatrix} A_1 & 0 & 0 & \cdots & 0 \\ 0 & A_2 & 0 & \cdots & 0 \\ \vdots & \vdots & & & \vdots \\ 0 & 0 & 0 & \cdots & 0 \\ 0 & 0 & 0 & \cdots & A_d \end{bmatrix},$$

where the A_k $(k = 1, \ldots, d)$ are $m_k \times m_k$ matrices given by

$$A_1 = P_1 P_2 \cdots P_d, \quad A_k = P_k \cdots P_d P_1 \cdots P_{k-1} \quad (k = 2, \ldots, d).$$

Observe that $(P_k)_{ij} = P\{X_{n+1} = j \mid X_n = i\}$, where $i \in G_{k-1}, j \in G_k$ and thus $(A_k)_{ij} = P\{X_{n+d} = j \mid X_n = i\}$ where $i, j \in G_{k-1}$. The A_k describe transitions from the states of G_{k-1} to the same set of states (after a further d steps) and, since the chain is irreducible, the A_k are irreducible (Theorem 5.4.5) and aperiodic. Consequently, the A_k are in fact transition matrices for regular (sub) Markov chains.

Applying Theorem 6.1.5 to the A_k, we deduce that A_k has a simple eigenvalue 1 with a corresponding $m_k \times 1$ nonzero eigenvector $\boldsymbol{\mu}_k = c_k \mathbf{e}$. Since $A_k \boldsymbol{\mu}_k = \boldsymbol{\mu}_k$, if we construct $m \times 1$ vectors $\mathbf{x}_k$ by adjoining an appropriate number of zeros on one or both sides of each $\boldsymbol{\mu}_k$, we determine linearly independent vectors $\mathbf{x}_1, \ldots, \mathbf{x}_d$ (since the only nonzero terms in $\mathbf{x}_k$ are those with indices $m_1 + \cdots + m_{k-1} + 1, \ldots, m_1 + \cdots + m_k$) such that

$$P^d \mathbf{x}_k = \mathbf{x}_k \quad (k = 1, \ldots, d).$$

Now let us define $\mathbf{u}_k = P^{d-k+1} \mathbf{x}_1$ $(k = 1, \ldots, d)$. Observe that $\mathbf{u}_1 = P^d \mathbf{x}_1 = \mathbf{x}_1$, and hence $\mathbf{u}_d = P \mathbf{u}_1$, $\mathbf{u}_{d-1} = P^2 \mathbf{u}_1 = P \mathbf{u}_d$ or generally $\mathbf{u}_k = P \mathbf{u}_{k+1}$ $(k = 2, \ldots, d-1)$. By successively generating $\mathbf{u}_1, \mathbf{u}_d, \mathbf{u}_{d-1}, \ldots, \mathbf{u}_2$ it is easy to verify that the nonzero terms in $\mathbf{u}_k$ are linearly independent. Furthermore, for $k = 1, \ldots, d$,

$$P^d \mathbf{u}_k = P^d P^{d-k+1} \mathbf{x}_1 = P^{d-k+1} P^d \mathbf{x}_1 = P^{d-k+1} \mathbf{x}_1 = \mathbf{u}_k.$$

It follows that if we restrict attention to the m_k-dimensional linear space obtained by considering only those components of $\mathbf{u}_k$ that lie in G_{k-1}, we obtain a right eigenvector with eigenvalue 1 for A_k.

Because the eigenvalue 1 has a simple multiplicity for A_k it follows that each $\mathbf{u}_k$ is a constant multiple of $\mathbf{x}_k$. The nonzero elements of $\mathbf{x}_k$ are each c_k and thus let us normalize $\mathbf{x}_k$ by taking $c_k = 1$. This implies $\sum_{k=1}^{d} \mathbf{x}_k = \mathbf{e}$. Consequently,

$$\sum_{k=1}^{d} \mathbf{u}_k = \left(\sum_{i=0}^{d-1} P^i \right) \mathbf{x}_1 = \mathbf{e}$$

to establish the result that $\mathbf{u}_k = \mathbf{x}_k = \mathbf{e}$ for $k = 1, 2, \ldots, d$.

Accordingly we deduce that $\mathbf{x}_1 = P\mathbf{x}_2$, $\mathbf{x}_2 = P\mathbf{x}_3, \ldots, \mathbf{x}_{d-1} = P\mathbf{x}_d$, $\mathbf{x}_d = P\mathbf{x}_1$.

Let $\omega = e^{2\pi i/d}$. Combining the above equations in the indicated manner, we obtain

$$P(\mathbf{x}_1 + \mathbf{x}_2 + \cdots + \mathbf{x}_d) = (\mathbf{x}_1 + \mathbf{x}_2 + \cdots + \mathbf{x}_d),$$

$$P(\mathbf{x}_1 + \omega \mathbf{x}_2 + \cdots + \omega^{d-1}\mathbf{x}_d) = \omega(\mathbf{x}_1 + \omega \mathbf{x}_2 + \cdots + \omega^{d-1}\mathbf{x}_d),$$

$$P(\mathbf{x}_1 + \omega^2 \mathbf{x}_2 + \cdots + \omega^{2(d-1)}\mathbf{x}_d) = \omega^2(\mathbf{x}_1 + \omega^2 \mathbf{x}_2 + \cdots + \omega^{2(d-1)}\mathbf{x}_d),$$

$$\vdots$$

$$P(\mathbf{x}_1 + \omega^{(d-1)}\mathbf{x}_2 + \cdots + \omega^{(d-1)^2}\mathbf{x}_d) = \omega^{(d-1)}(\mathbf{x}_1 + \omega^{(d-1)}\mathbf{x}_2 + \cdots + \omega^{(d-1)^2}\mathbf{x}_d).$$

The linear independence of the $\mathbf{x}_k$ ensures that none of the vectors appearing above are zero. These relations exhibit the property that the dth roots of unity are all eigenvalues of P with ω^k ($k = 0, 1, \ldots, d-1$) having a right eigenvector given by $\mathbf{x} \equiv \mathbf{x}_1 + \omega^k \mathbf{x}_2 + \cdots + \omega^{k(d-1)}\mathbf{x}_k$ such that $\mathbf{x}' = (\mathbf{e}', \omega^k \mathbf{e}', \ldots, \omega^{k(d-1)}\mathbf{e}')$. In particular, $\lambda = 1$ has a right eigenvector given by $\mathbf{e}$.

Suppose next that $P\mathbf{x} = \lambda \mathbf{x}$ for some nonzero $\mathbf{x}$. Then $P^d \mathbf{x} = \lambda^d \mathbf{x}$. If we write $\mathbf{x}' = (\mathbf{z}_1', \mathbf{z}_2', \ldots, \mathbf{z}_d')$ where $\mathbf{z}_k'$ is an $1 \times m_k$ vector, then $A_k \mathbf{z}_k = \lambda^d \mathbf{z}_k$, $k = 1, \ldots, d$. Since at least one of the $\mathbf{z}_k$ are nonzero and each A_k is regular, either $\lambda^d = 1$ or $|\lambda^d| < 1$. If $\lambda^d = 1$, then $A_k \mathbf{z}_k = \mathbf{z}_k$ so that there are constants $a_1, \ldots, a_d$ such that $\mathbf{z}_k = a_k \mathbf{\mu}_k$ and we see that

$$\mathbf{x} = \begin{bmatrix} a_1 \mathbf{\mu}_1 \\ \vdots \\ a_d \mathbf{\mu}_d \end{bmatrix} = a_1 \mathbf{x}_1 + \cdots + a_d \mathbf{x}_d.$$

Since $P\mathbf{x} = \lambda \mathbf{x}$, we require

$$a_1 \mathbf{x}_d + a_2 \mathbf{x}_1 + \cdots + a_d \mathbf{x}_{d-1} = \lambda a_1 \mathbf{x}_1 + \lambda a_2 \mathbf{x}_2 + \cdots + \lambda a_d \mathbf{x}_d.$$

From the linear independence of the $\mathbf{x}_k$ we have $a_1 = \lambda a_d$, $a_2 = \lambda a_1, \ldots$, $a_d = \lambda a_{d-1}$, and thus $\mathbf{x} = \lambda a_d(\mathbf{x}_1 + \lambda \mathbf{x}_2 + \cdots + \lambda^{d-1}\mathbf{x}_d)$.

This means that any eigenvector of a dth root of unity is, except for a constant multiple, one of the eigenvectors already constructed earlier. Our construction therefore gives all the possible eigenvalues with modules 1 and their associated eigenvectors. [The proof above is a modified version of that given by Karlin (1966, pp. 101–102).] □

THEOREM 6.1.7: Let P be the transition matrix of an absorbing Markov chain having r absorbing states so that

$$P = \begin{bmatrix} I & 0 \\ R & Q \end{bmatrix},$$

where I is of order r.

Then

(a) the eigenvalue $\lambda = 1$ occurs with multiplicity r, having associated with it r linearly independent right eigenvectors

$$\mathbf{x}_j = \begin{pmatrix} \boldsymbol{\alpha}_j \\ (I - Q)^{-1} R \boldsymbol{\alpha}_j \end{pmatrix}, \qquad j = 1, \ldots, r$$

(by choosing $\boldsymbol{\alpha}_1, \ldots, \boldsymbol{\alpha}_r$ as a linearly independent set of r-dim vectors) and r linearly independent left eigenvectors

$$\mathbf{y}_i' = (\mathbf{u}_i', \mathbf{0}'), \qquad i = 1, \ldots, r$$

(by choosing $\mathbf{u}_i', \ldots, \mathbf{u}_r'$ as a linearly independent set of r-dim vectors).

(b) The remaining eigenvalues λ are less than 1 in modulus with the right eigenvectors having the form $\mathbf{x} = \binom{0}{\beta}$, where $Q\beta = \lambda\beta$, and the left eigenvectors having the form $\mathbf{y}' = (\mathbf{u}', \mathbf{v}')$, where $\mathbf{v}'Q = \lambda\mathbf{v}'$ and $\mathbf{u}' = [1/(\lambda - 1)]\mathbf{v}'R$.

Proof: First observe that

$$\det(P - \lambda I) = \det \begin{bmatrix} I - \lambda I & 0 \\ R & Q - \lambda I \end{bmatrix} = (1 - \lambda)^r \det(Q - \lambda I).$$

From Corollary 6.1.1A, $Q^n \to 0$ and thus, by Theorem 4.5.4, $\rho(Q) < 1$, i.e., all the eigenvalues of Q are less than 1 in modulus; hence $\det(P - \lambda I) = 0$ has exactly r roots $\lambda = 1$.

Let $\mathbf{x} = \binom{\alpha}{\beta}$ be a right eigenvector associated with the eigenvalue λ. $P\mathbf{x} = \lambda\mathbf{x}$ implies that

$$\begin{bmatrix} I & 0 \\ R & Q \end{bmatrix} \begin{bmatrix} \alpha \\ \beta \end{bmatrix} = \lambda \begin{bmatrix} \alpha \\ \beta \end{bmatrix},$$

i.e., $\boldsymbol{\alpha} = \lambda\boldsymbol{\alpha}$ and $R\boldsymbol{\alpha} + Q\boldsymbol{\beta} = \lambda\boldsymbol{\beta}$.

If $\lambda = 1$, then $\boldsymbol{\alpha}$ is arbitrary and $R\boldsymbol{\alpha} = (I - Q)\boldsymbol{\beta}$, implying $\boldsymbol{\beta} = (I - Q)^{-1}R\boldsymbol{\alpha}$ since, by Corollary 6.1.1A, $I - Q$ is nonsingular. Observe that if $\boldsymbol{\alpha}$ is taken as $\mathbf{e}$, a vector of ones, then $\boldsymbol{\beta} = \mathbf{e}$ and hence $\mathbf{x} = \mathbf{e}$ in accordance with Theorem 6.1.4. The fact that we can find r linearly independent vectors $\boldsymbol{\alpha}_1, \ldots, \boldsymbol{\alpha}_r$ spanning the space of $r \times 1$ vectors ensures that $\lambda = 1$ has both algebraic and geometric multiplicity of r.

If $\lambda \neq 1$, then $\boldsymbol{\alpha} = \mathbf{0}$ so that $Q\boldsymbol{\beta} = \lambda\boldsymbol{\beta}$.

Let $\mathbf{y}' = (\mathbf{u}', \mathbf{v}')$ be a left eigenvector associated with the eigenvalue λ. $\mathbf{y}'P = \mathbf{y}'$ implies that

$$(\mathbf{u}', \mathbf{v}')\begin{bmatrix} I & 0 \\ R & Q \end{bmatrix} = \lambda(\mathbf{u}', \mathbf{v}'),$$

i.e., $\mathbf{u}' + \mathbf{v}'R = \lambda\mathbf{u}'$ and $\mathbf{v}'Q = \lambda\mathbf{v}'$.

If $\lambda = 1$, then $\mathbf{u}'$ is arbitrary with $\mathbf{v}'R = \mathbf{0}'$. Also, $\mathbf{v}'(I - Q) = \mathbf{0}'$ but since $(I - Q)^{-1}$ exists $\mathbf{v}' = \mathbf{0}'$.

If $\lambda \neq 1$, then $\mathbf{v}'Q = \lambda\mathbf{v}'$ and $\mathbf{u}' = [1/(\lambda - 1)]\mathbf{v}'R$. $\square$

From Theorems 6.1.4–6.1.7 we can make the following deductions:

(a)   The multiplicity of the eigenvalue $\lambda = 1$ determines the number of irreducible closed sets of states.

(b)   Any eigenvalue of P on the unit circle $|\lambda| = 1$ is a root of unity.

(c)   The dth roots of unity are eigenvalues of P if and only if P has an equivalence class of persistent states with period d.

(d)   The multiplicity of each dth root of unity is the number of irreducible subchains of period d.

EXAMPLE 6.1.5:   The transition matrix P of a Markov chain has the following eigenvalues on the unit circle:

$$\lambda_1 = 1, \quad \lambda_2 = 1, \quad \lambda_3 = 1, \quad \lambda_4 = -1, \quad \lambda_5 = -1, \quad \lambda_6 = i, \quad \lambda_7 = -i.$$

Identify the structure of the state space.

Since $\lambda = 1$ has multiplicity 3, there are 3 irreducible subchains: (i) periodic with period 4 ($\lambda_1 = 1, \lambda_4 = -1, \lambda_6 = i, \lambda_7 = -i$); (ii) periodic with period 2 ($\lambda_2 = 1, \lambda_5 = -1$); (iii) aperiodic ($\lambda_3 = 1$). $\square$

At the beginning of this section we mentioned the possibility of utilizing generating functions for determining the multistep transaction probabilities. The relevant procedure is based upon the following theorem.

THEOREM 6.1.8:   If

$$P_{ij}(s) = \sum_{n=0}^{\infty} p_{ij}^{(n)}s^n,$$

then

(a)
$$P_{ij}(s) = s \sum_k p_{ik} P_{kj}(s) + \delta_{ij}, \qquad (6.1.15)$$

(b)
$$P_{ij}(s) = s \sum_k P_{ik}(s) p_{kj} + \delta_{ij}. \qquad (6.1.16)$$

Proof: (a)   From Eq. (6.1.4)

$$p_{ij}^{(n+1)} = \sum_k p_{ik} p_{kj}^{(n)} \qquad (n \geq 0)$$

so that multiplying each side of this equation by s^n and summing over n gives

$$\frac{1}{s} [P_{ij}(s) - P_{ij}(0)] = \sum_k p_{ik} P_{kj}(s).$$

Equation (6.1.15) follows by noting $P_{ij}(0) = p_{ij}^{(0)} = \delta_{ij}$.

(b)   As above, using

$$p_{ij}^{(n+1)} = \sum_k p_{ik}^{(n)} p_{kj} \qquad (n \geq 0). \quad \square$$

Suppose the MC has m states. By a repeated use of Eqs. (6.1.15) we obtain a set of linear equations in $P_{1j}(s), P_{2j}(s), \ldots, P_{mj}(s)$ for fixed j, which can then be solved. Similarly, for fixed i Eqs. (6.1.16) yield relationships between $P_{i1}(s)$, $P_{i2}(s), \ldots, P_{im}(s)$. As an illustration you may wish to examine the two-state MC. Solving such equations is tedious and a more effective procedure is to use matrix generating functions.

THEOREM 6.1.9:   If $\mathbf{P}(s) = [\sum_{n=0}^{\infty} p_{ij}^{(n)} s^n]$, then for $|s| < 1$,
$$\mathbf{P}(s) = [I - sP]^{-1}.$$

Proof: Observe that

$$\mathbf{P}(s) = \sum_{n=0}^{\infty} [p_{ij}^{(n)}] s^n = \sum_{n=0}^{\infty} P^{(n)} s^n = \sum_{n=0}^{\infty} P^n s^n$$

$$= \sum_{n=0}^{\infty} (sP)^n = [I - sP]^{-1}$$

using Theorem 4.5.4 or Corollary 4.5.4A.

The convergence of the matrix power series follows for $|s| < 1$ from the fact that if $\lambda_1, \lambda_2, \ldots, \lambda_m$ are eigenvalues of P,

$$\rho(sP) = \max_{1 \leq i \leq m} |s\lambda_i| \qquad \text{[by Theorem 4.4.5(e)]}$$

$$= |s| \max_{1 \leq i \leq m} |\lambda_i|$$

$$\leq |s| < 1, \qquad \text{(by Theorem 6.1.4).}$$

Alternatively, $\lim_{n \to \infty} (sP)^n = \lim_{n \to \infty} s^n P^{(n)} = 0$ for $|s| < 1$ since $P^{(n)}$ is bounded by, say, $E = [1]$.

Another derivation of the theorem can be based on either Eq. (6.1.15) or Eq. (6.1.16):

$$\mathbf{P}(s) = \left[\sum_{n=0}^{\infty} p_{ij}^{(n)} s^n \right] = [P_{ij}(s)]$$

and thus

$$(1/s)[\mathbf{P}(s) - \mathbf{P}(0)] = \mathbf{P}(s)P.$$

Solving for $\mathbf{P}(s)$ with $\mathbf{P}(0) = I$ gives $\mathbf{P}(s) = [I - sP]^{-1}$. $\quad\square$

When the MC has a small number of states (usually ≤ 4) we have the following very useful procedure for finding $p_{ij}^{(n)}$:

(a)   Compute

$$[I - sP]^{-1} = \frac{1}{\det(I - sP)} \operatorname{adj}(I - sP).$$

(b)   The (i, j)th element of $[I - sP]^{-1}$ gives $P_{ij}(s)$.
(c)   The coefficient of s^n in $P_{ij}(s)$ gives $p_{ij}^{(n)}$.

Theorems 4.5.5, 4.5.6, and 4.5.7 also give some results and algorithms that aid in the computation of $[I - sP]^{-1}$.

EXAMPLE 6.1.6:   *Two-State Markov Chain.*   Obtain the results of Theorem 5.1.7 giving the n-step probabilities $p_{ij}^{(n)}$ of the MC with transition matrix

$$P = \begin{bmatrix} 1 - a & a \\ b & 1 - b \end{bmatrix} \quad (0 \leq a \leq 1, 0 \leq b \leq 1)$$

in the case $d = 1 - a - b \neq 1$.

It is easy to verify that $\det(I - Ps) = (1 - s)(1 - ds)$ and thus

$$[I - Ps]^{-1} = \frac{1}{(1 - s)(1 - ds)} \begin{bmatrix} 1 - (1 - b)s & as \\ bs & 1 - (1 - a)s \end{bmatrix}.$$

Now

$$P_{11}(s) = \frac{1 - (1 - b)s}{(1 - s)(1 - ds)} = \frac{1}{1 - d} \left\{ \frac{b}{1 - s} + \frac{a}{1 - ds} \right\},$$

leading to

$$p_{11}^{(n)} = \frac{1}{1-d}\{b + ad^n\}.$$

Also,

$$P_{12}(s) = \frac{as}{(1-s)(1-ds)} = \frac{1}{1-d}\left\{\frac{a}{1-s} - \frac{b}{1-ds}\right\}$$

giving

$$p_{12}^{(n)} = \frac{1}{1-d}\{a - bd^n\}$$

with analogous results for $p_{22}^{(n)}$ and $p_{21}^{(n)}$ by interchanging the roles of a and b.   □

EXAMPLE 6.1.7: Find the multistep transition probabilities for the Markov chain with transition matrix

$$P = \begin{bmatrix} 0 & 1 & 0 \\ q & 0 & p \\ 0 & 1 & 0 \end{bmatrix}$$

using the method of Theorem 6.1.9.
    Since

$$I - Ps = \begin{bmatrix} 1 & -s & 0 \\ -qs & 1 & -ps \\ 0 & -s & 1 \end{bmatrix},$$

$\det(I - Ps) = 1 - s^2$ and

$$[I - Ps]^{-1} = \frac{1}{(1-s)(1+s)} \begin{bmatrix} 1-ps^2 & s & ps^2 \\ qs & 1 & ps \\ qs^2 & s & 1-qs^2 \end{bmatrix},$$

yielding, via a partial fraction expansion,

$$[I - Ps]^{-1} = \begin{bmatrix} p & 0 & -p \\ 0 & 0 & 0 \\ -q & 0 & q \end{bmatrix} + \frac{1}{1-s} \begin{bmatrix} q/2 & \frac{1}{2} & p/2 \\ q/2 & \frac{1}{2} & p/2 \\ q/2 & \frac{1}{2} & p/2 \end{bmatrix}$$

$$+ \frac{1}{1+s} \begin{bmatrix} q/2 & -\frac{1}{2} & p/2 \\ -q/2 & \frac{1}{2} & -p/2 \\ q/2 & -\frac{1}{2} & p/2 \end{bmatrix}.$$

Extraction of the coefficient of s^n $(n = 0, 1, 2, \ldots)$ gives

$$P^{(0)} = \begin{bmatrix} 1 & 0 & 0 \\ 0 & 1 & 0 \\ 0 & 0 & 1 \end{bmatrix},$$

$$P^{(2n+1)} = \begin{bmatrix} 0 & 1 & 0 \\ q & 0 & p \\ 0 & 1 & 0 \end{bmatrix},$$

$$P^{(2n)} = \begin{bmatrix} q & 0 & p \\ 0 & 1 & 0 \\ q & 0 & p \end{bmatrix}.$$

These results could have been obtained by induction or the diagonaliza-tion technique. Concerning the latter procedure, note that the eigenvalues of P are 1, -1, and 0, showing the presence of a periodic (sub)chain with period 2. An examination of the transition graph, as given in Fig. 6.1.2, shows that the chain is in fact an irreducible periodic chain, which is a special case of the simple random walk with two reflecting barriers.   □

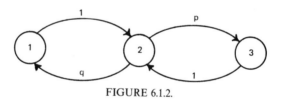

FIGURE 6.1.2.

In Lemma 4.5.7 it was established that

$$\det(I - Ps) = (1 - \lambda_1 s)(1 - \lambda_2 s) \cdots (1 - \lambda_m s),$$

where $\lambda_1, \lambda_2, \ldots, \lambda_m$ are the eigenvalues of P. Since every transition matrix has at least one eigenvalue $\lambda = 1$, we see that $\det(I - Ps)$ certainly has a factor $(1 - s)$. In fact, if the MC has r irreducible subchains this factor is repeated r times.

Because of the factoring of $\det(I - Ps)$ the standard technique for ex-tracting the coefficient of s^n in $P_{ij}(s)$ is via a partial fraction expansion. This procedure leads to the same result as that obtained by using the diag-onalization technique. The equivalence of the two methods can be estab-lished, and we give a proof for the case of a transition matrix with distinct eigenvalues.

THEOREM 6.1.10:   Let P be the transition matrix of a Markov chain with distinct eigenvalues $\lambda_1 = 1, \lambda_2, \ldots, \lambda_m$. Let $x_1, x_2, \ldots, x_m$ and $y'_1, y'_2, \ldots, y'_m$ be the corresponding right and left eigenvectors chosen so that $y'_i x_i = 1$ $(i = 1, 2, \ldots, m)$. Then for $|s| < 1$

$$[I - Ps]^{-1} = \sum_{i=1}^{m} \sum_{n=1}^{\infty} A_i \lambda_i^n s^n,$$

where for nonzero λ_i

$$A_i = \frac{\lambda_i^{m-1} \operatorname{adj}(I - \lambda_i^{-1} P)}{\prod_{j \neq i}(\lambda_i - \lambda_j)} = x_i y'_i.$$

Proof:

$$[I - Ps]^{-1} = \frac{\operatorname{adj}(I - Ps)}{\det(I - Ps)} = \frac{I + B_1 s + \cdots + B_{m-1} s^{m-1}}{(1 - \lambda_1 s) \cdots (1 - \lambda_m s)}$$

$$= \sum_{i=1}^{m} \frac{A_i}{1 - \lambda_i s}$$

by a partial fraction expansion, where

$$A_i = \lim_{s \to 1/\lambda_i} (1 - \lambda_i s)[I - Ps]^{-1}$$

$$= \frac{\lambda_i^{m-1} \operatorname{adj}(I - \lambda_i^{-1} P)}{\prod_{j \neq i}(\lambda_i - \lambda_j)}.$$

Extracting the coefficient of s^n from both sides of the expression for $[I - Ps]^{-1}$ above gives

$$P^{(n)} = \sum_{i=1}^{m} A_i \lambda_i^n.$$

The result will follow once we established the result that $A_i = x_i y'_i$.

Since the eigenvalues are all distinct, the eigenvalues have geometric multiplicity one so that any eigenvector u such that $Pu = \lambda_i u$ must be of the form $u = cx_i$ for some $c \neq 0$, where the $x_1, \ldots, x_m$ are a specified set of eigenvectors. Similarly, any v' such that $v'P = \lambda_i v'$ is of the form dy'_i for some $d \neq 0$.

Consider a fixed index i such that $\lambda_i \neq 0$. Then

$$(I - \lambda_i^{-1}P)\operatorname{adj}(I - \lambda_i^{-1}P) = I \det(I - \lambda_i^{-1}P) = 0.$$

Thus any column of $\operatorname{adj}(I - \lambda_i^{-1}P)$ may be taken as a right eigenvector corresponding to the eigenvalue λ_i and hence

$$\operatorname{adj}(I - \lambda_i^{-1}P) = [c_1 x_i, c_2 x_i, \ldots, c_m x_i] = x_i c',$$

where $\mathbf{c}' = (c_1, c_2, \ldots, c_m)$ for some such nonzero $c_1, \ldots, c_m$. Furthermore, since $\mathrm{adj}(I - \lambda_i^{-1}P)(I - \lambda_i^{-1}P) = 0$, any row of $\mathrm{adj}(I - \lambda_i^{-1}P)$ is a left eigenvector associated with λ_i so that $\mathbf{c}' = k\mathbf{y}_i'$ and

$$\mathrm{adj}(I - \lambda_i^{-1}P) = k\mathbf{x}_i\mathbf{y}_i'$$

for some nonzero k.

The eigenvalues of $I - \lambda_i^{-1}P$ are $\mu_j = 1 - \lambda_i^{-1}\lambda_j$ $(j = 1, 2, \ldots, m)$. Observe that $\mu_i = 0$ and that all the other μ_j are nonzero; thus, by Corollary 4.5.5A (with n taken as m and λ_j as μ_j),

$$\mathrm{tr}(\mathrm{adj}(I - \lambda_i^{-1}P)) = \prod_{j \neq i} (1 - \lambda_i^{-1}\lambda_j) = \frac{1}{\lambda^{m-1}} \prod_{j \neq i} (\lambda_i - \lambda_j).$$

Also, if $\mathbf{x}_i' = (x_{1i}, x_{2i}, \ldots, x_{mi})$ and $\mathbf{y}_i' = (y_{i1}, y_{i2}, \ldots, y_{im})$, then

$$\mathrm{tr}(\mathrm{adj}(I - \lambda_i^{-1}P)) = k\,\mathrm{tr}(\mathbf{x}_i\mathbf{y}_i') = k \sum_{l=1}^{m} x_{li}y_{il} = k\mathbf{y}_i'\mathbf{x}_i;$$

and thus if the sets of eigenvectors are chosen so that $\mathbf{y}_i'\mathbf{x}_i = 1$ we see that $k = \prod_{j \neq i}(\lambda_i - \lambda_j)/\lambda_i^{m-1}$ and hence $A_i = \mathbf{x}_i\mathbf{y}_i'$.   □

It is worthwhile pointing out that Theorem 6.1.10, besides establishing the equivalence between the two methods, also gives an alternative computational technique for the $A_i = \mathbf{x}_i\mathbf{y}_i'$ in terms of the eigenvalues and the $\mathrm{adj}(I - \lambda_i^{-1}P)$ so that once the eigenvalues (assumed to be distinct) and $\mathrm{adj}(I - Ps)$ have been computed we do not have to proceed any further with the computation of the eigenvectors, but instead use the expression for the A_i as given in the theorem.

Exercises 6.1

1. If

$$P = \begin{bmatrix} 0 & \frac{1}{2} & \frac{1}{2} \\ 1 & 0 & 0 \\ 1 & 0 & 0 \end{bmatrix}$$

show that, for $n \geq 1$,

$$P^{2n} = \begin{bmatrix} 1 & 0 & 0 \\ 0 & \frac{1}{2} & \frac{1}{2} \\ 0 & \frac{1}{2} & \frac{1}{2} \end{bmatrix}$$

and $P^{2n-1} = P$.

2. Show, using induction or otherwise that for the transition matrix

$$P = \begin{bmatrix} 0 & 1 & 0 & 0 \\ \frac{1}{2} & 0 & \frac{1}{2} & 0 \\ 0 & \frac{1}{2} & 0 & \frac{1}{2} \\ 0 & 0 & 1 & 0 \end{bmatrix},$$

$$P^{(n)} = \frac{1}{6} \begin{bmatrix} 1+u+2v+2w & 2-2u+2v-2w & 2+2u-2v-2w & 1-u-2v+2w \\ 1-u+v-w & 2+2u+v+w & 2-2u-v+w & 1+u-v-w \\ 1+u-v-w & 2-2u-v+w & 2+2u+v+w & 1-u+v-w \\ 1-u-2v+2w & 2+2u-2v-2w & 2-2u+2v-2w & 1+u+2v+2w \end{bmatrix},$$

where $u = (-1)^n$, $v = (\frac{1}{2})^n$, and $w = (-\frac{1}{2})^n$ [Gray (1967)].

3. If

$$P = \begin{bmatrix} a_1 & a_2 & a_3 \\ a_3 & a_1 & a_2 \\ a_2 & a_3 & a_1 \end{bmatrix}$$

prove that

$$P^{(n)} = \begin{bmatrix} a_{1n} & a_{2n} & a_{3n} \\ a_{3n} & a_{1n} & a_{2n} \\ a_{2n} & a_{3n} & a_{1n} \end{bmatrix},$$

where $a_{1n} + \omega a_{2n} + \omega^2 a_{3n} = (a_1 + a_2\omega + a_3\omega^2)^n$ and ω is the cube root of unity. In particular, if $a_1 = 0$ and $a_2 = a_3 = \frac{1}{2}$, show that if $a_{1n} \equiv a_n$, then $a_{2n} = a_{3n} = a_{n+1}$ and that $a_n = \frac{1}{3}[1 + \{(-1)^n/2^{n-1}\}]$.

4. For the Markov chain with transition matrix

$$P = \begin{bmatrix} 0 & 1 & 0 \\ 1-p & 0 & p \\ 0 & 1 & 0 \end{bmatrix},$$

find $P^{(2)}$ and $P^{(3)}$ and hence deduce a general expression for the n-step transition matrix $P^{(n)}$.

5. In a sequence of independent Bernoulli trials with probability of success S at each trial being p (and failure F being q, $p + q = 1$) we say that state 1 is observed at the nth trial if trials numbered $n - 1$ and n resulted in the sequence SS. Similarly, states 2, 3, and 4 stand for the sequences SF, FS, and FF on successive trials. Show that the process generated by

taking as state space $S = \{1, 2, 3, 4\}$ is a Markov chain. Find the transition matrix P, and the n-step transition matrix $P^{(n)}$ ($n \geq 1$). *Hint*: Find $P^{(2)}$ and $P^{(3)}$ and thence use induction [Feller (1968)].

6. (Generalization of Exercise 6.1.5). In the exercise above let the sequence of S's and F's be generated by Markov-dependent Bernoulli trials with transition matrix

$$\begin{bmatrix} 1 - a & a \\ b & 1 - b \end{bmatrix},$$

as described in Example 5.1.1. Label as states 1, 2, 3, and 4 the sequences SS, SF, FS, and FF on successive trials. Find the transition matrix P for the four-state Markov chain generated and find $P^{(n)}$ for $n \geq 1$. *Hint*:
(i)   Show by induction that $P^{(n)}$ has the form

$$\begin{bmatrix} a_{1n} & a_{2n} & a_{3n} & a_{4n} \\ b_{1n} & b_{2n} & b_{3n} & b_{4n} \\ a_{1n} & a_{2n} & a_{3n} & a_{4n} \\ b_{1n} & b_{2n} & b_{3n} & b_{4n} \end{bmatrix}$$

(ii)   Show that b_{1n}, b_{2n}, b_{3n}, and b_{4n} can be expressed in terms of a_{1n}, a_{2n}, a_{3n}, and a_{4n}.
(iii)   Establish relationships between these terms. In particular, show that b_{1n}, b_{2n}, b_{3n}, b_{4n} and a_{1n}, a_{2n}, a_{4n} can all be expressed in terms of the sequence $\{a_{3n}\}$.
(iv)   Show that $a_{3,n+1} - da_{3,n} = ab$ (where $d = 1 - a - b$) and hence solve for a_{3n} leading to an expression for P^n. Check that your results agree with those obtained for Exercise 6.1.5 (when $1 - a = b = p$).

7. Show that

$$\sum_{k=1}^{a-1} \sin(\pi kr/a) \sin(\pi ks/a) = \frac{a}{2} \delta_{rs} \qquad \text{for} \quad r, s = 1, \ldots, a - 1.$$

(*Hints*: $\sin\theta = \{e^{i\theta} - e^{-i\theta}\}/2i$ and $\sum_{k=1}^{a-1} x^k = \{x - x^a\}/(1 - x)$, provided $x \neq 1$.)

8. Let α_j ($j = 1, 2, \ldots, m$) be complex numbers. Show that if $\left|\sum_{j=1}^{m} \alpha_j\right| = \sum_{j=1}^{m} |\alpha_j|$, then $\alpha_j = |\alpha_j| e^{i\theta}$ ($j = 1, 2, \ldots, m$) for some real θ. (*Hint*: Show that the result is true for $m = 2$ and then use an inductive proof.)

9. If

$$P = \begin{bmatrix} I & 0 \\ R & Q \end{bmatrix}$$

is the transition matrix of an absorbing MC in canonical form show that

$$
\mathbf{P}(s) = \begin{bmatrix} \dfrac{1}{1-s}I & 0 \\[2ex] \dfrac{s}{1-s}(I - sQ)^{-1}R & (I - sQ)^{-1} \end{bmatrix}.
$$

10. Obtain the *n*-step transition probabilities for the transition matrix

$$
P = \begin{bmatrix} p & q & 0 \\ \tfrac{1}{2}q & p & \tfrac{1}{2}q \\ 0 & q & p \end{bmatrix},
$$

and for $0 < p < 1$ show that P^n tends to a limit as $n \to \infty$ [Gray (1967)].

11. The transition matrix for a cyclical four-state random walk is

$$
P = \begin{bmatrix} 0 & p & 0 & q \\ q & 0 & p & 0 \\ 0 & q & 0 & p \\ p & 0 & q & 0 \end{bmatrix},
$$

where $0 < p < 1$ and $p \neq q$. Show that the eigenvalues of P are ± 1 and $\pm i(p - q)$ $(i = \sqrt{-1})$ and hence prove that

$$
p_{jk}^{(n)} = \tfrac{1}{4}\{1 + (p - q)^n i^{k-j-n}\}\{1 + (-1)^{j+k+n}\}.
$$

Examine the behavior of $p_{jk}^{(n)}$ as $n \to \infty$ and discuss the limit of $\{p_k^{(n)}\}$ when the initial distribution is given by
   (i) $(1, 0, 0, 0)$,
   (ii) $(\tfrac{1}{6}, \tfrac{1}{3}, \tfrac{1}{3}, \tfrac{1}{6})$, and
   (iii) $(\tfrac{1}{4}, \tfrac{1}{4}, \tfrac{1}{4}, \tfrac{1}{4})$ [Gray (1967)].

12. The location of the eigenvalues of a transition matrix *on* the unit circle serve to identify the character of Markov chains. However, the locations of the eigenvalues *inside* the unit circle provide us with little information. Show that each of the following transition matrices has the same set of eigenvalues, yet one has no transient states, one has one transient state, and the other has two transient states.

$$
P = \begin{bmatrix} \tfrac{1}{2} & \tfrac{1}{2} & 0 \\ \tfrac{1}{2} & 0 & \tfrac{1}{2} \\ 0 & \tfrac{1}{2} & \tfrac{1}{2} \end{bmatrix}, \quad P_2 = \begin{bmatrix} \tfrac{1}{2} & \tfrac{1}{2} & 0 \\ 1 & 0 & 0 \\ \tfrac{1}{2} & 0 & \tfrac{1}{2} \end{bmatrix}, \quad P_3 = \begin{bmatrix} 1 & 0 & 0 \\ \tfrac{1}{2} & 0 & \tfrac{1}{2} \\ \tfrac{1}{2} & \tfrac{1}{2} & 0 \end{bmatrix}.
$$

13. Let

$$P = \begin{bmatrix} \frac{1}{4} & 0 & \frac{1}{4} & \frac{1}{4} & \frac{1}{4} \\ 0 & 1 & 0 & 0 & 0 \\ 0 & 0 & \frac{1}{3} & \frac{1}{3} & \frac{1}{3} \\ 0 & 0 & 0 & 1 & 0 \\ 0 & \frac{1}{4} & 0 & \frac{1}{4} & \frac{1}{2} \end{bmatrix}.$$

(1)   By relabeling the states express P in the canonical form of an absorbing MC with Q taken as

$$\begin{bmatrix} * & * & * \\ 0 & * & * \\ 0 & 0 & * \end{bmatrix}.$$

(2)   Using the result that

$$\begin{bmatrix} x & a & c \\ 0 & y & b \\ 0 & 0 & z \end{bmatrix} \begin{bmatrix} yz & -az & ab-cy \\ 0 & zx & -bx \\ 0 & 0 & xy \end{bmatrix} = xyzI,$$

find an expression for $\lim_{n \to \infty} P^{(n)}$.

14. Let

$$P = \begin{bmatrix} 0 & 0 & \frac{1}{2} & 0 & \frac{1}{2} \\ 0 & 1 & 0 & 0 & 0 \\ \frac{1}{4} & \frac{1}{4} & 0 & \frac{1}{4} & \frac{1}{4} \\ 0 & \frac{1}{2} & \frac{1}{2} & 0 & 0 \\ 0 & 0 & 0 & 0 & 1 \end{bmatrix}.$$

(i)   By relabeling the states express P in the canonical form of an absorbing MC with Q taken as

$$\begin{bmatrix} 0 & * & 0 \\ * & 0 & * \\ 0 & * & 0 \end{bmatrix}.$$

(ii)   Using the result that

$$\begin{bmatrix} 1 & -a & 0 \\ -b & 1 & -c \\ 0 & -d & 1 \end{bmatrix} \begin{bmatrix} 1-cd & a & ac \\ b & 1 & c \\ bd & d & 1-ab \end{bmatrix} = (1-ab-cd)I,$$

find an expression for $\lim_{n \to \infty} P^{(n)}$.

15. Let P be the transition matrix of an absorbing MC in canonical form with

$$Q = \begin{bmatrix} a_1 & a_2 & a_3 \\ 0 & b_1 & b_2 \\ 0 & 0 & c_1 \end{bmatrix}.$$

(i) If $Q^{(n)} = Q^n$, show that $Q^{(n)}$ has the form

$$Q^{(n)} = \begin{bmatrix} a_1^{(n)} & a_2^{(n)} & a_3^{(n)} \\ 0 & b_1^{(n)} & b_2^{(n)} \\ 0 & 0 & c_1^{(n)} \end{bmatrix}.$$

(ii) Show that $a_1^{(n)} = a_1^n$, $b_1^{(n)} = b_1^n$, and $c_1^{(n)} = c_1^n$.

(iii) Show also that

$$a_2^{(n)} = \frac{a_2(a_1^n - b_1^n)}{a_1 - b_1} \qquad (a_1 = b_1),$$

$$b_2^{(n)} = \frac{b_2(b_1^n - c_1^n)}{b_1 - c_1} \qquad (b_1 \neq c_1),$$

$$a_3^{(n)} = \frac{a_3(a_1^n - c_1^n)}{a_1 - c_1} + \frac{a_2 b_2}{a_1 - c_1} \left\{ \frac{a_1^n - b_1^n}{a_1 - b_1} - \frac{b_1^n - c_1^n}{b_1 - c_1} \right\}$$

$$(a_1 \neq b_1, b_1 \neq c_1, a_1 \neq c_1).$$

(iv) Find expressions for the elements of $Q^{(n)}$ when $a_1 = b_1 = c_1$.

6.2   Determination of the n-Step First Passage Time Probabilities $f_{ij}^{(n)}$

The importance of the $f_{ij}^{(n)}$ was exhibited in the classification of states as discussed in Section 5.2. In this section we present a variety of techniques for their evaluation given the transition matrix $P = [p_{ij}]$ of the MC.

Theorem 5.1.6 presented the main properties of the $f_{ij}^{(n)}$ that we restate for reference purposes:

(a) $f_{ij}^{(n)} = \sum_{j_1 \neq j} \cdots \sum_{j_{n-1} \neq j} p_{ij_1} p_{j_1 j_2} \cdots p_{j_{n-2} j_{n-1}} p_{j_{n-1} j} \qquad (n \geq 2)$

with $f_{ij}^{(1)} = p_{ij}.$ (6.2.1)

(b) $f_{ij}^{(n)} = \sum_{k \neq j} p_{ik} f_{kj}^{(n-1)} \qquad (n \geq 2).$ (6.2.2)

(c) $F^{(n)} = P(F^{(n-1)} - F_d^{(n-1)})$ $(n \geq 2)$ with $F^{(1)} = P$

where $F^{(n)} = [f_{ij}^{(n)}]$.   (6.2.3)

(d) $F^{(n)} = \sum_{i=0}^{n-1} P^{n-i} A_i,$

where A_i is a diagonal matrix satisfying the recurrence relationship

$$A_n + \sum_{i=0}^{n-1} D_{n-i} A_i = 0 \quad \text{with} \quad A_0 = I, \quad D_n = [P^n]_d \ (n \geq 1). \quad (6.2.4)$$

We have basically four methods we can use. In simple cases we can use direct probabilistic arguments or recursive techniques. In a more general setting, generating functions and matrix generating functions serve a useful role.

We examine each of these methods in turn. Firstly, when the transition matrix has a simple structure so that the number of possible transitions at any step is small it is often possible, by using the argument that led to Eq. (6.2.1), to write down the $f_{ij}^{(n)}$ without too much trouble. An example of this was given earlier in Section 5.1 when we considered the general two-state MC. As a further illustration we present a selection of examples.

EXAMPLE 6.2.1:   Suppose the MC has a transition matrix given by

$$P = \begin{bmatrix} 0 & p_1 & q_1 \\ q_2 & 0 & p_2 \\ p_3 & q_3 & 0 \end{bmatrix}.$$

From the transition graph, given in Fig. 6.2.1, elementary probabilistic arguments give

$$f_{11}^{(n)} = \begin{cases} 0, & n = 1; \\ p_1(p_2 q_3)^{m-1} q_2 + q_1(q_3 p_2)^{m-1} p_3, & n = 2m, m \geq 1; \\ p_1(p_2 q_3)^{m-1} p_2 p_3 + q_1(q_3 p_2)^{m-1} q_3 q_2, & n = 2m + 1, m \geq 1. \end{cases}$$

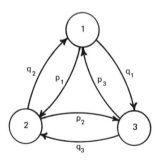

FIGURE 6.2.1.

Also,

$$f_{12}^{(n)} = \begin{cases} (q_1 p_3)^{m-1} q_1 q_3, & n = 2m, \ m \geq 1; \\ (q_1 p_3)^m p_1, & n = 2m + 1, \ m \geq 0; \end{cases}$$

and

$$f_{13}^{(n)} = \begin{cases} (p_1 q_2)^{m-1} p_1 p_2, & n = 2m, \ m \geq 1; \\ (p_1 q_2)^m q_1, & n = 2m + 1, \ m \geq 0; \end{cases}$$

with the other $f_{ij}^{(n)}$ following by symmetry.   □

EXAMPLE 6.2.2:   If

$$P = \begin{bmatrix} p_1 & q_1 & 0 \\ 0 & p_2 & q_2 \\ q_3 & 0 & p_3 \end{bmatrix},$$

then the MC has a transition graph of the form given in Fig. 6.2.2.

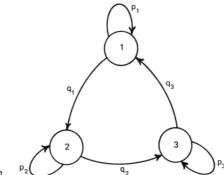

FIGURE 6.2.2.

By considering first returns to state 1,

$$f_{11}^{(n)} = \begin{cases} p_1, & n = 1; \\ 0, & n = 2; \\ \displaystyle\sum_{r=0}^{n-3} q_1 p_2^r q_2 p_3^{n-r-3} q_3 = \begin{cases} q_1 q_2 q_3 \left(\dfrac{p_3^{n-2} - p_2^{n-2}}{p_3 - p_2} \right), & n \geq 3, \ p_2 \neq p_3; \\ (n-2) q_1 q_2 q_3 p_3^{n-3}, & n \geq 2, \ p_2 = p_3. \end{cases} \end{cases}$$

The first passage distribution from state 1 to state 2 is a simple geometric distribution with

$$f_{12}^{(n)} = p_1^{n-1} q_1, \qquad n \geq 1,$$

while $f_{13}^{(n)}$ has a slightly more complicated form with

$$f_{13}^{(n)} = \sum_{r=0}^{n-2} p_1^r q_1 p_2^{n-r-2} q_3 = \begin{cases} q_1 q_2 \left(\dfrac{p_2^{n-1} - p_1^{n-1}}{p_2 - p_1} \right), & n \geq 2, \ p_1 \neq p_2; \\ (n-1) q_1 q_2 p_2^{n-2}, & n \geq 2, \ p_1 = p_2. \end{cases}$$

The other $f_{ij}^{(n)}$ follow by symmetry. □

Equations (6.2.3) form the basis of a recursive method for finding $F^{(n)} = [f_{ij}^{(n)}]$. An early illustration of this technique was given in the proof of Theorem 5.1.8 when the n-step first passage time matrix was derived for the general two-state MC. Let us apply this method to the canonical form of an absorbing MC.

THEOREM 6.2.1:   For an absorbing MC with transition matrix

$$P = \begin{bmatrix} I & 0 \\ R & Q \end{bmatrix}, \qquad F^{(n)} = \begin{bmatrix} F_1^{(n)} & F_2^{(n)} \\ F_3^{(n)} & F_4^{(n)} \end{bmatrix},$$

where

(a) $F_1^{(1)} = I, \qquad F_1^{(n)} = 0 \qquad (n \geq 2);$

(b) $F_2^{(n)} = 0 \qquad (n \geq 1);$

(c) $F_3^{(n)} = Q^{n-1} R \qquad (n \geq 1);$

(d) $F_4^{(n)} = Q[F_4^{(n-1)} - F_{4d}^{(n-1)}] \qquad (n \geq 2); \qquad F_4^{(1)} = Q.$

Proof: Since, by Eq. (6.2.3), $F^{(n)} = P[F^{(n-1)} - F_d^{(n-1)}]$, substitution with block forms yield

$$\begin{bmatrix} F_1^{(n)} & F_2^{(n)} \\ F_3^{(n)} & F_4^{(n)} \end{bmatrix} = \begin{bmatrix} I & 0 \\ R & Q \end{bmatrix} \begin{bmatrix} F_1^{(n-1)} - F_{1d}^{(n-1)} & F_2^{(n-1)} \\ F_3^{(n-1)} & F_4^{(n-1)} - F_{4d}^{(n-1)} \end{bmatrix} \qquad (n \geq 2).$$

By carrying out matrix multiplication and equating the respective blocks we obtain the following results.

(a) $F_1^{(n)} = F_1^{(n-1)} - F_{1d}^{(n-1)}$ $(n \geq 2)$ with $F_1^{(1)} = I$. Thus $F_1^{(2)} = I - I = 0$, which implies $F_1^{(n)} = 0$ for $n \geq 2$.

(b) $F_2^{(n)} = F_2^{(n-1)}$ and hence $= \cdots = F_2^{(1)} = 0$.

(c) $F_3^{(n)} = R(F_1^{(n-1)} - F_{1d}^{(n-1)}) + Q F_3^{(n-1)}$
$= Q F_3^{(n-1)}$ for $n \geq 2$ [using (a)] with $F_3^{(1)} = R$.
An inductive argument implies that $F_3^{(n)} = Q^{n-1} R$ for $n \geq 1$.

(d) $F_4^{(n)} = R F_2^{(n-1)} + Q(F_4^{(n-1)} - F_{4d}^{(n-1)})$, which leads to the stated result. □

It is possible to extend this theorem to obtain a general form for $F^{(n)}$ when the transition matrix P of a MC is expressed in its general canonical form (as given in Theorem 5.4.1).

THEOREM 6.2.2: Let

$$
P = \begin{bmatrix}
P_1 & 0 & \cdots & 0 & 0 \\
0 & P_2 & \cdots & 0 & 0 \\
\vdots & \vdots & & \vdots & \vdots \\
0 & 0 & \cdots & P_k & 0 \\
R_1 & R_2 & \cdots & R_k & Q
\end{bmatrix}
$$

be the transition matrix of a MC expressed in general canonical form. The *n*-step first passage matrix can be expressed in the following form:

$$
F^{(n)} = \begin{bmatrix}
F_1^{(n)} & 0 & \cdots & 0 & 0 \\
0 & F_2^{(n)} & \cdots & 0 & 0 \\
\vdots & \vdots & & \vdots & \vdots \\
0 & 0 & \cdots & F_k^{(n)} & 0 \\
F_{T1}^{(n)} & F_{T2}^{(n)} & \cdots & F_{Tk}^{(n)} & Q^{(n)}
\end{bmatrix},
$$

where

$$F_i^{(n)} = P_i(F_i^{(n-1)} - F_{id}^{(n-1)}) \quad (n \geq 2), \qquad F_i^{(1)} = P_i \quad (i = 1, \ldots, k);$$

$$F_{Ti}^{(n)} = R_i(F_i^{(n-1)} - F_{id}^{(n-1)}) + QF_{Ti}^{(n-1)} \quad (n \geq 2), \qquad F_{Ti}^{(1)} = R_i \quad (i = 1, \ldots, k);$$

$$Q^{(n)} = Q(Q^{(n-1)} - Q_d^{(n-1)}) \quad (n \geq 2), \qquad Q^{(1)} = Q. \quad \square$$

A formal proof of this theorem is omitted. An inductive argument leads directly to the stated results. It is worth observing that expressions for the $F_i^{(n)}$ are found by considering only the transitions within the equivalence class C_i governed by the transition matrix P_i, as to be expected. The $Q^{(n)}$ elements are found by considering only the transitions within the transient states. The only complication is the determination of $F_{Ti}^{(n)}$ dealing with first passages from the transient states to a state of the equivalence class C_i. If $P_i = I$, then $F_{Ti}^{(n)} = Q^{n-1}R_i$, as in Theorem 6.2.1.

It is interesting to compare the results of Theorems 6.1.1 and 6.2.1 concerning the determination of $P^{(n)}$ and $F^{(n)}$ when the underlying MC is an absorbing chain with transition matrix

$$
P = \begin{bmatrix}
I & 0 \\
R & Q
\end{bmatrix}.
$$

Let T denote the set of transient states and A denote the set of absorbing states.

Then from these theorems we obtain

$$[p_{ij}^{(n)}]_{i \in T, \, j \in T} = Q^n, \qquad\qquad n \geq 1,$$

$$[p_{ij}^{(n)}]_{i \in T, \, j \in A} = \left(\sum_{k=0}^{n-1} Q^k \right) R, \qquad n \geq 1,$$

$$[f_{ij}^{(n)}]_{i \in T, \, j \in A} = Q^{n-1} R, \qquad\qquad n \geq 1.$$

Thus, once Q^n is determined, using possibly a diagonalization procedure, we can find both the n-step transition probabilities and n-step first passage time probabilities from a transient state to an absorbing state. In particular, observe that for $i \in T$ and $j \in A$

$$p_{ij}^{(n)} = \sum_{k=0}^{n-1} \sum_{l \in T} p_{il}^{(k)} p_{lj}, \tag{6.2.5}$$

and

$$f_{ij}^{(n)} = \sum_{l \in T} p_{il}^{(n-1)} p_{lj}. \tag{6.2.6}$$

Note also that for $i \in T$ and $j \in A$,

$$f_{ij}^{(n)} = p_{ij}^{(n)} - p_{ij}^{(n-1)} \tag{6.2.7}$$

and

$$p_{ij}^{(n)} = \sum_{k=1}^{n} f_{ij}^{(k)}. \tag{6.2.8}$$

These last two results can be deduced directly from Eqs. (5.2.3) and (5.2.4).

In Example 6.1.4 we determined the multistep transition probabilities for the classical gambler's ruin model. Our derivation was, in effect, based upon an application of Eq. (6.2.5). Let us reconsider this same example but in relation to the derivation of the $f_{ij}^{(n)}$.

EXAMPLE 6.2.3: *Classical Gambler's Ruin Model.* In Example 6.1.4 we showed that such a model is represented as a Markov chain. By ordering the states as $\{0, a, 1, \ldots, a - 1\}$, where 0 and a are absorbing states and $1, 2, \ldots, a - 1$ are transient states, the transition matrix for the MC has the standard form of an absorbing MC given by Eq. (6.1.10). In particular, we showed [Eq. (6.1.11)] that for $r, s = 1, 2, \ldots, a - 1$,

$$p_{rs}^{(n)} = \frac{2^{n+1}}{a} p^{(n+s-r)/2} q^{(n-s+r)/2} \sum_{l=1}^{a-1} \cos^n(\pi l/a) \sin(\pi rl/a) \sin(\pi sl/a).$$

Now for $r \in T = \{1, 2, \ldots, a - 1\}$ it is easily seen, from Eq. (6.2.6) or Eqs. (6.2.7) and (6.1.12), that

$$f_{r0}^{(n)} = p_{r1}^{(n-1)}q \qquad \text{and} \qquad f_{ra}^{(n)} = p_{r, a-1}^{(n-1)}p.$$

Substitution gives

$$f_{r0}^{(n)} = \frac{2^n}{a} p^{(n-r)/2} q^{(n+r)/2} \sum_{l=1}^{a-1} \cos^{n-1}(\pi l/a) \sin(\pi r l/a) \sin(\pi l/a)$$

and

$$f_{ra}^{(n)} = \frac{2^n}{a} p^{(n+a-r)/2} q^{(n-a+r)/2} \sum_{l=1}^{a-1} \cos^{n-1}(\pi l/a) \sin(\pi r l/a) \sin(\overline{\pi a - 1}\, l/a).$$

By examining the expression for $f_{r0}^{(n)}$ in more detail we see that not all the terms in the summation need be evaluated.

Let $\alpha_{n,r,l} = \cos^{n-1}(\pi l/a) \sin(\pi r l/a) \sin(\pi l/a)$. Then, since $\cos(\pi - \theta) = -\cos\theta$, $\sin(r\pi - \theta) = (-1)^{r+1} \sin\theta$,

$$\alpha_{n,r,a-l} = (-1)^{n+r}\alpha_{n,r,l}.$$

Also, the term $l = a/2$ occurs only when a is even, in which case $\alpha_{n,r,a/2} = 0$. Consequently,

$$\sum_{l=1}^{a-1} \alpha_{n,r,l} = \sum_{1 \le l < a/2} \alpha_{n,r,l} + \sum_{a/2 < l \le a-1} \alpha_{n,r,l}$$

$$= \sum_{1 \le l < a/2} \{1 + (-1)^{n+r}\}\alpha_{n,r,l}$$

$$= \begin{cases} 0, & n - r \text{ odd}, \\ 2\sum_{1 \le l < a/2} \alpha_{n,r,l}, & n - r \text{ even}. \end{cases}$$

Thus, $f_{r0}^{(n)}$ is zero when $n - r$ is odd, and when $n - r$ is even

$$f_{r0}^{(n)} = \frac{2^{n+1}}{a} p^{(n-r)/2} q^{(n+r)/2} \sum_{1 \le l < a/2} \cos^{n-1}(\pi l/a) \sin(\pi r l/a) \sin(\pi l/a). \quad \square$$

Generating function techniques follow quite naturally when used in conjunction with the recursive methods suggested earlier. For each $i, j \in S$ we define $F_{ij}(s) = \sum_{n=0}^{\infty} f_{ij}^{(n)}s^n$.

THEOREM 6.2.3: For all $i, j \in S$, and $|s| < 1$,

$$F_{ij}(s) = sp_{ij} + s \sum_{k \ne j} p_{ik}F_{kj}(s). \qquad (6.2.9)$$

Proof: From Eq. (6.2.2),

$$F_{ij}(s) = \sum_{n=0}^{\infty} f_{ij}^{(n)} s^n = s p_{ij} + \sum_{n=2}^{\infty} \left(\sum_{k \neq j} p_{ik} f_{kj}^{(n-1)} \right) s^n$$

$$= s p_{ij} + \sum_{k \neq j} p_{ik} \sum_{n=2}^{\infty} f_{kj}^{(n-1)} s^n \qquad \text{(by Theorem 2.2.4)}$$

$$= s p_{ij} + s \sum_{k \neq j} p_{ik} F_{kj}(s). \qquad \square$$

EXAMPLE 6.2.4:   *Classical Gambler's Ruin Model.*   Let us reexamine Example 6.2.3 by deriving the $f_{i0}^{(n)}$ and $f_{ia}^{(n)}$ (for $i \in T$, the transient states) using generating functions and Theorem 6.2.3. From Eq. (6.2.9), $F_{i0}(s)$ the g.f. of the $f_{i0}^{(n)}$ satisfies

$$F_{i0}(s) = s p_{i0} + s \sum_{k \neq 0} p_{ik} F_{k0}(s),$$

and substituting for the transition probabilities (see Example 6.1.4) we obtain successively

$$F_{00}(s) = s,$$

$$F_{10}(s) = sq + spF_{20}(s),$$

$$F_{i0}(s) = sqF_{i-1,0}(s) + spF_{i+1,0}(s), \qquad 2 \leq i \leq a - 1,$$

$$F_{a0}(s) = sF_{a0}(s) \qquad\qquad \text{(which implies } F_{a0}(s) = 0\text{)}.$$

We are only interested in deriving $F_{i0}(s)$ for $i = 1, 2,, \ldots, a - 1$, and to this end let us define $U_i(s) = F_{i0}(s)$ for $i = 1, 2, \ldots, a - 1$ with $U_0(s) \equiv 1$ and $U_a(s) \equiv 0$. Then for $1 \leq i \leq a - 1$,

$$U_i(s) = sqU_{i-1}(s) + spU_{i+1}(s). \qquad (6.2.10)$$

By regarding s as an arbitrary constant, Eqs. (6.2.10) represent a system of difference equations that we can solve using the techniques of Section 2.9 (Theorem 2.9.1). A general solution is given by

$$U_i(s) = A(s)\lambda_1^i(s) + B(s)\lambda_2^i(s),$$

where $\lambda_1(s)$ and $\lambda_2(s)$ are roots of the auxiliary equation $ps\lambda^2(s) - \lambda(s) + qs = 0$, so that

$$\lambda_1(s) = \frac{1 + (1 - 4pqs^2)^{1/2}}{2ps}, \qquad \lambda_2(s) = \frac{1 - (1 - 4pqs^2)^{1/2}}{2ps}, \qquad (6.2.11)$$

where we take $0 < s < 1$ and the positive square root.

To satisfy the boundary conditions we require $A(s) + B(s) = 1$ and $A(s)\lambda_1^a(s) + B(s)\lambda_2^a(s) = 0$, whence

$$U_i(s) = \frac{\lambda_1^a(s)\lambda_2^i(s) - \lambda_1^i(s)\lambda_2^a(s)}{\lambda_1^a(s) - \lambda_2^a(s)}. \tag{6.2.12}$$

Since $\lambda_1(s)\lambda_2(s) = q/p$ this simplifies to give for $i = 1, 2, \ldots, a - 1$,

$$F_{i0}(s) = \left(\frac{q}{p}\right)^i \frac{\lambda_1^{a-i}(s) - \lambda_2^{a-i}(s)}{\lambda_1^a(s) - \lambda_2^a(s)}. \tag{6.2.13}$$

Concerning $F_{ia}(s)$, the generating function of the $f_{ia}^{(n)}$, a similar procedure as above gives

$$F_{0a}(s) = 0,$$

$$F_{ia}(s) = sqF_{i-1,a}(s) + spF_{i+1,a}(s), \qquad 1 \le i \le a - 2,$$

$$F_{a-1,a}(s) = sp + sqF_{a-2,a}(s),$$

$$F_{a,a}(s) = s.$$

Thus if we let $V_i(s) = F_{ia}(s)$ for $i = 1, 2, \ldots, a - 1$ with $V_0(s) \equiv 0$ and $V_a(s) \equiv 1$ we have for $1 \le i \le a - 1$,

$$V_i(s) = sqV_{i-1}(s) + spV_{i+1}(s),$$

the same form as Eq. (6.2.10) but with different boundary conditions. In this case the procedure above gives, for $i = 1, 2, \ldots, a - 1$,

$$F_{ia}(s) = \frac{\lambda_1^i(s) - \lambda_2^i(s)}{\lambda_1^a(s) - \lambda_2^a(s)}.$$

[Alternatively, this result can be derived from Eq. (6.2.13), since, by symmetry considerations, $f_{ia}^{(n)}$ can be found from $f_{i0}^{(n)}$ by replacing p, q, and i by q, p, and $a - i$.]

It is not at all obvious that the expressions for $F_{i0}(s)$ and $F_{ia}(s)$ are power series in s. By the binomial theorem we see that if $u = (1 - 4pqs^2)^{1/2}$,

$$\lambda_1^k(s) - \lambda_2^k(s) = (2ps)^{-k}[(1 + u)^k - (1 - u)^k]$$

$$= (2ps)^{-k}\left[\sum_{r=0}^k \binom{k}{r}\{1 - (-1)^r\}u^r\right]$$

$$= 2(2ps)^{-k}(1 - 4pqs^2)^{1/2}\sum_{m=0}^{[(k-1)/2]} \binom{k}{2m+1}(1 - 4pqs^2)^m$$

$$= s^{-k}(1 - 4pqs^2)^{1/2}P_k(s),$$

where $P_k(s)$ is an even polynomial of degree $k - 1$ when k is odd and of degree $k - 2$ when k is even. Thus, upon substitution,

$$F_{i0}(s) = \left(\frac{q}{p}\right)^i \frac{s^i P_{a-i}(s)}{P_a(s)}, \qquad F_{ia}(s) = \frac{s^{a-i} P_i(s)}{P_a(s)},$$

showing that both g.f.'s are the ratios of two polynomials whose degrees differ by at most 1. Consequently it is possible to derive explicit expressions for the probabilities using the method of partial fractions.

We can simply the calculations by introducing an auxiliary function $\alpha(s)$ defined by

$$1/2(pq)^{1/2}s = \cos \alpha(s). \qquad (6.2.14)$$

[For $0 < s < 1$ there correspond complex values of $\alpha(s)$, but this has no effect on the calculation.]

Reexpressing Eqs. (6.2.11) using $\alpha = \alpha(s)$, we obtain

$$\lambda_1(s) = (q/p)^{1/2}(\cos \alpha + i \sin \alpha) = (q/p)^{1/2}e^{i\alpha},$$

$$\lambda_2(s) = (q/p)^{1/2}(\cos \alpha - i \sin \alpha) = (q/p)^{1/2}e^{-i\alpha}.$$

Substitution in Eq. (6.2.12) yields for $i = 1, 2, \ldots, a - 1$,

$$F_{i0}(s) = \left(\frac{q}{p}\right)^{i/2} \frac{\sin(a - i)\alpha}{\sin a\alpha}.$$

We now make use of the classical identity

$$\frac{\sin a\alpha}{\sin \alpha} = \prod_{v=1}^{a-1} \left[\cos \alpha - \cos\left(\frac{\pi v}{a}\right)\right]$$

and note that the numbers $\cos(\pi v/a)$, $v = 1, \ldots, a - 1$ are real and distinct. Thus

$$F_{i0}(s) = \left(\frac{q}{p}\right)^{i/2} \frac{\sin(a - i)\alpha}{\sin \alpha} \prod_{v=1}^{a-1} \frac{1}{\cos \alpha - \cos(\pi v/a)}$$

$$= \left(\frac{q}{p}\right)^{i/2} \sum_{v=1}^{a-1} \frac{A_{iv}}{\cos \alpha - \cos(\pi v/a)}. \qquad (6.2.15)$$

Observe that

$$\frac{1}{\cos \alpha - \cos(\pi v/a)} = \frac{2(pq)^{1/2}s}{1 - 2(pq)^{1/2}s \cos(\pi v/a)}, \qquad (6.2.16)$$

and thus the partial fraction expansion in Eq. (6.2.15) is valid since there are no repeated factors.

To determine the A_{iv} we require

$$\frac{\sin(a-i)\alpha}{\sin\alpha} \equiv \sum_{v=1}^{a-1} A_{iv} \prod_{r \neq v} \left[\cos\alpha - \cos\left(\frac{\pi r}{a}\right) \right].$$

Setting $\alpha = \pi v/a$ we obtain

$$A_{iv} \prod_{r=1, r \neq v}^{a-1} \left[\cos\left(\frac{\pi v}{a}\right) - \cos\left(\frac{\pi r}{a}\right) \right] = \frac{\sin[\pi(a-i)v/a]}{\sin[\pi v/a]}.$$

But

$$\prod_{r=1, r \neq v}^{a-1} \left[\cos\left(\frac{\pi v}{a}\right) - \cos\left(\frac{\pi r}{a}\right) \right] = \lim_{\alpha \to \pi v/a} \frac{\sin a\alpha}{\sin\alpha(\cos\alpha - \cos(\pi v/a))}$$

$$= \frac{a(-1)^{v+1}}{\sin^2(\pi v/a)},$$

using L'Hospitals rule, and thus

$$A_{iv} = \frac{(-1)^{v+1}}{a} \sin\left(\frac{\pi v}{a}\right) \sin\left(\frac{\pi(a-i)v}{a}\right).$$

Finally, using Eqs. (6.2.15) and (6.2.16)

$$F_{i0}(s) = \left(\frac{q}{p}\right)^{i/2} \sum_{v=1}^{a-1} A_{iv} \sum_{n=1}^{\infty} 2^n p^{n/2} q^{n/2} \left[\cos\left(\frac{\pi v}{a}\right) \right]^{n-1} s^n,$$

which upon substitution of the A_{iv} and extraction of the coefficient of s^n yields

$$f_{i0}^{(n)} = \frac{2^n}{a} p^{(n-i)/2} q^{(n+i)/2} \sum_{v=1}^{a-1}$$

$$\times (-1)^{v+1} \sin\left(\frac{\pi v}{a}\right) \sin\left(\frac{\pi(a-i)v}{a}\right) \left[\cos\left(\frac{\pi v}{a}\right) \right]^{n-1}.$$

This is the same result as given in Example 6.2.3 since

$$(-1)^{v+1} \sin\left(\frac{\pi(a-i)v}{a}\right) = \sin\left(\frac{\pi iv}{a}\right).$$

A similar determination gives the required expression for $f_{ia}^{(n)}$.   □

EXAMPLE 6.2.5:   *Random Walk on the Nonnegative Integers with an Absorbing Barrier at the Origin.*   The previous example can be generalized to apply to random walks on the nonnegative integers with an absorbing barrier at 0. A particle starting at position $i > 0$ is eventually absorbed at the origin or else the random walk continues for ever. In terms of the gambler's ruin model, absorption corresponds to the ruin of a gambler with initial capital of i units playing against an infinitely rich opponent.

Let $F_{i0}(s)$ be the generating function of the $f_{i0}^{(n)}$, the probability that absorption takes place at the origin exactly at the nth trial starting from position i.

By taking the limit as $a \to \infty$ in Example 6.2.4 we have, from Eq. (6.2.12), for $i = 1, 2, \ldots$,

$$F_{i0}(s) = \lim_{a \to \infty} \frac{\lambda_1^a(s)\lambda_2^i(s) - \lambda_1^i(s)\lambda_2^a(s)}{\lambda_1^a(s) - \lambda_2^a(s)}$$

$$= \lambda_2^i(s) + \{\lambda_2^i(s) - \lambda_1^i(s)\} \lim_{a \to \infty} \frac{1}{\left(\frac{\lambda_1(s)}{\lambda_2(s)}\right)^a - 1}.$$

Observe that

$$\frac{\lambda_1(s)}{\lambda_2(s)} = \frac{1 + (1 - 4pqs^2)^{1/2}}{1 - (1 - 4pqs^2)^{1/2}} > 1 \quad \text{for} \quad 0 < s \le 1$$

and thus the last term is zero and

$$F_{i0}(s) = \lambda_2^i(s). \tag{6.2.17}$$

Alternatively this result could have been derived by observing that the $F_{i0}(s)$ [or, equivalently, the $U_i(s)$ for $i = 1, 2, \ldots$] satisfy Eq. (6.2.10) and hence must have a solution of the form

$$U_i(s) = A(s)\lambda_1^i(s) + B(s)\lambda_2^i(s).$$

However, for all $0 < s < 1$, $\lambda_1(s) > 1/2ps$, which implies that this solution is unbounded at infinity unless $A(s) = 0$. The unique boundary condition is now $U_0(s) = 1$ and Eq. (6.2.17) follows.

Observe that

$$F_{10}(s) = \lambda_2(s) = \frac{1 - (1 - 4pqs^2)^{1/2}}{2ps}$$

$$= \frac{1}{2ps} \sum_{n=1}^{\infty} \binom{1/2}{n}(-1)^{n+1}(4pq)^n s^{2n}$$

$$= \sum_{n=1}^{\infty} \frac{1}{n}\binom{2n-2}{n-1}p^{n-1}q^n s^{2n-1}$$

(cf. Example 3.3.2), so that

$$f_{10}^{(2n-1)} = \frac{1}{n}\binom{2n-2}{n-1}p^{n-1}q^n \quad (n \ge 1),$$

$$f_{10}^{(2n)} = 0 \quad (n \ge 0).$$

Furthermore, $F_{i0}(s) = [F_{10}(s)]^i$ so that the waiting time for absorption at the origin starting at position i can be regarded as the sum of i independent waiting times between the successive first passages through $i - 1$, $i - 2, \ldots, 1, 0$. In particular,

$$F_{ij}(s) = \lambda_2^{i-j}(s) \qquad \text{for} \quad i > j \geq 0.$$

This result can also be applied to first passage probabilities in an *unrestricted* random walk. If we move the origin to position i, then in a random walk on the entire line $f_{i0}^{(n)}$ is the probability that starting at the origin a first visit to the point $-i < 0$ takes place at the nth trial and can be regarded as the sum of i independent waiting times between the successive first passages through $-1, -2, \ldots, -i$. $\square$

EXAMPLE 6.2.6: *Random Walk on the Nonnegative Integers with a Reflecting Barrier at the Origin.*   Consider the random walk of Example 6.2.5 with the absorbing barrier at the origin replaced by a reflecting barrier. (This is commonly called a reflecting barrier at the origin, but in accordance with Example 5.1.2 it should be more correctly called a reflecting barrier at -1.) The effective state space is $S = \{0, 1, 2, \ldots\}$ with transition matrix given by

$$P = \begin{bmatrix} q & p & 0 & 0 & 0 & \cdots \\ q & 0 & p & 0 & 0 & \cdots \\ 0 & q & 0 & p & 0 & \cdots \\ 0 & 0 & q & 0 & p & \cdots \\ \vdots & \vdots & \vdots & \vdots & \vdots & \end{bmatrix}.$$

As in Example 6.2.4 we use Theorem 6.2.3 as our starting point to find expressions for the n-step first passage time probabilities. Let us consider finding the $f_{i0}^{(n)}$ via their generating functions $F_{i0}(s)$:

$$F_{00}(s) = sq + spF_{10}(s),$$

$$F_{10}(s) = sq + spF_{20}(s),$$

$$F_{i0}(s) = sqF_{i-1, 0}(s) + spF_{i+1, 0}(s), \qquad i = 2, 3, \ldots.$$

Using the same technique as in Examples 6.2.4 and 6.2.5, let $U_i(s) = F_{i0}(s)$ for $i = 1, 2, \ldots$. Then

$$U_i(s) = sqU_{i-1}(s) + spU_{i+1}(s) \qquad (i \geq 1),$$

with $U_0(s) = 1$. Using the same reasoning as in Example 6.2.5

$$U_i(s) = \lambda_2^i(s) \qquad \text{for} \quad i \geq 0,$$

leading to

$$F_{i0}(s) = \lambda_2^i(s) \qquad \text{for} \quad i \geq 1.$$

From the first boundary condition

$$
\begin{aligned}
F_{00}(s) &= sq + spF_{10}(s) \\
&= sq + sp\lambda_2(s) \\
&= sq + \tfrac{1}{2}[1 - (1 - 4pqs^2)^{1/2}].
\end{aligned}
$$

Thus the first passage time probabilities $f_{10}^{(n)}$ are the same as for the absorbing barrier case for $i \geq 1$ and for $i = 0$,

$$
f_{00}^{(1)} = q,
$$

$$
f_{00}^{(2n)} = \frac{1}{n}\binom{2n-2}{n-1}(pq)^n, \qquad n = 1, 2, \ldots,
$$

$$
f_{00}^{(2n+1)} = 0, \qquad\qquad n = 1, 2, \ldots. \quad \square
$$

Observe that for fixed values of j, Eq. (6.2.9) gives a linear relationship between the $F_{ij}(s)$ as i varies over S. Using matrix theory it is possible to obtain an explicit expression for $F_{ij}(s)$ when S is finite.

THEOREM 6.2.4:   Let P be the transition matrix of a finite MC. Then for $|s| < 1$, and each $i, j \in S$,

$$
F_{ij}(s) = s\mathbf{e}_i'[I - sP_j(0)]^{-1}P\mathbf{e}_j, \tag{6.2.18}
$$

where $P_j(0)$ is P with the jth column replaced by zeros.

Proof: Suppose $S = \{1, 2, \ldots, m\}$ and $P = [p_{ij}]$. Let

$$
\mathbf{F}_j'(s) \equiv (F_{1j}(s), F_{2j}(s), \ldots, F_{mj}(s)), \qquad \mathbf{p}_j' = (p_{1j}, p_{2j}, \ldots, p_{mj}).
$$

By expressing Eq. (6.2.9) in vector form we obtain

$$
\mathbf{F}_j(s) = s\mathbf{p}_j + sP_j(0)\mathbf{F}_j(s)
$$

and hence

$$
\mathbf{F}_j(s) = s[I - sP_j(0)]^{-1}\mathbf{p}_j.
$$

Equation (6.2.18) follows by noting that $F_{ij}(s) = \mathbf{e}_i'\mathbf{F}_j(s)$ and $\mathbf{p}_j = P\mathbf{e}_j$. The existence of the inverse follows since $\rho(sP_j(0)) < 1$, as in the proof of Theorem 6.1.8. [In fact, the theorem holds for $s = 1$ when P is irreducible since $0 \leq P_j(0) \leq P$ and $\rho(P_j(0)) < 1$ by Theorem 4.6.1(f).]

[Alternatively, let $\mathbf{f}_j^{(n)'} \equiv (f_{1j}^{(n)}, f_{2j}^{(n)}, \ldots, f_{mj}^{(n)})$. Then from Eq. (6.1.2) $\mathbf{f}_j^{(n)} = P_j(0)\mathbf{f}_j^{(n-1)}$ $(n \geq 2)$ with $\mathbf{f}_j^{(1)} = \mathbf{p}_j$, implying $\mathbf{f}_j^{(n)} = [P_j(0)]^{n-1}\mathbf{p}_j$ for $n \geq 1$. Then

$$
\begin{aligned}
\mathbf{F}_j(s) &= \sum_{n=0}^{\infty} \mathbf{f}_j^{(n)}s^n = s\left[\sum_{n=0}^{\infty} \{sP_j(0)\}^n\right]\mathbf{p}_j \\
&= s[I - sP_j(0)]^{-1}\mathbf{p}_j \qquad \text{(by Theorem 4.5.4).]} \quad \square
\end{aligned}
$$

If we have already obtained $P_{ij}(s)$ in deriving the n-step transition probabilities $p_{ij}^{(n)}$ (as in Theorem 6.1.8) we can use such a result to determine the generating function $F_{ij}(s)$, or vice versa, as shown by the following theorem.

THEOREM 6.2.5: For all $i, j \in S$ and $|s| < 1$,

$$F_{ij}(s) = \frac{P_{ij}(s) - \delta_{ij}}{P_{jj}(s)}, \tag{6.2.19}$$

$$P_{ij}(s) = \frac{F_{ij}(s)}{1 - F_{jj}(s)} + \delta_{ij}. \tag{6.2.20}$$

In particular,

$$F_{jj}(s) = 1 - \frac{1}{P_{jj}(s)}, \qquad P_{jj}(s) = \frac{1}{1 - F_{jj}(s)}.$$

Proof: Recall that in Theorem 5.2.4 we derived the following equations [Eqs. (5.2.3) and (5.2.4) respectively]: for $n \geq 1$,

$$p_{ij}^{(n)} = \sum_{k=1}^{n} f_{ij}^{(k)} p_{jj}^{(n-k)}$$

and

$$p_{ij}^{(n)} = f_{ij}^{(n)} + \sum_{k=1}^{n-1} f_{jj}^{(k)} p_{ij}^{(n-k)}.$$

When $i = j$ both of these above equations imply that

$$\{p_{jj}^{(n)}\} = \{f_{jj}^{(n)}\} * \{p_{jj}^{(n)}\} + \{\delta_{n0}\}.$$

Forming generating functions with the help of Theorem 2.5.2 gives

$$P_{jj}(s) = F_{jj}(s)P_{jj}(s) + 1$$

and Eqs. (6.2.19) and (6.2.20) both follow.

When $i \neq j$, Eq. (5.2.3) implies that $\{p_{ij}^{(n)}\} = \{f_{ij}^{(n)}\} * \{p_{jj}^{(n)}\}$ and hence that $P_{ij}(s) = F_{ij}(s)P_{jj}(s)$, leading to Eqs. (6.2.19) and (6.2.20). Alternatively Eq. (5.2.4) gives $\{p_{ij}^{(n)}\} = \{f_{ij}^{(n)}\} + \{f_{jj}^{(n)}\} * \{p_{ij}^{(n)}\}$. Thus $P_{ij}(s) = F_{ij}(s) + F_{jj}(s)P_{ij}(s)$, which is equivalent to Eq. (6.2.20).

[Note also that these results can be derived by considering the recurrent event "return to state j" and identifying $U(s) = P_{jj}(s)$, $V(s) = P_{ij}(s)$ (if $i \neq j$), $F(s) = F_{jj}(s)$ and $B(s) = F_{ij}(s)$ (if $i \neq j$), in accordance with the connections established in Section 5.2. Equations (6.2.19) and (6.2.20) are then just restatements of Eq. (3.3.2) (when $i = j$) and Eq. (3.5.4) (when $i \neq j$).]   □

We complete this section with a further technique for the derivation of the $f_{ij}^{(n)}$. In Section 6.1 we used matrix-generating functions to find the n-step transition probabilities. A similar approach can be used to find the n-step first passage time probabilities in finite-state MC's. This method is of use when the $p_{ij}^{(n)}$ have already been determined using Theorem 6.1.8.

THEOREM 6.2.6: Let P be the transition matrix of a finite-state MC and let $\mathbf{P}(s) = [I - sP]^{-1}$.

If $\mathbf{F}(s) = \left[\sum_{n=0}^{\infty} f_{ij}^{(n)} s^n \right]$, then, for $|s| < 1$,

$$\mathbf{F}(s) = [\mathbf{P}(s) - I][\mathbf{P}_d(s)]^{-1}. \qquad (6.2.21)$$

Proof: Observe that

$$\mathbf{F}(s) = \sum_{n=0}^{\infty} [f_{ij}^{(n)}] s^n = \sum_{n=0}^{\infty} F^{(n)} s^n$$

$$= sP + \sum_{n=2}^{\infty} s^n [P(F^{(n-1)} - F_d^{(n-1)})] \qquad \text{[using Eq. (6.2.3)]}$$

$$= sP + sP[\mathbf{F}(s) - \mathbf{F}_d(s)].$$

Then, since $[I - sP]^{-1}$ exists,

$$\mathbf{F}(s) = [I - sP]^{-1} sP[I - \mathbf{F}_d(s)] = [\mathbf{P}(s) - I][I - \mathbf{F}_d(s)], \qquad (6.2.22)$$

since

$$\mathbf{P}(s) - I = [I - sP]^{-1} - [I - sP]^{-1}[I - sP]$$
$$= [I - sP]^{-1}[I - I + sP]$$
$$= [I - sP]^{-1} sP.$$

Forming the matrices of diagonal elements from both sides of Eq. (6.2.22), we obtain

$$\mathbf{F}_d(s) = [\mathbf{P}_d(s) - I][I - \mathbf{F}_d(s)].$$

Solving for $\mathbf{F}_d(s)$ gives $\mathbf{F}_d(s) = I - [\mathbf{P}_d(s)]^{-1}$ and $I - \mathbf{F}_d(s) = [\mathbf{P}_d(s)]^{-1}$. Substitution into Eq. (6.2.21) leads to Eq. (6.2.21). Alternatively, since $\mathbf{F}(s) = [F_{ij}(s)]$, we have from Theorem 6.2.5

$$[F_{ij}(s)P_{jj}(s)] = [P_{ij}(s)] - [\delta_{ij}],$$

yielding $\mathbf{F}(s)\mathbf{P}_d(s) = \mathbf{P}(s) - I$, which immediately gives Eq. (6.2.21). $\square$

Exercises 6.2

1. Let

$$P = \begin{bmatrix} a_1 & a_2 & 0 \\ b_1 & 0 & b_3 \\ 0 & c_2 & c_3 \end{bmatrix}$$

be the transition matrix of a Markov chain. Find expressions for $f_{ij}^{(n)}$, $i, j = 1, 2, 3$, using

(a) direct probabilistic arguments,
(b) recursive arguments,
(c) matrix generating functions.

2. Let

$$P = \begin{bmatrix} p_{11} & p_{12} \\ p_{21} & p_{22} \end{bmatrix}$$

be a two-state Markov chain. Find expressions for the $f_{ij}^{(n)}$ using the methods of Theorems 6.2.4 and 6.2.6. In particular, show that

$$\mathbf{F}(s) = \begin{bmatrix} \dfrac{s(p_{11} - ds)}{1 - p_{22}s} & \dfrac{p_{12}s}{1 - p_{11}s} \\ \dfrac{p_{21}s}{1 - p_{22}s} & \dfrac{s(p_{22} - ds)}{1 - p_{11}s} \end{bmatrix},$$

where $d = 1 - p_{12} - p_{21}$. (Check your results with Theorem 5.1.8.)

3. Let

$$P = \begin{bmatrix} I & 0 \\ R & Q \end{bmatrix}$$

be the transition matrix of an absorbing MC in canonical form. Using the technique of Theorem 6.2.6, show that

$$\mathbf{F}(s) = \begin{bmatrix} sI & 0 \\ sQ(s)R & \{Q(s) - I\}\{Q_d(s)\}^{-1} \end{bmatrix},$$

where $Q(s) = (I - sQ)^{-1}$.

*4. Let $\{X_n, n \geq 0\}$ be a MC with state space S and transition matrix $P = [p_{ij}]$. For all $i, j \in S$, define

$$l_{ij}^{(n)} = P\{X_n = j, X_k \neq i \text{ for } k = 1, 2, \ldots, n - 1 | X_0 = i\} \qquad (n \geq 1),$$

i.e., $l_{ij}^{(n)}$ is the probability of going from $i \in S$ to $j \in S$ in n steps without revisiting state i in the meantime; it is an example of a *taboo* probability. Observe that $l_{ii}^{(n)} = f_{ii}^{(n)}$ ($n \geq 1$). Establish the following results:

(a) $l_{ij}^{(1)} = p_{ij}$, $l_{ij}^{(n+1)} = \sum\limits_{r \in S, r \neq i} l_{ir}^{(n)} p_{rj}$ $\qquad (n \geq 1)$.

(b) Using induction, show that for $n \geq 1$, with $l_{ij}^{(0)} \equiv 0$,

$$p_{ij}^{(n)} = \sum_{k=0}^{n} p_{ii}^{(k)} l_{ij}^{(n-k)}.$$

(c) $l_{ij}^{(n)} \leq p_{ij}^{(n)} (\leq 1)$; thus for $|s| < 1$ the generating functions $L_{ij}(s) \equiv \sum_{n=1}^{\infty} l_{ij}^{(n)} s^n$ are well defined; $i, j \in S$.

(d)   For $|s| < 1$,

$$P_{ii}(s) - 1 = P_{ii}(s) L_{ii}(s), \qquad P_{ij}(s) = P_{ii}(s) L_{ij}(s), \qquad i \neq j.$$

(e)   For $|s| < 1$, for all $i, j \in S$,

$$L_{ij}(s) = s \sum_r L_{ir}(s) p_{rj} + s p_{ij}(1 - L_{ii}(s)).$$

(f)   If $\mathbf{L}(s) = [L_{ij}(s)]$, then $\mathbf{L}(s) = [\mathbf{P}_d(s)]^{-1}[\mathbf{P}(s) - I]$.

5. Consider an *unrestricted* random walk on the integers $0, \pm 1, \pm 2, \ldots$, where $p_{i,i-1} = q$ and $p_{i,i+1} = p$ for all integers i $(p + q = 1)$. Using the method of Theorem 6.2.5 show that, for all i,

$$F_{ii}(s) = \sum_{n=1}^{\infty} f_{ii}^{(n)} s^n = 1 - (1 - 4pqs^2)^{1/2},$$

and hence find an expression for the $\{f_{ii}^{(n)}\}$.

6. Two players A and B with \$1 and \$2, respectively, agree to play not more than n equitable games staking \$1 at each game (i.e., one of the players either wins with probability $\frac{1}{2}$ or loses with probability $\frac{1}{2}$ at each game). Find the probability that A loses the series at the nth game. Show that the probability that A is ruined by the nth game is

$$\frac{2}{3} - \frac{3 + (-1)^n}{3 \cdot 2^{n+1}}.$$

7. Let $\{X_n\}$ $(n = 0, 1, \ldots)$ be a MC with state space $S = \{0, 1, 2, \ldots\}$. Suppose that state 0 is an absorbing state and that for all $j > 0$, $p_{jj} = p$ and $p_{j,j-1} = q$ where $p + q = 1$ $(0 < p < 1)$. Find $f_{j0}^{(n)}$, the probability that absorption occurs at the nth step $(j > 0)$.

6.3   Determination of the Reaching Probabilities f_{ij}

In Section 5.2 we defined f_{ij}, the probability that state j is ever reached from state i (in a finite number of steps) as $f_{ij} = \sum_{n=1}^{\infty} f_{ij}^{(n)}$ and the matrix of "reaching probabilities" as $F = [f_{ij}] = \sum_{n=1}^{\infty} F^{(n)}$.

In this section we examine techniques for finding the f_{ij}, or equivalently F, given the transition matrix $P = [P_{ij}]$ of the MC. The usual method is to attempt to solve the set of equations given earlier in Theorem 5.2.9, i.e., for all $i, j \in S$,

$$f_{ij} = p_{ij} + \sum_{k \neq j} p_{ik} f_{kj} \tag{6.3.1}$$

or, equivalently,

$$(I - P)F = P(I - F_d). \tag{6.3.2}$$

The generalization, given in Corollary 5.2.9A, is often useful, i.e., for all $n \geq 1$,

$$(I - P^n)F = \left(\sum_{k=1}^{n} P^k \right)(I - F_d). \tag{6.3.3}$$

In fact, Eq. (6.3.3) was used in Theorems 5.3.6 and 5.3.7 to show that for an irreducible MC

(a)   if all the states are persistent, then $f_{ij} = 1$ for all $i, j \in S$;

(b)   if all the states are transient, then for all $j \in S$ there exists an $i \neq j$ for which $f_{ij} < 1$.

If generating functions have been used to find $F_{ij}(s) = \sum_{n=0}^{\infty} f_{ij}^{(n)} s^n$, then Theorem 2.3.2(b) (the converse of Abel's convergence theorem) implies that

$$f_{ij} = \lim_{s \uparrow 1} F_{ij}(s) \equiv F_{ij}(1). \tag{6.3.4}$$

Such an approach, using Theorem 6.2.5, leads to the results of Theorem 5.2.5. In particular, using Theorem 2.5.3 we have that

$$f_{ij} \left(\sum_{n=0}^{\infty} p_{jj}^{(n)} \right) = \left(\sum_{n=1}^{\infty} p_{ij}^{(n)} \right), \tag{6.3.5}$$

provided one, and hence both, of $\sum_{n=0}^{\infty} p_{jj}^{(n)}$ and $\sum_{n=1}^{\infty} p_{ij}^{(n)}$ are finite. Furthermore, we obtain from Theorem 6.2.4, for finite MC's,

$$f_{ij} = \mathbf{e}_i' \lim_{s \uparrow 1} \left[I - sP_j(0) \right]^{-1} P\mathbf{e}_j$$

$(= \mathbf{e}_i'[I - P_j(0)]^{-1} P\mathbf{e}_j = 1$ when P is irreducible$)$.

When matrix generating functions have been used

$$F = \lim_{s \uparrow 1} \sum_{n=1}^{\infty} F^{(n)} s^n = \left[\lim_{s \uparrow 1} F_{ij}(s) \right] = \lim_{s \uparrow 1} \mathbf{F}(s),$$

and thus Theorem 6.2.6 implies that

$$F = \lim_{s \uparrow 1} (I - Ps)^{-1} Ps[\{(I - Ps)^{-1}\}_d]^{-1}.$$

This is useful only when $\mathbf{F}(s)$ has been completely determined. Care must be taken with the limit since $I - P$ is singular and hence $\lim_{s \uparrow 1} (I - Ps)^{-1}$ does not exist.

We shall examine some special MC's making use, primarily, of the connection between $P = [p_{ij}]$ and $F = [f_{ij}]$, as given by Eq. (6.3.1) or Eq. (6.3.2).

EXAMPLE 6.3.1:   *Two-State Markov Chain.*   The derivation of $F = [f_{ij}]$ for this MC was considered earlier in Theorem 5.2.3. If

$$P = \begin{bmatrix} 1 - a & a \\ b & 1 - b \end{bmatrix} \quad (0 \leq a \leq 1, 0 \leq b \leq 1),$$

then it can be shown that Eq. (6.3.2) is satisfied with

$$F = \begin{bmatrix} 1 - a\delta_{b0} & 1 - \delta_{a0} \\ 1 - \delta_{b0} & 1 - b\delta_{a0} \end{bmatrix},$$

where δ is the Kronecker delta.   □

THEOREM 6.3.1:   For an absorbing MC with transition matrix

$$P = \begin{bmatrix} I & 0 \\ R & Q \end{bmatrix}, \quad F = \begin{bmatrix} I & 0 \\ NR & (N - I)N_d^{-1} \end{bmatrix},$$

where $N = (I - Q)^{-1}$.

Proof: Firstly let us derive these results from first principles using the results of Theorem 6.2.1.

With appropriate partitioning,

$$F = \begin{bmatrix} F_1 & F_2 \\ F_3 & F_4 \end{bmatrix},$$

where

$$F_1 = \sum_{n=1}^{\infty} F_1^{(n)} = I,$$

$$F_2 = \sum_{n=1}^{\infty} F_2^{(n)} = 0,$$

$$F_3 = \sum_{n=1}^{\infty} F_3^{(n)} = \sum_{n=1}^{\infty} Q^{n-1}R = (I - Q)^{-1}R = NR$$

[where we have made use of Theorem 4.5.4 and the fact that $(I - Q)^{-1}$ exists as shown in Corollary 6.1.1A],

$$F_4 = \sum_{n=1}^{\infty} F_4^{(n)} = Q + \sum_{n=2}^{\infty} \{F_4^{(n-1)} - F_{4d}^{(n-1)}\} = Q + Q(F_4 - F_{4d}).$$

Thus $(I - Q)F_4 = Q(I - F_{4d})$,   implying $F_4 = (I - Q)^{-1}Q(I - F_{4d}) = NQ(I - F_{4d})$. But $N = (I - Q)^{-1}$ and thus $N(I - Q) = I$ or $N - I = NQ$. Therefore $F_4 = (N - I)(I - F_{4d}) = N - I - NF_{4d} + F_{4d}$. Taking the diagonal elements leads to the expression $F_{4d} = N_d - I - N_d F_{4d} + F_{4d}$ or

$I = N_d(I - F_{4d})$, implying $N_d^{-1} = I - F_{4d}$, and hence $F_4 = NQN_d^{-1} = (N - I)N_d^{-1}$.

As an alternative proof, consider solving Eq. (6.2.2) directly. Substitution of P and F as partitioned leads to

$$\begin{bmatrix} 0 & 0 \\ -R & I - Q \end{bmatrix}\begin{bmatrix} F_1 & F_2 \\ F_3 & F_4 \end{bmatrix} = \begin{bmatrix} I & 0 \\ R & Q \end{bmatrix}\begin{bmatrix} I - F_{1d} & 0 \\ 0 & I - F_{4d} \end{bmatrix}.$$

The $(1, 1)$ block gives $0 = I - F_{1d}$ and hence $F_{1d} = I$. ($F_1 = [f_{ij}]_{i \in A, j \in A}$, where A is the set of absorbing states and for such a set $f_{ij} = 0$ for $i \neq j$.) Consequently $F_1 = I$.

The $(1, 2)$ block gives no useful information.

The $(2, 1)$ block gives $-RF_1 + (I - Q)F_3 = R(I - F_{1d}) = 0$ and hence $F_3 = (I - Q)^{-1}RF_1 = NR$.

The $(2, 2)$ block gives $-RF_2 + (I - Q)F_4 = Q(I - F_{4d})$. ($F_2 = [f_{ij}]_{i \in A, j \in T}$, where T is the set of transient states. For such a transition it is obvious that $f_{ij} = 0$.) Thus $F_2 = 0$, implying

$$(I - Q)F_4 = Q(I - F_{4d}).$$

This leads to $F_4 = (N - I)N_d^{-1}$, as above. $\square$

As can be seen, the matrix $N \equiv (I - Q)^{-1}$ plays an important role for absorbing Markov chains and it often called the *fundamental matrix* of the absorbing Markov chain.

Let us extend this theorem to obtain a general form for F when the transition matrix P of a MC is expressed in its general canonical form.

THEOREM 6.3.2: Let

$$P = \begin{bmatrix} P_1 & 0 & \cdots & 0 & 0 \\ 0 & P_2 & \cdots & 0 & 0 \\ \vdots & \vdots & & \vdots & \vdots \\ 0 & 0 & \cdots & P_k & 0 \\ R_1 & R_2 & \cdots & R_k & Q \end{bmatrix}$$

be the transition matrix of a MC expressed in general canonical form. The matrix of "reaching probabilities" has the following form:

$$F = \begin{bmatrix} F_1 & 0 & \cdots & 0 & 0 \\ 0 & F_2 & \cdots & 0 & 0 \\ \vdots & \vdots & & \vdots & \vdots \\ 0 & 0 & \cdots & F_k & 0 \\ F_{T1} & F_{T2} & \cdots & F_{Tk} & F_{TT} \end{bmatrix},$$

where each F_i $(i = 1, \ldots, k)$ is a matrix of 1's, $F_i = E$;

$$F_{Ti} = NR_iE \qquad (i = 1, \ldots, k),$$

$$F_{TT} = (N - I)N_d^{-1},$$

where $N = (I - Q)^{-1}$, the fundamental matrix of the MC. □

Proof: The form of $F = \sum_{n=1}^{\infty} F^{(n)}$ follows from Theorem 6.2.2. From this same theorem we deduce that if $F_i = \sum_{n=1}^{\infty} F_i^{(n)}$, then $(I - P_i)F_i = P_i(I - F_{id})$. But P_i is the transition matrix of an irreducible subchain and, hence, by Theorem 5.3.6, $F_i = E$. Further, $F_{Ti} = \sum_{n=1}^{\infty} F_{Ti}^{(n)}$, and from Theorem 6.2.2, $F_{Ti} = R_i + R_i(F_i - F_{id}) + QF_{Ti}$, leading to $F_{Ti} = NR_iE$.

Finally, $F_{TT} = \sum_{n=1}^{\infty} Q^{(n)} = Q + Q(F_{TT} - F_{TTd})$, leading to $F_{TT} = (N - I)N_d^{-1}$, as in the proof of Theorem 6.3.1. □

Note that if $F_{Ti} = [f_{rs}]$, then f_{rs} is the probability that state $s \in C_i$, the ith equivalence class, is ever reached from $r \in T$, the transient states. From above, $F_{Ti} = NR_iE$, so that for $r \in T$, $s \in C_i$, $f_{rs} = \sum_{t \in C_i} (NR_i)_{rt}$, which does not depend on s. Of course, if $s \in C_i$ is ever reached so are all the states of C_i since C_i is an irreducible set MC (cf. Theorem 5.3.6); and thus for each $s \in C_i$, $f_{rs} = f_{rC_i}$, *the probability that the equivalence class C_i is ever reached from $r \in T$.*

COROLLARY 6.3.2A: Under the conditions of Theorem 6.3.2, if

$$F_C = [f_{rC_i}]_{r \in T, i \in \{1, \ldots, k\}}$$

and

$$R_C = [p_{rC_i}]_{r \in T, i \in \{1, \ldots, k\}}$$

where $p_{rC_i} = \sum_{t \in C_i} p_{rt}$, then $F_C = NR_C$.

Proof: Since

$$R_i = [p_{rs}]_{r \in T, s \in C_i} \qquad \text{and} \qquad N = [n_{rs}]_{r \in T, s \in T},$$

then, for $r \in T$,

$$f_{rC_i} = \sum_{t \in C_i} \left(\sum_{j \in T} n_{rj} p_{jt} \right) = \sum_{j \in T} n_{rj} \left(\sum_{t \in C_i} p_{jt} \right)$$

$$= \sum_{j \in T} n_{rj} p_{rC_i},$$

implying that $F_C = NR_C$.

Alternatively, if in the canonical form of P we replace each P_i by 1 and R_i by the column vector $[p_{rC_i}]_{r \in T}$, we have reduced the canonical form to that of an absorbing MC and consequently we can use the results of Theorem 6.3.1 to obtain the required result $F_C = NR_C$.

Furthermore, observe that direct from Eq. (6.3.1) if $r \in T$, $s \in C_i$,

$$f_{rC_i} = f_{rs} = p_{rs} + \sum_{t \in C_i, t \ne s} p_{rt} f_{ts} + \sum_{t \in C_j, j \ne i} p_{rt} f_{ts} + \sum_{t \in T} p_{rt} f_{ts}$$

$$= p_{rs} + \sum_{t \in C_i, t \ne s} p_{rt} \cdot 1 + \sum_{t \in C_j, j \ne i} p_{rt} \cdot 0 + \sum_{t \in T} p_{rt} f_{tC_i},$$

i.e.,

$$f_{rC_i} = p_{rC_i} + \sum_{t \in T} p_{rt} f_{tC_i}. \tag{6.3.6}$$

Equation (6.3.6) implies that $F_C = R_C + QF_C$, leading to $F_C = NR_C$.   □

When we are faced with an infinite MC we cannot use the matrix solution and we need to examine the solutions of Eq. (6.3.6) directly. We shall examine this case later. Related to this is the problem of determining h_r, *the probability that the MC remains forever in the transient states T if it starts in $r \in T$.*

*THEOREM 6.3.3:   For any MC let

$$h_r^{(n)} = P\{X_n \in T \,|\, X_0 = r\} \qquad (n \ge 1, r \in T),$$
$$h_r = P\{X_n \in T \text{ for all } n \,|\, X_0 = r\} \qquad (r \in T).$$

Then

(a) $h_r^{(n+1)} = \sum_{s \in T} p_{rs} h_s^{(n)},$

(b) $h_r = \lim_{n \to \infty} h_r^{(n)}$ exists,

(c) $h_r = \sum_{s \in T} p_{rs} h_s,$

(d) h_r is given by the *maximal* solution of

$$x_r = \sum_{s \in T} p_{rs} x_s, \tag{6.3.7}$$

where $0 \le x_r \le 1$ for all $r \in T$.

Proof:

(a) $h_r^{(n+1)} = \sum_{k \in T} P\{X_{n+1} = k \,|\, X_0 = r\} = \sum_{k \in T} p_{rk}^{(n+1)}$

$$= \sum_{k \in T} \left[\sum_{s \in T} p_{rs} p_{sk}^{(n)} \right] \qquad \text{(by Theorem 6.1.1),}$$

$$= \sum_{s \in T} \left[\sum_{k \in T} p_{rs} p_{sk}^{(n)} \right] \qquad \text{(by Theorem 2.2.4),}$$

$$= \sum_{s \in T} p_{rs} h_s^{(n)}.$$

(b) Define $h_r^{(0)} = 1$ for all r. From this it follows that $h_r^{(1)} = \sum_{s \in T} p_{rs} \leq h_r^{(0)}$. Using induction, it can be shown that for all states $r \in T$, $h_r^{(n+1)} \leq h_r^{(n)}$ (Exercise 6.3.9). Hence, since $0 \leq h_r^{(n)} \leq 1$ and the sequence is monotone nonincreasing, $\lim_{n \to \infty} h_r^{(n)}$ exists. Obviously this limit is h_r.

(c) Since $|p_{rs} h_s^{(n)}| \leq p_{rs}$ and $\sum_{s \in T} p_{rs} < \infty$, application of Theorem 2.2.5, the dominated convergence theorem, to (a) above yields $h_r = \sum_{s \in T} p_{rs} h_s$.

(d) If the system of Eqs. (6.3.7) had a unique solution we would have a convenient way of finding $\{h_r\}_{r \in T}$. Unfortunately, there are, in general, many solutions to this system of equations, so additional information must be used in order to find the solution corresponding to the $\{h_r\}$. (There is always at least one solution, i.e., $x_r = 0$ for all $r \in T$.)

Let $\{x_r\}$ be any solution of the system $x_r = \sum_{s \in T} p_{rs} x_s$ with $0 \leq x_r \leq 1$. We use induction to show that for all $r \in T$ and all n, $x_r \leq h_r^{(n)}$. Firstly, $0 \leq x_r \leq h_r^{(0)} = 1$ and thus

$$x_r = \sum_{s \in T} p_{rs} x_s \leq \sum_{s \in T} p_{rs} h_s^{(0)} = h_r^{(1)}.$$

Consequently if $x_r \leq h_r^{(n)}$ for all $r \in T$,

$$x_r = \sum_{s \in T} p_{rs} x_s \leq \sum_{s \in T} p_{rs} h_s^{(n)} = h_r^{(n+1)}.$$

This implies that $x_r \leq h_r^{(n)}$ for all n and since $h_r^{(n)} \downarrow h_r$ we can deduce that $x_r \leq h_r$. Hence among all the solutions of Eqs. (6.3.7) with $0 \leq x_r \leq 1$, the $\{h_r\}$ solution is the maximal solution. $\square$

*COROLLARY 6.3.3A: (a) If the MC is finite, $h_r \equiv 0$ for all $r \in T$.

(b) If the MC is infinite, $h_r = 0$ for all $r \in T$ if and only if Eqs. (6.3.7) have no solution $\{x_r\}$ $(0 \leq x_r \leq 1)$, except $x_r = 0$.

*Proof: (a) If $\mathbf{h}_T$ is the column vector of h_r, then $\mathbf{h}_T = Q\mathbf{h}_T$ where $Q = [p_{rs}]_{r \in T, s \in T}$, i.e., $(I - Q)\mathbf{h}_T = \mathbf{0}$. But $(I - Q)^{-1}$ exists by Corollary 6.1.1A and hence $\mathbf{h}_T \equiv \mathbf{0}$.

(b) If $h_r = 0$ for all $r \in T$, then there is no other solution x_r with $0 \leq x_r \leq 1$ since $x_r \leq h_r = 0$. Conversely, if $x_r = 0$ is the only solution with $0 \leq x_r \leq 1$, then $h_r = 0$ for all $r \in T$. $\square$

Thus, for a MC with infinite state space, the probability of remaining forever in the transient states is zero (regardless of the initial state) if and only if the null solution is the only *bounded* solution of the homogeneous equations (6.3.7).

Let us return to the problem of determining f_{rC_i} when the MC is infinite. For the sake of convenience let $f_{rC_i} \equiv g_r$, so that Eq. (6.3.6) can be written as

$$g_r = \sum_{s \in T} p_{rs} g_s + p_{rC_i}.$$

This means that Eq. (6.3.6) is in effect a nonhomogeneous equation with Eq. (6.3.7) as its homogeneous component.

*COROLLARY 6.3.3B: A necessary and sufficient condition for a (bounded) solution f_{rC_i} of Eq. (6.3.6) to be unique is that $h_r = 0$ for all $r \in T$.

Proof: Suppose Eq. (6.3.6) has two solutions, say $g_r^{(1)}$ and $g_r^{(2)}$. Then $g_r^{(1)} - g_r^{(2)}$ is a solution of Eq. (6.3.7). Conversely, if Eq. (6.3.7) has a solution h_r, then $g_r + h_r$ is a solution of Eq. (6.3.6). The result now follows. (The result is always true for finite MC's with the solution given by Corollary 6.3.2A.) □

The states of an infinite irreducible MC are either all transient or all persistent. Using Theorem 6.3.3, we can derive a procedure for determining the nature of such MC's. Without loss of generality the state space of the MC can be taken as the nonnegative integers.

*THEOREM 6.3.4: An irreducible MC with state space $S = \{0, 1, 2, \ldots\}$ is transient if and only if the system of equations

$$h_r = \sum_{s=1}^{\infty} p_{rs} h_s, \qquad r = 1, 2, \ldots \tag{6.3.8}$$

possesses a bounded solution $\{h_r\}$ that is not identically zero.

Proof: From Theorems 5.3.6 and 5.3.7 the MC is persistent iff $f_{j0} = 1$ for $j = 1, 2, \ldots$. Now consider a new MC with the same state space but with transition probabilities the same as the original MC for $i = 1, 2, \ldots$ and $j = 0, 1, 2, \ldots$ but equal to 1 for $i = 0$, $j = 0$ and 0 for $i = 0$, $j = 1, 2, \ldots$. Thus in this new chain state 0 is absorbing and the remaining states $\{1, 2, \ldots\}$ are transient, since by the irreducibility of the original chain state 0 can be reached from every other state. Further, f_{j0} is also the probability of absorption in state 0 in the new chain starting in state j so that $1 - f_{j0}$ is the probability that the MC will remain indefinitely in the (transient) states $\{1, 2, \ldots\}$ starting at j. By Corollary 6.3.3A a necessary and sufficient condition for $1 - f_{j0} > 0$ for some $j = 1, 2, \ldots$ is that there exists a bounded solution to Eq. (6.3.8) that is not identically equal to zero. However, the original MC is transient iff $1 - f_{j0} > 0$ for some $j = 1, 2, \ldots$ and the proof is complete. □

We present a summary of the basic methods and results that we have available for determining the reaching probabilities f_{ij} in a given MC (see Table 6.3.1).

TABLE 6.3.1

TECHNIQUES FOR FINDING f_{ij}

Initial state	Final state	Result or method
i arbitrary	j arbitrary	(a) $f_{ij} = \sum_{n=1}^{\infty} f_{ij}^{(n)}$
		(b) $f_{ij} = p_{ij} + \sum_{k \neq j} p_{ik} f_{kj}$
		(c) $f_{ij} = \lim_{s \uparrow 1} F_{ij}(s)$
		[Eqs. (6.3.1), (6.3.4)]
i persistent	j persistent	$f_{ij} = 1, \quad i \leftrightarrow j$
		$f_{ij} = 0, \quad i \nleftrightarrow j$
		[Theorems 5.3.5, 5.3.6]
i persistent	j transient	$f_{ij} = 0$
		[Theorem 5.3.5]
i transient	j persistent	(a) $F = NR, \quad F_{Ti} = NR_i E, \quad F_C = NR_C$
		[Theorems 6.3.1, 6.3.2; Corollary 6.3.2A]
		(b) $f_{ij} = \sum_{k \in T} p_{ik} f_{kj} + \sum_{k \in C} p_{ik}$
		where C is equivalence class containing j.
		[Eq. (6.3.6)]
i transient	j transient	(a) $F_{TT} = (N - I)N_d^{-1}$
		[Theorems 6.3.1, 6.3.2]
		(b) $f_{ij} = \sum_{k \in T, k \neq j} p_{ik} f_{kj}$
		[Eq. (6.3.1)]
		(c) $f_{ij} = \sum_{n=1}^{\infty} p_{ij}^{(n)} \Big/ \sum_{n=0}^{\infty} p_{jj}^{(n)}$
		Eq. (6.3.5)]

EXAMPLE 6.3.2:   *Classical Gambler's Ruin Model.*   In Example 6.1.4 we showed that the transition matrix of the MC associated with this model can be expressed in canonical form as an absorbing MC with two absorbing states 0 and a, and transient states $1, 2, \ldots a - 1$ [Eq. (6.1.10)].

Let us first focus attention on determining f_{i0}, the probability that the gambler is ultimately ruined (absorbed at 0) when he starts with i units.

A direct application of Eq. (6.3.1) gives

$$f_{i0} = p_{i0} + \sum_{k \neq 0} p_{ik} f_{k0},$$

implying that

$$f_{i0} = pf_{i+1,\,0} + qf_{i-1,\,0} \qquad (1 < i < a - 1),$$
$$f_{10} = pf_{20} + q \qquad (i = 1),$$
$$f_{a-1,\,0} = qf_{a-2,\,0} \qquad (i = a - 1).$$

If we define $f_{0,0} \equiv 1$ and $f_{a,0} \equiv 0$, then

$$pf_{i+2,\,0} - f_{i+1,\,0} + qf_{i0} = 0 \qquad (i = 0, 1, \ldots, a - 2).$$

This is a system of homogeneous difference equations that can be solved using the techniques of Theorem 2.9.1. The general solution is given by

$$f_{i0} = \begin{cases} A\lambda_1^i + B\lambda_2^i & (\lambda_1 \neq \lambda_2), \\ (A + Bi)\lambda_1^i & (\lambda_1 = \lambda_2), \end{cases}$$

where λ_1 and λ_2 are roots of the auxiliary equation

$$p\lambda^2 - \lambda + q = (p\lambda - q)(\lambda - 1) = 0,$$

implying that $\lambda_1 = 1$ and $\lambda_2 = q/p$.

If $p \neq q$, the roots are distinct and thus $f_{i0} = A + B(q/p)^i$. A and B are determined from the boundary conditions

$$f_{00} = A + B = 1, \qquad f_{a0} = A + B\left(\frac{q}{p}\right)^a = 0,$$

giving

$$f_{i0} = \frac{\left(\dfrac{q}{p}\right)^i - \left(\dfrac{q}{p}\right)^a}{1 - \left(\dfrac{q}{p}\right)^a} \qquad (i = 1, 2, \ldots, a - 1). \tag{6.3.9}$$

If $p = q$, the roots are equal and thus $f_{i0} = A + Bi$. A and B must satisfy the boundary conditions $f_{00} = A = 1$, $f_{a0} = A + Ba = 0$, giving

$$f_{i0} = 1 - i/a \qquad (i = 1, 2, \ldots, a - 1). \tag{6.3.10}$$

We can carry through the same exercise to find an expression for f_{ia} but consider interchanging the roles of the gambler and the opponent:

f_{ia} = prob. gambler eventually wins starting with i units

     = prob. opponent is eventually ruined starting with $a - i$ units.

Since

prob. opponent wins 1 unit = prob. gambler loses 1 unit = q

and

prob. opponent loses 1 unit = prob. gambler wins 1 unit = p,

to find f_{ia} interchange p and q and replace i with $a - i$ in f_{i0}. This leads to

$$
f_{ia} = \begin{cases} \dfrac{\left(\dfrac{q}{p}\right)^i - 1}{\left(\dfrac{q}{p}\right)^a - 1} & (i = 1, \ldots, a - 1;\ p \neq q), \\[4mm] \dfrac{i}{a} & (i = 1, \ldots, a - 1;\ p = q), \end{cases}
$$

from which we see that $f_{i0} + f_{ia} = 1$ (which, of course, is a consequence of Theorem 6.3.3).

An alternative method is to use the results derived in Example 6.2.4 concerning the g.f. $F_{i0}(s)$. From Eqs. (6.2.13) and (6.3.4)

$$
f_{i0} = \left(\frac{q}{p}\right)^i \lim_{s \uparrow 1} \frac{\lambda_1^{a-i}(s) - \lambda_2^{a-i}(s)}{\lambda_1^a(s) - \lambda_2^a(s)}. \tag{6.3.11}
$$

Now

$$
\lambda_1(1) = \frac{1 + (1 - 4pq)^{1/2}}{2p} = \frac{1 + |p - q|}{2p}
$$

and

$$
\lambda_2(1) = \frac{1 - (1 - 4pq)^{1/2}}{2p} = \frac{1 - |p - q|}{2p}.
$$

If $p > q$, then $\lambda_1(1) = 1$, $\lambda_2(1) = q/p$, whereas if $p < q$, then $\lambda_1(1) = q/p$, $\lambda_2(1) = 1$. In both these cases the results of Eq. (6.3.9) follow from Eq. (6.3.11). When $p = q$, it is easily verified that

$$
F_{i0}(s) = \left(\frac{q}{p}\right)^i \frac{s P_{a-i}(s)}{P_a(s)},
$$

where $P_k(s) = [1 + (1 - s^2)^{1/2}]^k - [1 - (1 - s^2)^{1/2}]^k$. Since $P_k(1) = 0$ and $P_k^{(1)}(1) = k$, an application of L'Hospitals rule shows that $F_{i0}(1)$ is the expression f_{i0} given by Eq. (6.3.10).

When the number of states is small we can use the results of Theorem 6.3.1 to find f_{ij} for general i, j. For example, suppose $a = 4$. Then

$$P = \begin{bmatrix} I & 0 \\ R & Q \end{bmatrix} \quad \text{where} \quad Q = \begin{bmatrix} 0 & p & 0 \\ q & 0 & p \\ 0 & q & 0 \end{bmatrix} \quad \text{and} \quad R = \begin{bmatrix} q & 0 \\ 0 & 0 \\ 0 & p \end{bmatrix}.$$

Now

$$I - Q = \begin{bmatrix} 1 & -p & 0 \\ -q & 1 & -p \\ 0 & -q & 1 \end{bmatrix}$$

and

$$N = (I - Q)^{-1} = \frac{1}{1 - 2pq} \begin{bmatrix} 1 - pq & p & p^2 \\ q & 1 & p \\ q^2 & q & 1 - pq \end{bmatrix}.$$

$$F_3 = \begin{bmatrix} f_{10} & f_{14} \\ f_{20} & f_{24} \\ f_{30} & f_{34} \end{bmatrix} = NR = \frac{1}{1 - 2pq} \begin{bmatrix} q(1 - pq) & p^3 \\ q^2 & p^2 \\ q^3 & p(1 - pq) \end{bmatrix}.$$

$$F_4 = \begin{bmatrix} f_{11} & f_{12} & f_{13} \\ f_{21} & f_{22} & f_{23} \\ f_{31} & f_{32} & f_{33} \end{bmatrix} = (N - I)N_d^{-1} = \frac{1}{1 - pq} \begin{bmatrix} pq & p(1 - pq) & p^2 \\ q & 2pq(1 - pq) & p \\ q^2 & q(1 - pq) & pq \end{bmatrix}.$$

The expressions given in F_3 for f_{i0} and f_{ia} can be shown to be equivalent to those derived earlier.   □

EXAMPLE 6.3.3:   *Random Walk on the Nonnegative Integers with an Absorbing Barrier at the Origin.*   Using the results of Example 6.2.5, for $i = 1, 2, \ldots,$

$$f_{i0} = \lim_{s \uparrow 1} F_{i0}(s) = \lim_{s \uparrow 1} \lambda_2^i(s) = \lim_{s \uparrow 1} \left[\frac{1 - (1 - 4pqs^2)^{1/2}}{2ps} \right]^i.$$

It is easily verified that

$$f_{i0} = \left[\frac{1 - |p - q|}{2p} \right]^i = \begin{cases} 1, & p \leq \tfrac{1}{2}, \\ \left(\dfrac{q}{p} \right)^i, & p > \tfrac{1}{2}. \end{cases}$$

This example can be regarded as a model for a gambler playing an infinitely rich adversary. Thus if $p \leq \tfrac{1}{2}$, then the gambler will go broke with prob. 1; whereas if $p > \tfrac{1}{2}$, there is a positive probability that the game will not finish and the gambler's fortune will increase indefinitely.

Alternatively, we can take the limit as $a \to \infty$ in Eqs. (6.3.9) and (6.3.10) to obtain the above results. □

EXAMPLE 6.3.4: *Random Walk on the Nonnegative Integers with a Reflecting Barrier at the Origin.* Using the results of Examples 6.2.6 and 6.3.3, it is easily seen that for $i = 1, 2, \ldots$,

$$f_{i0} = \begin{cases} 1, & p \leq \tfrac{1}{2}, \\ \left(\dfrac{q}{p}\right)^i, & p > \tfrac{1}{2}, \end{cases}$$

and that

$$f_{00} = q + pf_{10} = \begin{cases} 1, & p \leq \tfrac{1}{2}, \\ 2q, & p > \tfrac{1}{2}. \end{cases}$$

Since the chain is irreducible we can conclude that all the states are persistent if $p \leq q$ and transient if $p > q$. □

EXAMPLE 6.3.5: *Unrestricted Random Walk.* This model was originally considered in Example 6.1.1 where we showed if $S = \{0, \pm 1, \pm 2, \ldots\}$ and $p_{i,i+1} = p$, $p_{i,i-1} = q$, $p_{ij} = 0$ for $j \neq i - 1, i + 1$ for all $i \in S$ with $0 < p < 1$, then the MC is irreducible and that all the states are persistent if $p = \tfrac{1}{2}$ and transient if $p \neq \tfrac{1}{2}$.

This implies $f_{ii} = 1$ if $p \neq \tfrac{1}{2}$. We obtain expressions for f_{ii} in the transient case and investigate the determination of the f_{ij} for $i \neq j$.

For all $i, j \in S$ it was shown that

$$p_{ij}^{(n)} = \binom{n}{\dfrac{n+j-i}{2}} p^{(n+j-i)/2} q^{(n-j+i)/2}.$$

Thus if k is a nonnegative integer, then

$$p_{i,i+k}^{(n)} = p_{0,k}^{(n)} = \binom{n}{\dfrac{n+k}{2}} p^{(n+k)/2} q^{(n-k)/2}$$

and

$$p_{i,i-k}^{(n)} = p_{0,-k}^{(n)} = \binom{n}{\dfrac{n-k}{2}} p^{(n-k)/2} q^{(n+k)/2},$$

where, of course, the binomial coefficient is defined only if $n + k$ (or equivalently $n - k$) is an even integer with $n \geq k$. Thus by setting $n - k = 2l$ we obtain

$$P_{0,k}(s) = \sum_{n=k}^{\infty} p_{0,k}^{(n)} s^n = (ps)^k \sum_{l=0}^{\infty} \binom{2l+k}{l}(pqs^2)^l$$

and

$$P_{0,-k}(s) = \sum_{n=k}^{\infty} p_{0,-k}^{(n)} s^n = (qs)^k \sum_{l=0}^{\infty} \binom{2l+k}{l}(pqs^2)^l.$$

Since $P_{ij}(s) = P_{0,j-i}(s)$ for all $i, j \in S$, we can deduce, using Theorem 6.2.5, that

$$F_{ij}(s) = \frac{P_{0,j-i}(s) - \delta_{ij}}{P_{00}(s)}, \qquad (6.3.12)$$

from which we can obtain expressions for the f_{ij} using Eq. (6.3.4). Observe that $P_{0,-k}(s)$ is of the same form as $P_{0,k}(s)$ but with p and q interchanged. This means that f_{ij} with $i > j$ can be obtained from the f_{ij} with $j > i$ by interchanging p and q. From Example 3.3.2 we obtain the result that

$$P_{00}(s) = \sum_{l=0}^{\infty} \binom{2l}{l}(pqs^2)^l = (1 - 4pqs^2)^{-1/2},$$

and thus Eq. (6.3.12) implies that for all $i \in S$

$$F_{ii}(s) = 1 - (1 - 4pqs^2)^{1/2}$$

and hence

$$f_{ii} = \begin{cases} 1, & p = \frac{1}{2}, \\ 1 - |p - q|, & p \neq \frac{1}{2}. \end{cases}$$

In general it is not an easy matter to find closed form expressions for $P_{0,k}(s)$ [or equivalently $P_{0,-k}(s)$]. We leave it as an exercise to show that for $l \geq 0$

$$\binom{2l+1}{l} = \frac{1}{2}\binom{2l+2}{l+1},$$

$$\binom{2l+2}{l} = \frac{1}{2}\binom{2l+4}{l+2} - \binom{2l+2}{l+1},$$

$$\binom{2l+3}{l} = \frac{1}{2}\binom{2+6}{l+3} - \frac{3}{2}\binom{2l+4}{l+2}$$

$$\vdots$$

[These can be derived by using the following relationship concerning binomial coefficients: $\binom{x}{r-1} + \binom{x}{r} = \binom{x+1}{r}$.] Thence using the expression for

$P_{00}(s)$ above we have

$$\sum_{l=0}^{\infty} \binom{2l+2}{l+1} x^l = \frac{1}{x}[(1-4x)^{-1/2} - 1],$$

$$\sum_{l=0}^{\infty} \binom{2l+4}{l+2} x^l = \frac{1}{x^2}[(1-4x)^{-1/2} - 1 - 2x],$$

$$\sum_{l=0}^{\infty} \binom{2l+6}{l+3} x^l = \frac{1}{x^3}[(1-4x)^{-1/2} - 1 - 2x - 6x^2].$$

Using these summations results and the expressions for the binomial coefficients, we can deduce that

$$P_{0,1}(s) = \frac{1}{2qs}\{1 - 4pqs^2)^{-1/2} - 1\},$$

$$P_{0,2}(s) = \frac{1}{2q^2s^2}\{(1 - 2pqs^2)(1 - 4pqs^2)^{-1/2} - 1\},$$

$$P_{0,3}(s) = \frac{1}{2q^3s^3}\{(1 - 3pqs^2)(1 - 4pqs^2)^{-1/2} - 1 + pqs^2\}.$$

From Eq. (6.3.11) we obtain

$$F_{i,\,i+1}(s) = \frac{1}{2qs}\{1 - (1 - 4pqs^2)^{1/2}\},$$

$$F_{i,\,i+2}(s) = \frac{1}{2q^2s^2}\{1 - 2pqs^2 - (1 - 4pqs^2)^{1/2}\},$$

$$F_{i,\,i+3}(s) = \frac{1}{2q^3s^3}\{1 - 3pqs^2 - (1 - pqs^2)(1 - 4pqs^2)^{1/2}\}.$$

By considering separately the cases $p = \frac{1}{2}$, $p < \frac{1}{2}$, and $p > \frac{1}{2}$ it can be easily seen that

$$f_{i,\,i+1} = \frac{1}{2q}\{1 - |p - q|\} = \begin{cases} 1, & p \geq \frac{1}{2}, \\ \dfrac{p}{q}, & p < \frac{1}{2}; \end{cases}$$

$$f_{i,\,i+2} = \frac{1}{2q^2}\{1 - 2pq - |p - q|\} = \begin{cases} 1, & p \geq \frac{1}{2}, \\ \dfrac{p^2}{q^2}, & p < \frac{1}{2}; \end{cases}$$

$$f_{i,\,i+3} = \frac{1}{2q^3}\{1 - 3pq - (1 - pq)|p - q|\} = \begin{cases} 1, & p \geq \frac{1}{2}, \\ \dfrac{p^3}{q^3}, & p < \frac{1}{2}. \end{cases}$$

We can make some interesting observations from these results. Note that for $k = 1, 2, 3$,

$$f_{i,i+k} = (f_{i,i+1})^k = \begin{cases} 1, & p \geq \frac{1}{2}, \\ \left(\dfrac{p}{q}\right)^k, & p < \frac{1}{2}. \end{cases}$$

In fact it is easily verified that for these values of k

$$F_{i,i+k}(s) = [F_{i,i+1}(s)]^k.$$

A moment's reflection should convince the reader that these results are generally true since any first passage from i to $i + k$ can be regarded as the sum of k independent first passages from $i + j$ to $i + j + 1$ $(j = 0, 1, \ldots, k - 1)$. In a similar manner

$$F_{i,i-k}(s) = [F_{i,i-1}(s)]^k$$

and

$$f_{i,i-k} = (f_{i,i-1})^k = \begin{cases} \left(\dfrac{q}{p}\right)^k, & p > \frac{1}{2}, \\ 1, & p \leq \frac{1}{2}. \end{cases}$$

The results of this example exemplify the statement of Theorem 5.3.7 that when an irreducible MC consists of transient states (in this case $p \neq \frac{1}{2}$) then for all $j \in S$ there exists an $i \neq j$ for which $f_{ij} < 1$ (e.g., $f_{j-1,j}$ when $p < \frac{1}{2}$ or $f_{j+1,j}$ when $p > \frac{1}{2}$). Of course, we can still have some $f_{ij} = 1$ in the transient case, but there is no general theory that tells us which such i, j.   □

EXAMPLE 6.3.6:   *Finite Birth and Death Chain.*   Consider the MC with state space $S = \{0, 1, \ldots, a\}$ and transition matrix given by

$$\begin{bmatrix} r_0 & p_0 & 0 & 0 & \cdots & & & 0 \\ q_1 & r_1 & p_1 & 0 & \cdots & & & 0 \\ 0 & q_2 & r_2 & p_2 & \cdots & & & 0 \\ \vdots & \vdots & \vdots & \vdots & & & & \vdots \\ & & & & \cdots & q_{a-1} & r_{a-1} & p_{a-1} \\ 0 & 0 & 0 & 0 & \cdots & 0 & q_a & r_a \end{bmatrix},$$

where we assume $q_i > 0$ and $p_i > 0$ for $i = 1, \ldots, a - 1$. We examine the determination of f_{i0} for $i = 0, 1, \ldots, a$. Using Eq. (6.3.1) we obtain

$$f_{00} = r_0 + p_0 f_{10},$$

$$f_{10} = q_1 + r_1 f_{10} + p_1 f_{20},$$

$$f_{i0} = q_i f_{i-1,0} + r_i f_{i0} + p_i f_{i+1,0} \qquad (i = 2, \ldots, a - 1)$$

$$f_{a0} = q_a f_{a-1,0} + r_a f_{a0}.$$

By defining

$$x_0 = 1 - f_{00} \qquad x_1 = 1 - f_{10},$$

$$x_i = f_{i-1,0} - f_{i,0} \qquad (i = 2, \ldots, a);$$

we obtain from the above equations

$$q_i x_i = p_i x_{i+1} \qquad (i = 1, 2, \ldots, a - 1),$$

with $x_0 = p_0 x_1$ and $q_a x_a = 0$.

If $q_a \neq 0$, then it is easily seen that $x_i = 0$ for $i = 0, \ldots, a$ and thus $f_{i0} = 1$ for $i = 0, \ldots, a$. (Note that if, in addition $p_0 \neq 0$, the MC is irreducible and finite and hence cannot contain any transient states and this result follows from Theorem 5.3.6.)

Thus in what follows we assume that $q_a = 0$ or equivalently $r_a = 1$. This implies that state a is absorbing and hence $f_{a0} = 0$.

It can be easily seen that

$$\sum_{i=1}^{k} x_i = 1 - f_{k0} \qquad (k = 1, \ldots, a),$$

and that if $\rho_0 \equiv 1$,

$$\rho_i \equiv \frac{q_1 \cdots q_i}{p_1 \cdots p_i} \qquad (i = 1, 2, \ldots, a - 1),$$

$$x_i = \rho_{i-1} x_1 \qquad (i = 1, \ldots, a).$$

This leads to the result that

$$1 - f_{k0} = \left(\sum_{i=0}^{k-1} \rho_i \right) x_1 \qquad (k = 1, \ldots, a). \tag{6.3.13}$$

By taking $k = a$ we obtain an expression for x_1, i.e.,

$$x_1 = 1 \bigg/ \sum_{i=0}^{a-1} \rho_i,$$

from which we can deduce that

$$f_{k0} = 1 - \left(\sum_{i=0}^{k-1} \rho_i \right) \bigg/ \left(\sum_{i=0}^{a-1} \rho_i \right) \qquad (k = 1, \ldots, a) \tag{6.3.14}$$

and

$$f_{00} = 1 - p_0 \bigg/ \left(\sum_{i=0}^{a-1} \rho_i \right). \tag{6.3.15}$$

[The reader may wish to verify that the results of Example 6.3.2, the classical gambler's ruin model, can be obtained as a special case when $r_0 = 1$, $p_0 = 0$, $q_i = q$, $r_i = 0$, and $p_i = p$ $(i = 1, \ldots, a - 1)$.] □

*EXAMPLE 6.3.7:   *Infinite Birth and Death Chain.*   Let us now consider a birth and death chain on the nonnegative integers, which is irreducible. Thus $S = \{0, 1, 2, \ldots\}$ with transition probabilities $p_{i, i+1} = p_i > 0$ $(i \geq 0)$; $p_{i,i} = r_i \geq 0$ $(i \geq 0)$; and $p_{i, i-1} = q_i > 0$ $(i \geq 1)$. Such an infinite chain will either be transient or persistent.

If we follow through the procedure of Example 6.3.6 we obtain the same equations (6.3.13) for $k \geq 1$. It seems reasonable to assume that the probabilities f_{k0} may be obtained by taking the limit of Eqs. (6.3.14) and (6.3.15) as $a \to \infty$: Let $\rho \equiv \sum_{i=0}^{\infty} \rho_i$.

If $\rho < \infty$, then

$$f_{k0} = \begin{cases} 1 - p_0/\rho & (k = 0), \\ 1 - \left(\sum_{i=0}^{k-1} \rho_i\right)\Big/\rho & (k = 1, 2, \ldots). \end{cases}$$

If $\rho = \infty$, then $f_{k0} = 1$ $(k = 1, 2, \ldots)$.

To justify these results we can prove that an irreducible birth and death chain on $\{0, 1, 2, \ldots\}$ is persistent iff

$$\rho = \sum_{i=0}^{\infty} \frac{q_1 \cdots q_i}{p_1 \cdots p_i} = \infty.$$

Using Theorem 6.3.4 we consider the solution of the equations

$$h_i = \sum_{j=1}^{\infty} p_{ij} h_j \qquad \text{for} \quad i = 1, 2, \ldots.$$

These equations lead to

$$h_1 = r_1 h_1 + p_1 h_2,$$
$$h_i = q_i h_{i-1} + r_i h_i + p_{i, i+1}, \qquad i = 2, 3, \ldots.$$

Thus

$$p_i(h_{i+1} - h_i) = q_i(h_i - h_{i-1}), \qquad i = 2, 3, \ldots,$$

and hence, as in the derivation of Eq. (6.3.13) for $i = 2, 3, \ldots$,

$$h_{i+1} - h_i = \frac{q_i \cdots q_2}{p_i \cdots p_2}(h_2 - h_1).$$

This means that for $k = 2, \ldots$,

$$h_{k+1} - h_1 = \left[1 + \sum_{i=2}^{k} \frac{q_i \cdots q_2}{p_i \cdots p_2}\right](h_2 - h_1).$$

The first equation implies that, since $p_1 > 0$,

$$h_2 - h_1 = (q_1/p_1)h_1,$$

and that the general solution $\{h_k, \; k = 1, 2, \ldots\}$ is given, up to an undetermined constant h_1, by

$$h_k = h_1 \sum_{i=0}^{k-1} \rho_i, \qquad k = 1, 2, \ldots.$$

Thus a bounded, nonnegative solution $\{h_k\}$ exists iff $\sum_{i=0}^{\infty} \rho_i < \infty$, which is thus the necessary and sufficient condition for a transient MC. Equivalently, the chain is persistent iff the series diverges.

This, of course, establishes the fact that the $f_{k0} = 1$ for $k = 1, 2, \ldots$ when $\rho = \infty$, since in this case $f_{ij} = 1$ for all $i, j \in S$ by Theorem 5.3.6. A rigorous derivation of the results for f_{k0} when $\rho < \infty$ is outside the scope of this text (see Chung, 1960, p. 69). □

Exercises 6.3

1. In the terminology of Theorem 6.2.4, show that when P is irreducible:

$$\mathbf{e}_i'[I - P_j(0)]^{-1} P \mathbf{e}_j = 1 \qquad \text{for all} \quad i, j \in S.$$

2. Show, using Eqs. (6.2.4), that

$$F = \left(\sum_{n=1}^{\infty} P^n \right) \left(\sum_{i=0}^{\infty} A_i \right),$$

where

$$\left(\sum_{i=0}^{\infty} A_i \right) \left(\sum_{n=0}^{\infty} D_n \right) = I, \qquad D_n = [P^n]_d,$$

so that

$$F \left(\sum_{n=0}^{\infty} D_n \right) = \left(\sum_{n=1}^{\infty} P^n \right),$$

which can also be obtained from Eq. (6.3.5).

3. Let $P = [p_{ij}]$ be the transition matrix of a Markov chain with finite state space S. Define $A = [a_{ij}]$, where

$$a_{ij} = \begin{cases} p_{ij}, & i \neq r, \text{ all } j \in S; \\ 0, & i = r, j \neq r; \\ 1, & i = r, j = r; \end{cases}$$

so that A is formed from P by replacing the rth row by e'_r. Let $A^n = [a_{ij}^{(n)}]$ $(n = 1, 2, 3, \ldots)$ with $[a_{ij}^{(0)}] = I$. Show that

(a) $a_{rr}^{(n)} = 1$ $(n = 0, 1, 2, \ldots)$;
(b)   for $i \neq r$, $f_{ir}^{(n)} = a_{ir}^{(n)} - a_{ir}^{(n-1)}$ $(n = 1, 2, \ldots)$;
(c)   for $i \neq r$, $f_{ir} = \lim_{n \to \infty} a_{ir}^{(n)}$
[*Hint*: For (a) and (b) use induction.] [Based on Pearl (1973).]

4. Suppose a gambler with initial capital k plays against an infinitely rich adversary and wins two units or loses one unit with probabilities p and q, respectively ($p + q = 1$). What is the gambler's chance of being ruined? (In the course of your solution, two cases $q > 2p$ and $q \leq 2p$ will emerge.) [Bailey (1964).]

5. Consider a finite population (of fixed size N) of individuals of possible types A and B undergoing the following growth process. At instants of time $t_1 < t_2 < t_3 < \cdots$ one individual dies and is immediately replaced by another type A or B. If just before a replacement time t_n there are j A's and $N - j$ B's present, the probability that an A individual dies is $j\mu_1/c_j$ and that a B individual dies is $(N - j)\mu_2/c_j$, where $c_j = \mu_1 j + \mu_2(N - j)$ and $\mu_1 \neq \mu_2$. Also, it is assumed that the replacement is type A with probability j/N and type B with probability $(N - j)/N$. Given that there are k A's to begin with show that the probability that all the A's die out is [Karlin (1966)]

$$\frac{(\mu_1/\mu_2)^N - (\mu_1/\mu_2)^k}{(\mu_1/\mu_2)^N - 1}.$$

6. The gambler's ruin problem (as considered in Example 6.3.2) can be generalized so that the chances of winning and losing one unit are p and q, respectively, and the chance of a tie or draw is $r(p + q + r = 1)$. Derive the probability of ruin (i.e., ever reaching 0) given the gambler's initial capital is k and his opponent has $a - k$. (The cases $p \neq q$ and $p = q$ will need to be considered separately.)

7. A series of independent games is played by A and B, whose respective probabilities of winning any single game are p and $q(= 1 - p)$. If the series is won by the first player to establish a lead of two games, show that the probability of A winning the series is

$$p^2/(p^2 + q^2).$$

8. Examine the case of a simple random walk (see Example 5.1.2) that starts at position i. The probabilities of moving one step to the right or left are p and q, respectively, $x = a$ is an absorbing barrier ($\beta = 1$),

and $x = 0$ is an elastic barrier $(0 < \alpha < 1)$. What is the probability of absorption at $x = 0$ (i.e., of ever reaching $x = 0$ starting from $x = i$)?

9. With the notation introduced in Theorem 6.3.3, show that $h_r^{(n)}$ is monotone nonincreasing as $n \to \infty$.

10. Consider a population kept constant in size by the selection of N individuals in each successive generation. A particular gene assuming the forms A and a has $2N$ representatives; if at the nth generation A occurs j times then a occurs $2N - j$ times. In this case we say that the population is in state j $(0 \leq j \leq 2N)$ at time n. Assuming random mating, the composition of the following generation is determined by $2N$ Bernoulli trials in which the A-gene has probability $j/2N$. We have therefore an MC with transition probabilities

$$p_{jk} = \binom{2N}{k} \left(\frac{j}{2N}\right)^k \left(1 - \frac{j}{2N}\right)^{2N-k}, \qquad j, k = 0, 1, \ldots, 2N.$$

There are two absorbing states $j = 0, 2N$. Verify that the probability that the population ultimately consists only of aa-individuals (i.e., absorption in 0) starting in state j is given by

$$f_{j0} = 1 - j/2N.$$

(This result is plausible in that at the moment when the A- and a-genes are in the proportion $j : 2N - j$ their survival chances should be in the same ratio.) [Feller (1968), attributed to Malécot (1944).]

Chapter 7

Markov Chains in Discrete Time— Limiting Behavior

7.1 Limiting and Stationary Distributions

In Section 5.2, using recurrent event theory, we obtained asymptotic expressions for the $p_{ij}^{(n)}$. In this section we consider special types of Markov chains (MC's) and examine the relevance of these asymptotic results to the n-step transition matrix $P^{(n)}$, the n-step absolute probabilities $p_j^{(n)}$, and the n-step absolute probability vector $\mathbf{p}^{(n)'}$.

We use the following notation whenever the limits exist:

$$\lim_{n \to \infty} p_{ij}^{(n)} \equiv \pi_{ij}, \qquad \lim_{n \to \infty} P^{(n)} \equiv \Pi = [\pi_{ij}],$$

$$\lim_{n \to \infty} p_j^{(n)} = p_j, \qquad \lim_{n \to \infty} \mathbf{p}^{(n)'} = \mathbf{p}' = (p_j)'.$$

To motivate some of the more general results we give a brief survey of the relevant results for the two-state Markov chain.

EXAMPLE 7.1.1: *Two-State Markov Chain.* Let

$$P = \begin{bmatrix} p_{11} & p_{12} \\ p_{21} & p_{22} \end{bmatrix} = \begin{bmatrix} 1-a & a \\ b & 1-b \end{bmatrix} \qquad (0 \le a \le 1, 0 \le b \le 1),$$

$$d = 1 - p_{12} - p_{21} = 1 - a - b \qquad (|d| \le 1),$$

and

$$\mathbf{p}^{(0)'} = (p_1^{(0)}, p_2^{(0)}) = (\alpha, \beta).$$

From Theorem 5.1.7 and its associated corollary (Corollary 5.1.7A) we can make the following observations.

(a)   If $|d| < 1$, then

$$P^{(n)} = \frac{1}{1-d} \begin{bmatrix} b + ad^n & a - ad^b \\ b - bd^n & a + bd^n \end{bmatrix} \quad (n \geq 1)$$

so that

$$\Pi = \frac{1}{1-d} \begin{bmatrix} b & a \\ b & a \end{bmatrix} = \frac{1}{p_{12} + p_{21}} \begin{bmatrix} p_{21} & p_{12} \\ p_{21} & p_{12} \end{bmatrix}.$$

Also,

$$\mathbf{p}^{(n)'} = \left(\frac{b + (\alpha a - \beta b)d^n}{1-d}, \frac{a - (\alpha a - \beta b)d^n}{1-d} \right) \quad (n \geq 1),$$

so that

$$\mathbf{p}' = \left(\frac{b}{1-d}, \frac{a}{1-d} \right) = \left(\frac{p_{21}}{p_{12} + p_{21}}, \frac{p_{12}}{p_{12} + p_{21}} \right).$$

If, in addition, we have $a > 0$ and $b > 0$ (i.e., $0 < a < 1$, $0 < b < 1$), Theorem 5.2.3 assures us that the MC is finite, irreducible, aperiodic, and consists solely of persistent nonnull states, i.e., the MC is regular. For such MC's, as we shall prove later,

(i) Π exists with π_{ij} depending only on j, i.e., $\pi_{ij} = \pi_j$.
(ii)   A limiting distribution $\mathbf{p}'$ exists that does not depend on $\mathbf{p}^{(0)'}$.
(iii) If $\mathbf{p}^{(0)'} = \mathbf{p}'$, then $\mathbf{p}^{(n)'} = \mathbf{p}'$ for all $n \geq 1$.

(b)   If $d = 1$, then

$$P^{(n)} = \begin{bmatrix} 1 & 0 \\ 0 & 1 \end{bmatrix} \quad (n \geq 1)$$

so that

$$\Pi = \begin{bmatrix} 1 & 0 \\ 0 & 1 \end{bmatrix}.$$

Also,

$$\mathbf{p}^{(n)'} = (\alpha, \beta) = \mathbf{p}^{(0)'},$$

so that

$$\mathbf{p}' = (\alpha, \beta) = \mathbf{p}^{(0)'}.$$

In this case the MC has absorbing states (as also in the case $|d| < 1$ with $a = 0$ or $b = 0$). For absorbing MC's we shall show that

(i) Π exists but π_{ij} will in general depend on both i and j.

(ii)   A family of limiting distributions $\mathbf{p}'$ exists that will depend on $\mathbf{p}^{(0)'}$.

(c)   If $d = -1$, then

$$P^{(n)} = \begin{cases} \begin{bmatrix} 1 & 0 \\ 0 & 1 \end{bmatrix} & (n \text{ even}), \\[12pt] \begin{bmatrix} 0 & 1 \\ 1 & 0 \end{bmatrix} & (n \text{ odd}), \end{cases}$$

with

$$\mathbf{p}^{(n)'} = \begin{cases} (\alpha, \beta) & (n \text{ even}), \\ (\beta, \alpha) & (n \text{ odd}). \end{cases}$$

In this case the MC is irreducible and periodic, and in general for such MC's we shall show that

(i)   A limiting matrix Π does not exist, but $P^{(n)}$ ultimately oscillates between a number of possible forms.

(ii)   A limiting distribution $\mathbf{p}'$ does not exist independent of $\mathbf{p}^{(0)'}$. However, with a suitable choice of initial probability vectors a limiting distribution $\mathbf{p}'$ can be found—a "pseudolimiting distribution" [e.g., if $\mathbf{p}^{(0)'} = (\frac{1}{2}, \frac{1}{2})$, then $\mathbf{p}' = (\frac{1}{2}, \frac{1}{2})$]. $\square$

We digress to introduce the concept of *stationarity*, which is intimately connected with limiting distributions.

DEFINITION 7.1.1:   A probability distribution $\{v_j\}$ is called a *stationary distribution* for the MC with transition matrix $P = [p_{ij}]$ if, for all $j \in S$,

$$v_j = \sum_{i \in S} v_i p_{ij} \qquad \left(\text{with } \sum_{j \in S} v_j = 1 \right).$$

The vector $\mathbf{v}' = (v_j)'$ $(j \in S)$ is called a *stationary probability vector* for the MC and thus $\mathbf{v}'$ must satisfy the equation

$$\mathbf{v}' = \mathbf{v}'P \qquad \text{with } \mathbf{v}'e = 1. \quad \square$$

THEOREM 7.1.1:   Let $\mathbf{v}'$ be a stationary probability vector for the MC with transition matrix P.

(a) $\mathbf{v}' = \mathbf{v}'P^n$ for all $n \geq 1$ (or equivalently, $v_j = \sum_{i \in S} v_i p_{ij}^{(n)}$ for all $n \geq 1$).

(b)   If $\mathbf{p}^{(0)'} = \mathbf{v}'$, then $\mathbf{p}^{(n)'} = \mathbf{v}'$ for all $n \geq 1$, and thus $\mathbf{p}' = \lim_{n \to \infty} \mathbf{p}^{(n)'}$ exists and is given by $\mathbf{v}'$.

(c) If $\mathbf{p}^{(n)'} = \mathbf{p}^{(0)'}$ for all $n \geq 1$, then $\mathbf{p}^{(0)'}$ is a stationary probability vector.

(d) If $\mathbf{p}' = \lim_{n \to \infty} \mathbf{p}^{(n)'}$ exists, then $\mathbf{p}'$ is a stationary probability vector.

Proof: (a) Assume $\mathbf{v}' = \mathbf{v}'P^n$ for some $n = k$ (certainly true for $n = 1$). Then

$$\mathbf{v}'P^{k+1} = \mathbf{v}'P^kP = \mathbf{v}'P = \mathbf{v},$$

and thus the result is true for $n = k + 1$ and hence generally by induction.

(b) From Theorem 5.1.3(c) and (a) above, for all $n \geq 1$,

$$\mathbf{p}^{(n)'} = \mathbf{p}^{(0)'}P^n = \mathbf{v}P^n = \mathbf{v}'.$$

(c) From Theorem 5.1.3(d), since $\mathbf{p}^{(n)'} = \mathbf{p}^{(n-1)'}P$ for all $n \geq 1$, $\mathbf{p}^{(0)'} = \mathbf{p}^{(1)'} = \mathbf{p}^{(0)'}P$, which implies $\mathbf{p}^{(0)'}$ is a stationary probability vector.

(d) Suppose $\mathbf{p}' = \lim_{n \to \infty} \mathbf{p}^{(n)'}$ exists. Then from Theorem 5.1.3(d)

$$\mathbf{p}' = \lim_{n \to \infty} (\mathbf{p}^{(n-1)'}P) = \left(\lim_{n \to \infty} \mathbf{p}^{(n-1)'} \right)P = \mathbf{p}'P,$$

showing that $\mathbf{p}'$ is a stationary probability vector.

The interchange of the limiting operation and matrix multiplication is valid for finite matrices by Theorem 4.5.1. In the countably infinite case the result still holds true by virtue of Theorem 2.2.5 since in fact we are interchanging a limit and a summation, i.e., in element form

$$p_j = \lim_{n \to \infty} \sum_{i \in S} p_i^{(n-1)}p_{ij} = \sum_{i \in S} \lim_{n \to \infty} p_i^{(n-1)}p_{ij} = \sum_{i \in S} p_ip_{ij}$$

since

$$\left| p_i^{(n-1)}p_{ij} \right| \leq p_i^{(n-1)} \quad \text{and} \quad \sum_{i \in S} p_i^{(n-1)} = 1. \quad \square$$

Observe that if the initial distribution is a stationary distribution, then the absolute distribution of X_n, which is specified by the n-step absolute probability vector $\mathbf{p}^{(n)'}$, does not depend on n. In other words, $\mathbf{p}^{(n)'}$ does not change with time and thence the MC is not evolving in time but is *stationary* or in *statistical equilibrium*.

EXAMPLE 7.1.1 (Continued): *Two-State Markov Chain.* Let $\mathbf{v}' = (v_1, v_2)$ be a stationary probability vector for a general two-state MC. Then

$$(v_1, v_2) = (v_1, v_2) \begin{bmatrix} p_{11} & p_{12} \\ p_{21} & p_{22} \end{bmatrix}$$

and thus

$$v_1 = v_1p_{11} + v_2p_{21},$$
$$v_2 = v_1p_{12} + v_2p_{22}$$

together with

$$1 = v_1 + v_2.$$

Note that one of the first two equations above is redundant [a consequence of $\det(I - P) = 0$]. It is an easy matter to see that

(a) if $p_{12} + p_{21} \neq 0$ (i.e., $-1 \le d < 1$), there exists *a unique* stationary vector

$$\mathbf{v}' = \left(\frac{p_{21}}{p_{12} + p_{21}}, \frac{p_{12}}{p_{12} + p_{21}} \right);$$

(b) if $p_{12} + p_{21} = 0$ (i.e., $d = 1$), there is *no unique* stationary vector but that $\mathbf{v}' = (v_1, v_2)$ is a stationary probability vector for all v_1, v_2 such that $v_1 + v_2 = 1$.

Observe that if

$$\mathbf{p}^{(0)'} = \left(\frac{p_{21}}{p_{12} + p_{21}}, \frac{p_{12}}{p_{12} + p_{21}} \right) = \left(\frac{b}{1 - d}, \frac{a}{1 - d} \right)$$

is substituted in the expressions for $\mathbf{p}^{(n)'}$ given in Example 7.1.1 we see that $\mathbf{p}^{(n)'} = \mathbf{p}^{(0)'}$ for the cases $-1 \le d < 1$, which is a consequence of Theorem 7.1.1(b).

It is also easy to verify, using the results of Example 7.1.1, that for $\mathbf{p}^{(n)'}$ to be independent of n for $-1 \le d \le 1$ then either $\alpha a - \beta b = 0$ or $d = 0$. The first condition implies that

$$\mathbf{p}^{(0)'} = (\alpha, \beta) = \left(\frac{b}{1 - d}, \frac{a}{1 - d} \right) \qquad \text{for} \quad -1 \le d < 1,$$

in accordance with Theorem 7.1.1(c). The second condition implies that

$$\mathbf{p}^{(n)'} = \left(\frac{b}{1 - d}, \frac{a}{1 - d} \right) = (b, a) \qquad \text{for all } n \ge 1 \quad \text{for } any \ \mathbf{p}^{(0)'}.$$

This is still in accordance with Theorem 7.1.1(c), although it should be remarked that this is a feature of all MC's that consist of independent trials (in contrast to the usual assumed first-order dependency).

Note also that for $-1 < d < 1$, $\mathbf{p}' = \lim_{n \to \infty} \mathbf{p}^{(n)'}$ exists and is given by

$$\mathbf{p}' = \left(\frac{b}{1 - d}, \frac{a}{1 - d} \right)$$

the unique stationary probability vector, in accordance with Theorem 7.1.1(d). □

Let us now restrict attention to *regular* MC's and establish the results alluded to in Example 7.1.1(a).

THEOREM 7.1.2:   Let P be the transition matrix of a regular MC with state space $S = \{1, 2, \ldots, m\}$. Then

(a)
$$\lim_{n \to \infty} P^{(n)} = \Pi = \begin{bmatrix} \pi_1 & \pi_2 & \cdots & \pi_m \\ \pi_1 & \pi_2 & \cdots & \pi_m \\ \vdots & \vdots & \ddots & \vdots \\ \pi_1 & \pi_2 & \cdots & \pi_m \end{bmatrix},$$

where $\pi_j \equiv 1/\mu_j > 0$ with μ_j the mean recurrence time of state j. Thus if $\boldsymbol{\pi}' \equiv (\pi_1, \pi_2, \ldots, \pi_m)$, then $\Pi = \mathbf{e}\boldsymbol{\pi}'$.

(b)   For *any* initial probability vector $\mathbf{p}^{(0)'}$, $\lim_{n \to \infty} \mathbf{p}^{(n)'} = \boldsymbol{\pi}'$.

(c)   The vector $\boldsymbol{\pi}'$ is the *unique* stationary probability vector for the MC and thus the limiting distribution is stationary.

(d)   If $\mathbf{p}^{(0)'} = \boldsymbol{\pi}'$, then $\mathbf{p}^{(n)'} = \boldsymbol{\pi}'$ for all $n \geq 1$.

(e) Π satisfies the equation $\Pi P = P\Pi = \Pi$. Further, any matrix A such that $AP = PA = A$ is a scalar multiple of Π. In particular, if $A\mathbf{e} = \mathbf{e}$, then $A = \Pi$.

Proof: (a)   Since state j is ergodic for all $j \in S$, Theorem 5.2.8(b) implies that

$$\lim_{n \to \infty} p_{ij}^{(n)} = f_{ij}/\mu_j \qquad \text{for all} \quad i, j \in S.$$

Furthermore, since the MC is irreducible, Theorem 5.3.6 implies that $f_{ij} = 1$ for all $i, j \in S$. Thus

$$\lim_{n \to \infty} P^{(n)} = \left[\lim_{n \to \infty} p_{ij}^{(n)} \right] = [1/\mu_j],$$

where $\mu_j < \infty$ and $\pi_j \equiv 1/\mu_j > 0$ since each state j is nonnull by Theorem 5.3.10.

(b)   From Theorem 5.1.3(c),

$$\lim_{n \to \infty} \mathbf{p}^{(n)'} = \lim_{n \to \infty} (\mathbf{p}^{(0)'} P^{(n)}) = \mathbf{p}^{(0)'} \left(\lim_{n \to \infty} P^{(n)} \right)$$
$$= \mathbf{p}^{(0)'} \Pi = \mathbf{p}^{(0)'} \mathbf{e}\boldsymbol{\pi}'$$
$$= \boldsymbol{\pi}',$$

since $\mathbf{p}^{(0)'}$ is a probability vector.

(c)   From (a) above, Theorem 5.1.5, and the finiteness of the MC,

$$\pi_j = \lim_{n \to \infty} p_{ij}^{(n)} = \lim_{n \to \infty} \sum_{k=1}^{m} p_{ik}^{(n-1)} p_{kj},$$
$$= \sum_{k=1}^{m} \lim_{n \to \infty} p_{ik}^{(n-1)} p_{kj} = \sum_{k=1}^{m} \pi_k p_{kj}.$$

Furthermore,

$$\sum_{j=1}^{m} \pi_j = \sum_{j=1}^{m} \lim_{n \to \infty} p_{ij}^{(n)} = \lim_{n \to \infty} \sum_{j=1}^{m} p_{ij}^{(n)} = 1,$$

so that $\{\pi_j\}$ is a stationary distribution and π' is a stationary probability vector. To show the uniqueness let $\mathbf{v}'$ be any other probability vector such that $\mathbf{v}' = \mathbf{v}'P$. Then by (b) above, if $\mathbf{p}^{(0)'} = \mathbf{v}'$, then $\lim_{n \to \infty} \mathbf{p}^{(n)'} = \lim_{n \to \infty} \mathbf{v}'P^n = \pi'$. But by Theorem 7.1.1(a), since $\mathbf{v}'$ is stationary, $\mathbf{v}'P^n = \mathbf{v}'$ for all $n \geq 1$ and hence $\mathbf{v}' = \pi'$.

(d)   This follows from Theorem 7.1.1(b).

(e)   It is easy to establish that $\Pi = \Pi P = P\Pi$. From Theorem 5.1.2(c)

$$P^{(n+1)} = P^{(n)}P = PP^{(n)},$$

and since $\lim_{n \to \infty} P^{(n)}$ exists and the MC is finite, using Theorem 4.5.1,

$$\lim_{n \to \infty} P^{(n+1)} = \left(\lim_{n \to \infty} P^{(n)} \right) P = P \left(\lim_{n \to \infty} P^{(n)} \right),$$

i.e., $\Pi = \Pi P = P\Pi$.

Now let A be any matrix such that $AP = PA = A$. Let the ith row of A be $\mathbf{y}_i'$. Since $AP = A$, we have that $\mathbf{y}_i'P = \mathbf{y}_i'$ so that by (c) $\mathbf{y}_i' = a_i\pi'$ for some a_i, and hence if $A = [a_{ij}]$ we have $a_{ij} = a_i\pi_j$. Furthermore, if the jth column of A is $\mathbf{x}_j$, then, since $PA = A$, we have that $P\mathbf{x}_j = \mathbf{x}_j$ and thus by Theorem 6.1.5 $\mathbf{x}_j = b_j\mathbf{e}$. This implies that $a_{ij} = b_j$. Hence for all $i, j = 1, 2, \ldots, m$, $a_{ij} = a_i\pi_j = b_j$. In particular, $a_i\pi_j = a_j\pi_j$ and since $\pi_j > 0$ we have that $a_i = a_j$ for all i, j and hence $a_i = a$ and $A = [a\pi_j] = a\Pi$, i.e., A is a scalar multiple of Π. Also $A\mathbf{e} = a\Pi\mathbf{e} = a\mathbf{e}\pi'\mathbf{e} = a\mathbf{e}$, which equals $\mathbf{e}$ iff $a = 1$. $\square$

Thus for regular MC's the limiting distribution is the stationary distribution, and hence to find the limiting distribution we need only solve the system of linear equations:

$$\pi_j = \sum_{i=1}^{m} \pi_i p_{ij} \quad \text{and} \quad \sum_{j=1}^{m} \pi_j = 1. \tag{7.1.1}$$

EXAMPLE 7.1.2:   Consider a MC with state space $S = \{1, 2, 3\}$ and transition matrix

$$P = \begin{bmatrix} r & 1-r & 0 \\ q & r & p \\ 0 & 1-r & r \end{bmatrix} \quad (0 < r < 1, p > 0, q > 0).$$

It is easily seen that this MC is regular and thus the MC has a unique stationary distribution $\{\pi_i, i = 1, 2, 3\}$ where the π_i satisfy the stationary equations (7.1.1). Substitution yields

$$\pi_1 = \pi_1 r + \pi_2 q,$$

$$\pi_2 = \pi_1(1 - r) + \pi_2 r + \pi_3(1 - r),$$

$$\pi_3 = \pi_2 p + \pi_3 r,$$

subject to the condition that $\pi_1 + \pi_2 + \pi_3 = 1$.

We need only take any two of the stationary equations together with the normalizing condition to obtain the solution. Since

$$\pi_2 = \frac{(1 - r)}{q}\pi_1, \qquad \pi_3 = \frac{p}{(1 - r)}\pi_2 = \frac{p}{q}\pi_1,$$

we require

$$\pi_1\left(1 + \frac{1 - r}{q} + \frac{p}{q}\right) = 1,$$

which leads to the unique solution

$$\pi_1 = \frac{q}{2(1 - r)}, \qquad \pi_2 = \frac{1}{2}, \qquad \pi_3 = \frac{p}{2(1 - r)}. \quad \square$$

Although $\lim_{n \to \infty} p_{ij}^{(n)} = 1/\mu_j$, we have not given any indication as to how fast this convergence takes place.

EXAMPLE 7.1.1 (Continued):   *Two-State Markov Chain.*   When $|d| < 1$ and

$$P = \begin{bmatrix} p_{11} & p_{12} \\ p_{21} & p_{22} \end{bmatrix} = \begin{bmatrix} 1 - a & a \\ b & 1 - b \end{bmatrix}$$

it is easily verified that

$$P^{(n)} - \Pi = \frac{1}{1 - d}\begin{bmatrix} ad^n & -ad^n \\ -bd^n & bd^n \end{bmatrix},$$

and thus $|p_{ij}^{(n)} - \pi_j| \le \alpha\beta^n$, where

$$\alpha = \max(a, b)/(1 - d) = \max(p_{12}, p_{21})/(p_{12} + p_{21}) > 0,$$

$$\beta = |d| \qquad (0 \le \beta < 1).$$

Such a form of convergence is known as *geometric ergodicity* and occurs in any regular MC as the following theorem shows. $\square$

THEOREM 7.1.3:   If P is the transition matrix of a regular MC and $\pi_j = 1/\mu_j$, then there exist constants $\alpha > 0$ and $0 \le \beta < 1$ such that

$$|p_{ij}^{(n)} - \pi_j| \le \alpha\beta^n \qquad \text{for all} \quad n \ge 1.$$

Proof: Let $\lambda_1 = 1, \lambda_2, \ldots, \lambda_m$ be the eigenvalues of P. Since P is regular, Theorem 6.1.5 shows that $|\lambda_i| < 1$ for $i = 2, \ldots, m$. Rather than prove the theorem in full generality, we make the assumption that P is diagonalizable. Under such a restriction we can use Theorem 6.1.3 to express P^n as

$$P^n = \sum_{l=1}^{m} \lambda_l^n \mathbf{x}_l \mathbf{y}_l',$$

where $\mathbf{x}_1, \mathbf{x}_2, \ldots, \mathbf{x}_m$ are linearly independent right eigenvectors corresponding to the eigenvalues $\lambda_1, \lambda_2, \ldots, \lambda_m$ of P and $\mathbf{y}_1', \mathbf{y}_2', \ldots, \mathbf{y}_m'$ are the corresponding linearly independent left eigenvectors chosen so that $\mathbf{y}_i' \mathbf{x}_j = \delta_{ij}$ $(i, j = 1, 2, \ldots, m)$.

Thus

$$P^n = \mathbf{x}_1 \mathbf{y}_1' + \sum_{l=2}^{m} \lambda_l^n A_l, \tag{7.1.2}$$

where $A_l = \mathbf{x}_l \mathbf{y}_l' = [a_{ij}^{(l)}]$. Since $|\lambda_l| < 1$ for $l = 2, \ldots, m$, it is easy to see that $|\lambda_l|^n$ tends to 0 as $n \to \infty$ and hence

$$\lim_{n \to \infty} P^n = \mathbf{x}_1 \mathbf{y}_1'.$$

Observe that $\mathbf{x}_1$ satisfies $P\mathbf{x}_1 = \mathbf{x}_1$, and thus from Theorem 6.1.4 we can take $\mathbf{x}_1 = \mathbf{e}$ where $\mathbf{e}' = (1, 1, \ldots, 1)$. Also, $\mathbf{y}_1'$ satisfies $\mathbf{y}_1' P = \mathbf{y}_1'$ so that $\mathbf{y}_1'$ is a multiple of the stationary probability vector, which is unique for regular MC's by Theorem 7.1.2(c). In order that $\mathbf{y}_1' \mathbf{x}_1 = 1$ we take $\mathbf{y}_1' = \boldsymbol{\pi}' = (\pi_1, \pi_2, \ldots, \pi_m)$. Thus

$$\lim_{n \to \infty} P^n = \mathbf{e}\boldsymbol{\pi}',$$

as found in Theorem 7.1.2(a).

Extracting the (i, j)th element from Eq. (7.1.2) leads to

$$p_{ij}^{(n)} - \pi_j = \sum_{l=2}^{m} \lambda_l^n a_{ij}^{(l)},$$

and thus

$$|p_{ij}^{(n)} - \pi_j| \leq \sum_{l=2}^{m} |\lambda_l^n| |a_{ij}^{(l)}| \leq \beta^n \sum_{l=2}^{m} |a_{ij}^{(l)}| \leq \alpha\beta^n,$$

where

$$0 \leq \beta \leq \max_{l=2,\ldots,m} |\lambda_l| < 1$$

and

$$\alpha = \max_{i,j=1,2,\ldots,m} \sum_{l=2}^{m} |a_{ij}^{(l)}| > 0.$$

The assumption of diagonalizability of P can be removed by basing the proof upon Theorem 4.4.10.

Alternatively, simpler proofs can be given by bounding the elements in each column of P by maximal and minimal elements and using Theorem 5.3.12 that for a regular MC we can find an N such that for all $n \geq N$ P^n has no zero elements. The details to this approach are given by Parzen (1962, pp. 270–273) or Bhat (1972, pp. 85–88). In particular, it can be shown that the results of this theorem hold with

$$\alpha = (1 - 2\varepsilon_N)^{-1} \quad \text{and} \quad \beta = (1 - 2\varepsilon_N)^{1/N},$$

where ε_N is the smallest element of P^N. $\quad\square$

Let us now consider a finite MC that contains transient states, T. In Theorem 5.2.8 we saw that for all $i, j \in T$, $\lim_{n \to \infty} p_{ij}^{(n)} = 0$. The following theorem shows that in this case the convergence also occurs at a geometric rate.

THEOREM 7.1.4: For any finite Markov chain with a set of transient states T, there exists an $\alpha > 0$ and $0 < \beta < 1$ such that

$$p_{ij}^{(n)} \leq \alpha \beta^n \quad \text{for all} \quad i, j \in T.$$

Proof: Suppose that the MC starts in $i \in T$. Since there is a positive probability that the process will eventually leave T, let n_i be the least possible number of steps for this to occur.

Let $n_0 = \max_{i \in T} n_i$. Observe that n_0 is finite since each n_i is less than or equal to the number of transient states.

Furthermore, there is a state $j_i \in \bar{T}$, the persistent states such that $p_{ij_i}^{(n_0)} > 0$ (since once the process leaves T after n_i steps it will remain in a closed set of persistent states $\in \bar{T}$ for the subsequent steps and hence will be in some state $j_i \in \bar{T}$ after n_0 steps).

Let $p = \min_{i \in T} p_{ij_i}^{(n_0)}$. Then $0 < p < 1$. Now

$$p_{iT}^{(n_0)} = \sum_{j \in T} p_{ij}^{(n_0)} = 1 - \sum_{j \in \bar{T}} p_{ij}^{(n_0)} \leq 1 - p \quad \text{for all} \quad i \in T.$$

Furthermore, for all $i, j \in T$

$$p_{ij}^{(n_0 k)} = \sum_{r_1 \in T} \cdots \sum_{r_{k-1} \in T} p_{ir_1}^{(n_0)} p_{r_1 r_2}^{(n_0)} \cdots p_{r_{k-1} j}^{(n_0)},$$

and thus

$$p_{iT}^{(n_0 k)} = \sum_{j \in T} p_{ij}^{(n_0 k)} \leq (1 - p)^k, \quad k = 0, 1, 2, \ldots,$$

and

$$p_{iT}^{(n_0 k + l)} = \sum_{j \in T} \sum_{r \in T} p_{ir}^{(n_0 k)} p_{rj}^{(l)} \leq (1 - p)^k \quad (0 \leq l < n_0).$$

Thus, obviously for all $i, j \in T$,

$$p_{ij}^{(n_0 k + l)} \le (1 - p)^k, \qquad k = 0, 1, 2, \ldots, \quad 0 \le l < n_0.$$

Let $n = n_0 k + l$. Then

$$p_{ij}^{(n)} \le (1 - p)^{(n - l)/n_0} \le (1 - p)^{n/n_0 - 1} = \alpha \beta^n,$$

with $\alpha = (1 - p)^{-1} > 0$ and $\beta = (1 - p)^{1/n_0}$. [Proof based on Bhat (1972, p. 67).]   □

In Example 7.1.1(b) we stated some general results that hold for *absorbing* MC's. We establish these results in the next theorem, which parallels Theorem 7.1.2 for regular MC's.

THEOREM 7.1.5:   Let P be the transition matrix of an absorbing MC with canonical form

$$P = \begin{bmatrix} I & 0 \\ R & Q \end{bmatrix}.$$

Then

(a)
$$\lim_{n \to \infty} P^{(n)} = \Pi = \begin{bmatrix} I & 0 \\ NR & 0 \end{bmatrix},$$

where $N = (I - Q)^{-1}$, the fundamental matrix of the absorbing MC.

(b)   A limiting distribution exists but it depends on the initial probability vector. If $\mathbf{p}^{(0)'} = (\mathbf{p}_{\bar{T}}^{(0)'}, \mathbf{p}_T^{(0)'})$, then

$$\lim_{n \to \infty} \mathbf{p}^{(n)'} = (\mathbf{p}_{\bar{T}}^{(0)'} + \mathbf{p}_T^{(0)'} NR, \mathbf{0}').$$

(c)   A unique stationary distribution does not exist. In fact, $\pi' = (\pi_{\bar{T}}', \mathbf{0}')$ is a stationary probability vector for any $\pi_{\bar{T}}'$ such that $\pi' e = 1$.

(d)   If $\mathbf{p}^{(0)'} = (\pi_{\bar{T}}', \mathbf{0}')$, then $\mathbf{p}^{(n)'} = (\pi_{\bar{T}}', \mathbf{0}')$ for all $n \ge 1$.

(e) Π satisfies the equation $\Pi P = P \Pi = \Pi$. Further, any matrix A such that $AP = PA = A$ must be of the form

$$A = \begin{bmatrix} A_1 & 0 \\ NRA_1 & 0 \end{bmatrix},$$

where A_1 is an arbitrary submatrix.

Proof: (a)   The result follows immediately from Corollary 6.1.1A where we established that

$$P^{(n)} = \begin{bmatrix} I & 0 \\ N(I - Q^n)R & Q^n \end{bmatrix}$$

and that $\lim_{n \to \infty} Q^n = 0$.

(b)   Let $\mathbf{p}^{(0)\prime} = (\mathbf{p}_{\bar{T}}^{(0)\prime}, \mathbf{p}_{T}^{(0)\prime})$. Then, for all $n \geq 1$,

$$\mathbf{p}^{(n)\prime} = \mathbf{p}^{(0)\prime} P^{(n)} = (\mathbf{p}_{\bar{T}}^{(0)\prime}, \mathbf{P}_{T}^{(0)\prime}) \begin{bmatrix} I & 0 \\ N(I - Q^n)R & Q^n \end{bmatrix}$$

$$= (\mathbf{p}_{\bar{T}}^{(0)\prime} + \mathbf{p}_{T}^{(0)\prime} N(I - Q^n)R, \ \mathbf{p}_{T}^{(0)\prime} Q^n),$$

and thus

$$\lim_{n \to \infty} \mathbf{p}^{(n)\prime} = (\mathbf{p}_{\bar{T}}^{(0)\prime} + \mathbf{p}_{T}^{(0)\prime} NR, \ \mathbf{0}\prime).$$

(c)   Suppose $\pi\prime = (\pi_{\bar{T}}\prime, \pi_{T}\prime)$ is stationary. Then $\pi\prime P = \pi\prime$, i.e.,

$$(\pi_{\bar{T}}\prime, \pi_{T}\prime) \begin{bmatrix} I & 0 \\ R & Q \end{bmatrix} = (\pi_{\bar{T}}\prime, \pi_{T}\prime).$$

Equating components on carrying out the block multiplication yields

$$\pi_{\bar{T}}\prime + \pi_{T}\prime R = \pi_{\bar{T}}\prime, \qquad \pi_{T}\prime Q = \pi_{T}\prime.$$

These equations give $\pi_{T}\prime R = \mathbf{0}\prime$ and $\pi_{T}\prime(I - Q) = \mathbf{0}\prime$. But since $(I - Q)^{-1}$ exists the last equation implies that $\pi_{T}\prime = \mathbf{0}\prime$, which also satisfies the first equation. Thus $\pi\prime = (\pi_{\bar{T}}\prime, \mathbf{0}\prime)$ is stationary for any $\pi_{\bar{T}}\prime$.

(d)   This follows from the proof of part (c) by putting $\mathbf{p}_{\bar{T}}^{(0)\prime} = \pi_{\bar{T}}\prime$ and $\mathbf{p}_{T}^{(0)\prime} = \mathbf{0}\prime$.

(e)   Let

$$A = \begin{bmatrix} A_1 & A_2 \\ A_3 & A_4 \end{bmatrix}.$$

Block multiplication to determine the A_1, A_2, A_3, and A_4, such that $AP = PA = A$, yields

$$\begin{bmatrix} A_1 + A_2 R & A_2 Q \\ A_3 + A_4 R & A_4 Q \end{bmatrix} = \begin{bmatrix} A_1 & A_2 \\ RA_1 + QA_3 & RA_2 + QA_4 \end{bmatrix}$$

$$= \begin{bmatrix} A_1 & A_2 \\ A_3 & A_4 \end{bmatrix}.$$

Observe that $A_2 Q = A_2$ and $A_4 Q = A_4$ and hence $A_2(I - Q) = 0$ and $A_4(I - Q) = 0$. Since $I - Q$ is nonsingular, this implies that $A_2 = 0$ and $A_4 = 0$. With this observation we have that $A_3 = RA_1 + QA_3$ and hence that $A_3 = NRA_1$. Consequently any matrix so partitioned with A_1 arbitrary, $A_2 = 0$, $A_3 = NRA_1$, and $A_4 = 0$ satisfies the equation $AP = PA = A$. By taking $A_1 = I$, A becomes Π and hence $\Pi P = P\Pi = \Pi$. $\square$

Observe that Π is fully determined once NR is known. Furthermore, from Theorem 7.1.5(a), for $i \in T$ and $j \in \bar{T}$,

$$\pi_{ij} = \lim_{n \to \infty} p_{ij}^{(n)} = (NR)_{ij}$$

$$= \text{Probability that starting in } i \in T \text{ the MC will be absorbed in } j \in \bar{T}$$

$$= f_{ij}, \quad \text{(as also seen in Theorem 6.3.1)}.$$

Theorem 7.1.5(e) gives us an alternative method for determining such probabilities by solving the linear equations $P\Pi = \Pi$, as follows:

Suppose that there are r absorbing states and t transient states and thus $S = \bar{T} \cup T$ where $\bar{T} = \{1, \ldots, r\}$, $T = \{r + 1, \ldots, r + t\}$. From Theorem 7.1.5(a), Π can be expressed as

$$\Pi = [\pi_1, \ldots, \pi_r, 0, \ldots, 0],$$

where for $j = 1, \ldots, r$,

$$\pi_j = \begin{bmatrix} \mathbf{e}_j \\ \mathbf{f}_j \end{bmatrix} \quad \text{with} \quad \begin{array}{ll} \mathbf{e}_j' = (0, \ldots, 0, 1, 0, \ldots, 0) & \text{(an } 1 \times r \text{ vector)}, \\ \mathbf{f}_j' = (f_{r+1, j}, \ldots, f_{r+t, j}) & \text{(an } 1 \times t \text{ vector)}. \end{array}$$

Since $P\Pi = \Pi$ we have $P\pi_j = \pi_j$ for $j = 1, \ldots, r$ and hence

$$\begin{bmatrix} I & 0 \\ R & Q \end{bmatrix} \begin{bmatrix} \mathbf{e}_j \\ \mathbf{f}_j \end{bmatrix} = \begin{bmatrix} \mathbf{e}_j \\ \mathbf{f}_j \end{bmatrix},$$

leading to

$$R\mathbf{e}_j + Q\mathbf{f}_j = \mathbf{f}_j.$$

Thus

$$(I - Q)\mathbf{f}_j = R\mathbf{e}_j \quad (j = 1, \ldots, r),$$

a system of linear equations that can be solved for $\mathbf{f}_j (= NR\mathbf{e}_j)$ leading to the required expressions for the π_{ij} or f_{ij}.

EXAMPLE 7.1.3: *Gambler's Ruin Model.* We have seen earlier in Example 6.1.4 that this model leads to an absorbing MC with absorbing barriers at 0 and a. With the transition matrix as given in Example 6.1.4 it is easily seen that the stationary probabilities $\{\pi_i, i = 0, 1, \ldots, a\}$ must satisfy the equations

$$\pi_0 = \pi_0 + q\pi_1,$$

$$\pi_1 = q\pi_2,$$

$$\pi_i = p\pi_{i-1} + q\pi_{i+1}, \quad i = 2, \ldots, a - 2,$$

$$\pi_{a-1} = p\pi_{a-2}$$

$$\pi_a = p\pi_{a-1} + \pi_a.$$

Thus, $\pi_1 = \pi_2 = \cdots = \pi_{a-1} = \pi_{a-2} = 0$ with π_0 and π_a arbitrarily chosen so that $\pi_0 + \pi_a = 1$; and hence $\pi_0 = \alpha$, $\pi_a = 1 - \alpha$ is a stationary distribution for any α $(0 \leq \alpha \leq 1)$ in accordance with Theorem 7.1.5(d). $\quad\square$

Let us now examine the case of finite *irreducible periodic* Markov Chains.

THEOREM 7.1.6: Let P be the transition matrix of a finite irreducible periodic MC with period d and canonical form given by

$$P = \begin{bmatrix} 0 & P_1 & 0 & \cdots & 0 \\ 0 & 0 & P_2 & \cdots & 0 \\ \vdots & \vdots & \vdots & \ddots & \vdots \\ 0 & 0 & 0 & \cdots & P_{d-1} \\ P_d & 0 & 0 & \cdots & 0 \end{bmatrix}.$$

Then

(a) $\lim_{n\to\infty} P^{(n)}$ does not exist but $\lim_{k\to\infty} P^{(kd+l)} = P^l B$, $(0 \leq l < d)$, where

$$B = \begin{bmatrix} \Pi_1 & 0 & \cdots & 0 \\ 0 & \Pi_2 & \cdots & 0 \\ \vdots & \vdots & \ddots & \vdots \\ 0 & 0 & \cdots & \Pi_d \end{bmatrix},$$

with $\Pi_i = \lim_{k\to\infty} A_i^k$, $A_1 = P_1 \cdots P_d$, $A_i = P_i \cdots P_d P_1 \cdots P_{i-1}$ $(i = 2, \ldots, d)$.
The A_i are transition matrices of regular MC's with unique stationary probability vectors π_i' and $\Pi_i = \mathbf{e}\pi_i'$.

(b) A finite irreducible periodic MC has a *unique* stationary probability vector $\pi' = (1/d)(\pi_1', \pi_2', \ldots, \pi_d')$. If $\pi' = (\pi_1, \pi_2, \ldots, \pi_m)$, then $\pi_j = 1/\mu_j > 0$.

(c) Let $\Pi \equiv \mathbf{e}\pi'$. If A is any matrix such that $AP = PA = A$, then A must be a scalar multiple of Π. In particular if $A\mathbf{e} = \mathbf{e}$, then $A = \Pi$. Furthermore,

$$\Pi = \frac{1}{d} \sum_{l=0}^{d-1} P^l B.$$

(d) Given an initial probability vector $\mathbf{p}^{(0)'}$ such that $\mathbf{p}^{(0)'}B = \mathbf{p}^{(0)'}PB = \cdots = \mathbf{p}^{(0)'}P^{d-1}B$, then a limiting distribution exists (a "pseudolimiting" distribution), which is given by the stationary distribution π'. (The family of such initial distributions is nonempty since $\mathbf{p}^{(0)'} = \pi'$ certainly satisfies the required conditions.)

Proof: (a) Firstly $\lim_{n \to \infty} P^n$ does not exist by virtue of Theorems 4.5.2. and 6.1.6. There are eigenvalues on the unit circle besides $\lambda = 1$, contrary to the conditions for convergence of the sequence $\{P^n\}$. However, in in the proof of Theorem 5.4.5 we showed that

$$P^{(d)} = \begin{bmatrix} A_1 & 0 & \cdots & 0 \\ 0 & A_2 & \cdots & 0 \\ \vdots & \vdots & \ddots & \vdots \\ 0 & 0 & \cdots & A_d \end{bmatrix}$$

and thus

$$P^{(kd+l)} = P^l \begin{bmatrix} A_1^k & 0 & \cdots & 0 \\ 0 & A_2^k & \cdots & 0 \\ \vdots & \vdots & \ddots & \vdots \\ 0 & 0 & \cdots & A_d^k \end{bmatrix} \quad (0 \le l < d),$$

where the A_i are as given in the statement of the theorem. Furthermore, since P is irreducible, we can deduce that the A_i are irreducible [Theorem 5.4.5(c)] and aperiodic (cf. the proof of Theorem 6.1.6) and hence the A_i are transition matrices of regular MC's. Thus, from Theorem 7.1.2(a) $\lim_{k \to \infty} A_i^k = \Pi_i$, where each row of Π_i is the unique probability vector π_i' satisfying the equation $\pi_i' = \pi_i' A_i$. The result now follows.

(b) Let $\pi' = (\mathbf{u}_1', \mathbf{u}_2', \dots \mathbf{u}_d')$ be a stationary probability vector for the periodic MC where $\mathbf{u}_i$ represents the stationary probabilities associated with the states in G_{i-1}. The transitions from G_{i-1} to G_i are governed by the subtransition matrix P_i. Now $\pi' = \pi' P$ and hence

$$(\mathbf{u}_1', \mathbf{u}_2', \dots, \mathbf{u}_d') = (\mathbf{u}_1', \mathbf{u}_2', \dots, \mathbf{u}_d') \begin{bmatrix} 0 & P_1 & 0 & \cdots & 0 \\ 0 & 0 & P_2 & \cdots & 0 \\ \vdots & \vdots & \vdots & \ddots & \vdots \\ 0 & 0 & 0 & \cdots & P_{d-1} \\ P_d & 0 & 0 & \cdots & 0 \end{bmatrix}$$

$$= (\mathbf{u}_d' P_d, \mathbf{u}_1' P_1, \dots, \mathbf{u}_{d-1}' P_{d-1}).$$

Equating subvectors, we obtain $\mathbf{u}_1' = \mathbf{u}_d' P_d$, $\mathbf{u}_i' = \mathbf{u}_{i-1}' P_{i-1}$ $(i = 2, \dots, d)$. Thus

$$\pi' = (\mathbf{u}_1', \mathbf{u}_1' P_1, \mathbf{u}_1' P_1 P_2, \dots, \mathbf{u}_1' P_1 P_2 \cdots P_{d-1}). \tag{7.1.3}$$

To find $\mathbf{u}_1'$ note that $\mathbf{u}_1' = \mathbf{u}_d' P_d = \mathbf{u}_1' P_1 P_2 \cdots P_{d-1} P_d$, and thus $\mathbf{u}_1' = \mathbf{u}_1' A_1$.

In part (a) we saw that π_1' is the unique probability vector satisfying the equation $\pi_1' = \pi_1' A_1$ and thus $\mathbf{u}_1' = c\pi_1'$ for some $c > 0$. From Eq. (7.1.3)

$$\pi' = c(\pi_1', \pi_1' P_1, \pi_1' P_1 P_2, \ldots, \pi_1' P_1 P_2 \cdots P_{d-1})$$

Now

$$\pi'\mathbf{e} = 1 = c(\pi_1'\mathbf{e} + \pi_1' P_1\mathbf{e} + \cdots + \pi_1' P_1 P_2 \cdots P_{d-1}\mathbf{e}),$$

where $\mathbf{e}$ is a vector of 1's of appropriate order. Since

$$\pi_1' P_1 P_2 \cdots P_{d-1}\mathbf{e} = \pi_1' P_1 \cdots P_{d-2}\mathbf{e} = \cdots = \pi_1'\mathbf{e} = 1,$$

it is evident that $1 = cd$, implying that $c = 1/d$. Observe that $\pi_1' P_1 = \pi_1' A_1 P_1 = \pi_1' P_1 P_2 \cdots P_d P_1 = \pi_1' P_1 A_2$, but π_2' is the unique stationary probability vector satisfying $\pi_2' = \pi_2' A_2$ so that $\pi_1' P_1 = \pi_2'$ (since $\pi_1' P_1$ is also a probability vector). Similarly we can show that $\pi_{i+1}' = \pi_1' P_1 \cdots P_i$, $(i = 1, \ldots, d-1)$ and hence $\pi' = (1/d)(\pi_1', \pi_2', \ldots, \pi_d')$.

Suppose A_j is of order m_j and thus the states of the MC can be labeled as i_j where $j = 1, 2, \ldots, d$, $i = 1, \ldots, m_j$, $(\sum_1^d m_j = m)$. From part (a)

$$\left(\lim_{k \to \infty} P^{(kd)}\right)_{i_j i_j} = \left(\lim_{k \to \infty} A_j^k\right)_{ii} = (\Pi_j)_{ii} = (\pi_j')_i.$$

Also,

$$\left(\lim_{k \to \infty} P^{(kd)}\right)_{i_j i_j} = \lim_{k \to \infty} p_{i_j i_j}^{(kd)} = \frac{d}{\mu_{i_j}} (> 0)$$

from Theorem 5.2.7(c). Now the i_jth element of π' is given by

$$\frac{1}{d}(\pi_j')_i = \frac{1}{d}\frac{d}{\mu_{i_j}} = \frac{1}{\mu_{i_j}}$$

and hence the result follows by relabeling the states as $1, 2, \ldots, m$.

(c) The proof of the first part follows along similar lines to the proof of Theorem 7.1.2(e). Let A be any matrix satisfying $A = PA = AP$. Then the rows of A must be multiples of the unique stationary probability vector π' and by Theorem 6.1.6 the columns must be multiples of the vector $\mathbf{e}$, the only possible right eigenvectors of the simple eigenvalue $\lambda = 1$. This implies $A = a\Pi$, as in the proof of Theorem 7.1.2(e). Take $A = (1/d)\sum_{l=0}^{d-1} P^l B$. We show that $PA = AP = A$ and hence since $A\mathbf{e} = \mathbf{e}$ we deduce that $A = \Pi$.

$$PA = \frac{1}{d}\sum_{l=0}^{d-1} P^{l+1}B = A + \frac{1}{d}(P^d - I)B$$

where

$$(P^d - I)B = \begin{bmatrix} A_1 - I & 0 & \cdots & 0 \\ 0 & A_2 - I & \cdots & 0 \\ \vdots & \vdots & \ddots & \vdots \\ 0 & 0 & \cdots & A_d - I \end{bmatrix} \begin{bmatrix} \Pi_1 & 0 & \cdots & 0 \\ 0 & \Pi_2 & \cdots & 0 \\ \vdots & \vdots & \ddots & \vdots \\ 0 & 0 & \cdots & \Pi_d \end{bmatrix}$$

$$= \begin{bmatrix} A_1\Pi_1 - \Pi_1 & 0 & \cdots & 0 \\ 0 & A_2\Pi_2 - \Pi_2 & \cdots & 0 \\ \vdots & \vdots & \ddots & \vdots \\ 0 & 0 & \cdots & A_d\Pi_d - \Pi_d \end{bmatrix} = 0,$$

since, by (a),

$$A_i\Pi_i - \Pi_i = A_i \mathbf{e}\pi_i' - \mathbf{e}\pi_i' = \mathbf{e}\pi_i' - \mathbf{e}\pi_i' = 0.$$

This implies that $PA = A$. Furthermore, observe that $PA = AP$ if and only if

$$\frac{1}{d} \sum_{l=0}^{d-1} P^l PB = \frac{1}{d} \sum_{l=0}^{d-1} P^l BP,$$

which is satisfied if $PB = BP$, i.e., if

$$\begin{bmatrix} 0 & P_1\Pi_2 & 0 & \cdots & 0 \\ 0 & 0 & P_2\Pi_3 & \cdots & 0 \\ \vdots & \vdots & \vdots & \ddots & \vdots \\ 0 & 0 & 0 & \cdots & P_{d-1}\Pi_d \\ P_d\Pi_1 & 0 & 0 & \cdots & 0 \end{bmatrix}$$

$$= \begin{bmatrix} 0 & \Pi_1 P_1 & 0 & \cdots & 0 \\ 0 & 0 & \Pi_2 P_2 & \cdots & 0 \\ \vdots & \vdots & \vdots & \ddots & \vdots \\ 0 & 0 & 0 & \cdots & \Pi_{d-1}P_d \\ \Pi_d P_d & 0 & 0 & \cdots & 0 \end{bmatrix}.$$

Now

$$P_i\Pi_{i+1} = P_i\mathbf{e}\pi_{i+1}' = \mathbf{e}\pi_{i+1}' = \Pi_{i+1} \qquad (\Pi_1 \text{ if } i = d)$$

and

$$\Pi_i P_i = \mathbf{e}\pi_i' P_i = \mathbf{e}\pi_1' P_1 \cdots P_{i-1} P_i \qquad \text{[from the proof of (b)]}$$
$$= \mathbf{e}\pi_{i+1}' = \Pi_{i+1} \qquad\qquad\qquad (\Pi_1 \text{ if } i = d).$$

Thus $P_i\Pi_{i+1} = \Pi_i P_i = \Pi_{i+1}$ and $PB = BP$ and hence the required conclusion.

(d)   Let $\mathbf{p}^{(0)'}$ be any initial probability vector and because

$$\lim_{k \to \infty} \mathbf{p}^{(kd+l)'} = \lim_{k \to \infty} \mathbf{p}^{(0)'} P^{(kd+l)} = \mathbf{p}^{(0)'} P^l \lim_{k \to \infty} P^{(kd)}$$

$$= \mathbf{p}^{(0)'} P^l B,$$

if $\mathbf{p}^{(0)'} B = \mathbf{p}^{(0)'} PB = \cdots = \mathbf{p}^{(0)'} P^{d-1} (= \mathbf{p}'$ say), then $\lim_{n \to \infty} \mathbf{p}^{(n)'}$ exists (and $= \mathbf{p}')$.

Since $\mathbf{p}'$ assumes the common value of d equal vectors it must also assume the value of their mean and hence

$$\mathbf{p}' = \mathbf{p}^{(0)'} \left(\frac{1}{d} \sum_{l=0}^{d-1} P^l \right) B = \mathbf{p}^{(0)'} \Pi \qquad \text{[by part (c)]}$$

$$= \mathbf{p}^{(0)'} \mathbf{e}\pi' = \pi'.$$

Thus the "pseudolimiting" distribution is the stationary distribution.   □

EXAMPLE 7.1.4:   *A Four-State, Period-Two Markov Chain.*   This example is included to illustrate some of the main features of Theorem 7.1.6. We consider a MC with transition matrix given by

$$P = \begin{bmatrix} 0 & P_1 \\ P_2 & 0 \end{bmatrix} \quad \text{where } P_1 = \begin{bmatrix} 1 & 0 \\ \frac{2}{3} & \frac{1}{3} \end{bmatrix} \quad \text{and} \quad P_2 = \begin{bmatrix} \frac{1}{3} & \frac{2}{3} \\ 0 & 1 \end{bmatrix}.$$

Thus

$$P^2 = \begin{bmatrix} A_1 & 0 \\ 0 & A_2 \end{bmatrix} \quad \text{where} \quad A_1 = P_1 P_2 = \begin{bmatrix} \frac{1}{3} & \frac{2}{3} \\ \frac{2}{9} & \frac{7}{9} \end{bmatrix}$$

$$\text{and} \quad A_2 = P_2 P_1 = \begin{bmatrix} \frac{7}{9} & \frac{2}{9} \\ \frac{2}{3} & \frac{1}{3} \end{bmatrix}.$$

The transition graph of the MC is given in Fig. 7.1.1, so that P is the transition matrix of a four-state, periodic, period-two MC with subclasses $G_0 = \{1, 2\}$ and $G_1 = \{3, 4\}$.

A_1 and A_2 are irreducible aperiodic transition matrices with unique stationary probability vectors given by $\pi_1' = (\frac{1}{4}, \frac{3}{4})$ and $\pi_2' = (\frac{3}{4}, \frac{1}{4})$.

The unique stationary probability vector of the periodic MC is given by $\pi' = \frac{1}{2}(\pi_1', \pi_2') = (\frac{1}{8}, \frac{3}{8}, \frac{3}{8}, \frac{1}{8})$, which can also be verified directly.

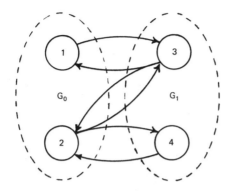

FIGURE 7.1.1.

Observe that

$$\lim_{k \to \infty} P^{(kd+l)} = \begin{cases} B = \begin{bmatrix} \Pi_1 & 0 \\ 0 & \Pi_2 \end{bmatrix} & (l = 0), \\[2ex] PB = \begin{bmatrix} 0 & P_1\Pi_2 \\ P_2\Pi_1 & 0 \end{bmatrix} = \begin{bmatrix} 0 & \Pi_2 \\ \Pi_1 & 0 \end{bmatrix} & (l = 1), \end{cases}$$

where

$$\Pi_1 = e\pi_1' = \begin{bmatrix} \frac{1}{4} & \frac{3}{4} \\ \frac{1}{4} & \frac{3}{4} \end{bmatrix},$$

$$\Pi_2 = e\pi_2' = \begin{bmatrix} \frac{3}{4} & \frac{1}{4} \\ \frac{3}{4} & \frac{1}{4} \end{bmatrix}.$$

Furthermore, the unique matrix Π such that $P\Pi = \Pi P = \Pi$ is given by

$$\Pi = \begin{Bmatrix} e\pi' \\ \frac{1}{2}(B + PB) \end{Bmatrix} = \begin{bmatrix} \frac{1}{8} & \frac{3}{8} & \frac{3}{8} & \frac{1}{8} \\ \frac{1}{8} & \frac{3}{8} & \frac{3}{8} & \frac{1}{8} \\ \frac{1}{8} & \frac{3}{8} & \frac{3}{8} & \frac{1}{8} \\ \frac{1}{8} & \frac{3}{8} & \frac{3}{8} & \frac{1}{8} \end{bmatrix}.$$

Let $\mathbf{p}^{(0)'} = (p_1, p_2, p_3, p_4)$ be an initial probability vector such that $\mathbf{p}^{(0)'}B = \mathbf{p}^{(0)'}PB$, i.e.,

$$\mathbf{p}^{(0)'}B = (p_1, p_2, p_3, p_4) \begin{bmatrix} \frac{1}{4} & \frac{3}{4} & 0 & 0 \\ \frac{1}{4} & \frac{3}{4} & 0 & 0 \\ 0 & 0 & \frac{3}{4} & \frac{1}{4} \\ 0 & 0 & \frac{3}{4} & \frac{1}{4} \end{bmatrix}$$

$$= (\tfrac{1}{4}(p_1 + p_2), \tfrac{3}{4}(p_1 + p_2), \tfrac{3}{4}(p_3 + p_4), \tfrac{1}{4}(p_3 + p_4)),$$

and

$$\mathbf{p}^{(0)'}PB = (p_1, p_2, p_3, p_4)\begin{bmatrix} 0 & 0 & \frac{3}{4} & \frac{1}{4} \\ 0 & 0 & \frac{3}{4} & \frac{1}{4} \\ \frac{1}{4} & \frac{3}{4} & 0 & 0 \\ \frac{1}{4} & \frac{3}{4} & 0 & 0 \end{bmatrix}$$

$$= (\tfrac{1}{4}(p_3 + p_4), \tfrac{3}{4}(p_3 + p_4), \tfrac{3}{4}(p_1 + p_2), \tfrac{1}{4}(p_1 + p_2)).$$

Equating terms implies that $p_1 + p_2 = p_3 + p_4 \, (= \tfrac{1}{2})$. Hence if

$$\mathbf{p}^{(0)'} = (p_1, \tfrac{1}{2} - p_1, \tfrac{1}{2} - p_4, p_4),$$

then $\lim_{n \to \infty} \mathbf{p}^{(n)'}$ exists and is given by $\boldsymbol{\pi}' = (\tfrac{1}{8}, \tfrac{3}{8}, \tfrac{3}{8}, \tfrac{1}{8})$ for all p_1, p_4 $(0 \le p_1 \le \tfrac{1}{2}, 0 \le p_4 \le \tfrac{1}{2})$.

Note that this family of initial probability vectors also includes the stationary probability vector as a member (when $p_1 = \tfrac{1}{8}$, $p_4 = \tfrac{1}{8}$).   □

We have seen that in general $\lim_{n \to \infty} P^n$ need not exist (cf. Theorem 7.1.6). However, for any Markov chain, a Cesaró limit of $\{P^n\}$ always exists.

THEOREM 7.1.7:   For any Markov chain

$$\lim_{n \to \infty} \frac{1}{n} \sum_{k=1}^{n} P^k \qquad \text{exists.}$$

Proof: To say that the above limit exists means that

$$\lim_{n \to \infty} \left(\frac{1}{n}\right) \sum_{k=1}^{n} p_{ij}^{(k)}$$

exists for every $i, j \in S$.

Let $P_{ij}(s) = \sum_{n=0}^{\infty} p_{ij}^{(n)} s^n$ be the generating function of the $p_{ij}^{(n)}$. By Theorem 2.3.5, since $p_{ij}^{(n)} \ge 0$, the required limit will exist and will assume the value π_{ij} if $\lim_{s \uparrow 1}(1 - s)P_{ij}(s)$ exists and equals π_{ij}.

Now, when $i = j$, $\lim_{s \uparrow 1}(1 - s)P_{jj}(s)$ exists [by Theorem 3.3.7 by identifying $U(s)$ with $P_{jj}(s)$ in reference to the recurrent event "return to state j"] and assumes the value zero except when j is nonnull persistent, in which case the limit is $1/\mu_j$.

When $i \ne j$, we have from Theorem 6.2.5,

$$\lim_{s \uparrow 1}(1 - s)P_{ij}(s) = \lim_{s \uparrow 1}(1 - s)P_{jj}(s)F_{ij}(s)$$

$$= \lim_{s \uparrow 1}(1 - s)P_{jj}(s) \lim_{s \uparrow 1} F_{ij}(s)$$

$$= f_{ij} \lim_{s \uparrow 1}(1 - s)P_{jj}(s),$$

which exists by virtue of the $i = j$ case.   □

The following corollaries follow from the proof of the theorem.

COROLLARY 7.1.7A: For any MC if $P_{jj}(s) = \sum_{n=0}^{\infty} p_{jj}^{(n)} s^n$, then

$$\lim_{s \uparrow 1} (1 - s) P_{jj}(s) = u_j = \begin{cases} \dfrac{1}{\mu_j} & \text{if } j \text{ is persistent nonnull,} \\ 0 & \text{if } j \text{ is persistent null or transient.} \end{cases}$$

Furthermore,

$$\lim_{s \uparrow 1} (1 - s) P_{jj}(s) = u_j \qquad \text{iff} \qquad \lim_{n \to \infty} \frac{1}{n} \sum_{k=1}^{n} p_{jj}^{(k)} = u_j. \quad \square$$

COROLLARY 7.1.7B: If

$$\lim_{n \to \infty} \frac{1}{n} \sum_{k=1}^{n} p_{jj}^{(k)} = u_j \qquad \text{for some } j \in S,$$

then

$$\lim_{n \to \infty} \frac{1}{n} \sum_{k=1}^{n} p_{ij}^{(k)} = f_{ij} u_j \qquad \text{for all } i (\neq j) \in S. \quad \square$$

COROLLARY 7.1.7C: For all $i, j \in S$

$$\lim_{n \to \infty} \frac{1}{n} \sum_{k=1}^{n} p_{ij}^{(k)} = \begin{cases} \dfrac{f_{ij}}{\mu_j} & \text{if } j \text{ is persistent nonnull,} \\ 0 & \text{otherwise.} \quad \square \end{cases}$$

COROLLARY 7.1.7D: If the MC is irreducible, then

$$\lim_{n \to \infty} \frac{1}{n} \sum_{k=1}^{n} P^k = \Pi = [\pi_{ij}],$$

where

$$\pi_{ij} = \begin{cases} \dfrac{1}{\mu_j} & \text{if the states are persistent nonnull,} \\ 0 & \text{if the states are persistent null,} \\ 0 & \text{if the states are transient.} \quad \square \end{cases}$$

Let us now restrict attention to irreducible MC's and consider the relationships between their limiting and stationary distributions.

Firstly we are in a position to summarize the results for *finite irreducible* MC's. Such MC's contain only persistent nonnull states (Theorem 5.3.10) and thus the MC is either regular or periodic irreducible. A comparison of Theorems 7.1.2, 7.1.6, and Corollary 7.1.7D together with some observations from Theorems 6.1.5, 6.1.6, and 4.6.1 (the Perron–Frobenius theorem) lead to the following summary.

THEOREM 7.1.8:   Let P be the transition matrix of a *finite irreducible* MC. Then

(a) $$\lim_{n\to\infty} \frac{1}{n} \sum_{k=1}^{n} P^k = \Pi = \mathbf{e}\boldsymbol{\pi}';$$

(b) $\lim_{n\to\infty} P^n = \Pi = \mathbf{e}\boldsymbol{\pi}'$ iff the MC is aperiodic; $\lim_{n\to\infty} P^n$ does not exist if the MC is periodic.

(c) $\boldsymbol{\pi}'$ is the *unique* positive stationary probability vector and if $\boldsymbol{\pi}' = (\pi_1, \pi_2, \ldots, \pi_m)$, then $\pi_j = 1/\mu_j$, the mean recurrence time of state j.

Furthermore, P has a *simple* eigenvalue $\lambda = 1$ and, apart from multiplication by a scalar constant, $\mathbf{e}$ is the *unique* right eigenvector of $\lambda = 1$ and $\boldsymbol{\pi}'$ is the *unique* left eigenvector of $\lambda = 1$. $\square$

When the MC is *infinite* and *irreducible* there are three cases to consider. All the states are either persistent nonnull, persistent null, or transient. The following theorem summarizes the key results for such chains.

THEOREM 7.1.9:   Let P be the transition matrix of an *infinite irreducible* MC. Then

(a) $$\lim_{n\to\infty} \frac{1}{n} \sum_{k=1}^{n} P^k = \Pi = [\pi_{ij}],$$

where

$$\pi_{ij} = \begin{cases} \dfrac{1}{\mu_j} & \text{if the states are persistent nonnull,} \\ 0 & \text{otherwise.} \end{cases}$$

(b) $\lim_{n\to\infty} P^n$ does not exist if the states are persistent nonnull and periodic; otherwise $\lim_{n\to\infty} P^n = \Pi$.

(c)   The MC has a stationary distribution if and only if the states are persistent nonnull, in which case the stationary distribution $\{\pi_j\}$ $(j \in S)$ is unique with $\pi_j = 1/\mu_j > 0$.

Proof: (a)   Part (a) is a restatement of Corollary 7.1.7D.

(b)   Part (b) is an application of Theorem 5.2.8.

(c)   Firstly, let a stationary distribution $\{v_j\}$ exist. From Theorem 7.1.1(a)

$$v_j = \sum_{i \in S} v_i p_{ij}^{(n)}.$$

Multiplying by s^n and adding over $n = 0, 1, \ldots$, we obtain

$$\frac{v_j}{1-s} = \sum_{i \in S} v_i P_{ij}(s)$$

(the interchange of summation being valid for $|s| < 1$). From Theorem 6.2.5, since $P_{ij}(s) = F_{ij}(s)P_{jj}(s)$ $(i \neq j)$ and $P_{jj}(s) = F_{jj}(s)P_{jj}(s) + 1$, we obtain

$$\frac{v_j}{1-s} = \sum_{i \neq j} v_i F_{ij}(s)P_{jj}(s) + v_j F_{jj}(s)P_{jj}(s) + v_j,$$

leading to

$$sv_j = (1-s)P_{jj}(s) \sum_{i \in S} v_i F_{ij}(s).$$

Using Corollary 7.1.7A by taking the limit as $s \uparrow 1$ we see that

$$v_j = u_j \sum_{i \in S} v_i f_{ij}. \tag{7.1.4}$$

The interchange of the limit as $s \uparrow 1$ and the summation is valid by Theorems 2.2.3(a) and 2.2.2 since for $|s| \leq 1$

$$|v_i F_{ij}(s)| \leq v_i f_{ij} \leq v_i \qquad \text{where} \quad \sum_{i \in S} v_i = 1.$$

In addition this implies from Eq. (7.1.4) that $v_j \leq u_j$. From Corollary 7.1.7(a), $u_j = 0$ in the transient and persistent null cases and hence $v_j = 0$, which is impossible. Thus $u_j > 0$ for at least one j and hence the persistent nonnull case occurs so that $u_j > 0$ for all j. In the persistent nonnull case we have from Theorem 5.3.6 $f_{ij} = 1$ for all $i, j \in S$ and thus from Eq. (7.1.4)

$$v_j = u_j = 1/\mu_j \qquad (j \in S).$$

This proves the "only if" statement and that the stationary distribution is unique with $\pi_j = 1/\mu_j$.

Next let the chain be persistent nonnull so that $u_j = \pi_j > 0$. From the fact that $\sum_{j \in S} p_{ij}^{(n)} = 1$ it follows that

$$\frac{1}{1-s} = \sum_{n=0}^{\infty} s^n \sum_{j \in S} p_{ij}^{(n)} = \sum_{j \in S} \sum_{n=0}^{\infty} p_{ij}^{(n)} s^n = \sum_{j \in S} P_{ij}(s)$$

with the interchange of summations being valid by Theorem 2.2.4 for $|s| < 1$.

Once again we use the aforementioned results of Theorem 6.2.5 separating out the cases $i = j$ and $i \neq j$ to obtain

$$s = (1-s) \sum_{j \in S} F_{ij}(s)P_{jj}(s).$$

Without loss of generality we can take $S = \{1, 2, \ldots\}$ so that

$$s = (1 - s) \sum_{j=1}^{m} F_{ij}(s)P_{jj}(s) + (1 - s) \sum_{j=m+1}^{\infty} F_{ij}(s)P_{jj}(s)$$

and hence for $0 < s < 1$ and any finite $m \geq 1$

$$s \geq \sum_{j=1}^{m} (1 - s)P_{jj}(s)F_{ij}(s).$$

Taking the limit as $s \uparrow 1$ we obtain, using Corollary 7.1.7A, that

$$1 \geq \sum_{j=1}^{m} u_j f_{ij}$$

and hence also that

$$1 \geq \sum_{j \in S} u_j f_{ij}.$$

Thus, since $f_{ij} = 1$ (Theorem 5.3.6), we have that

$$\sum_{j \in S} u_j \leq 1. \tag{7.1.5}$$

From Theorem 6.1.8 [Eq. (6.1.16)]

$$P_{ij}(s) = s \sum_{k \in S} P_{ik}(s)p_{kj} + \delta_{ij}$$

so that for $i \neq j$, using Theorem 6.2.5 once again, we obtain

$$F_{ij}(s)P_{jj}(s) = s \sum_{k \in S} F_{ik}(s)P_{kk}(s)p_{kj} + sp_{ij}.$$

Multiplication by $(1 - s)$ $(0 < s < 1)$ and truncation with any finite $m \geq 1$ gives

$$(1 - s)F_{ij}(s)P_{jj}(s) \geq s \sum_{k=1}^{m} F_{ik}(s)(1 - s)P_{kk}(s)p_{kj}.$$

Taking the limit as $s \uparrow 1$ as before yields

$$f_{ij}u_j \geq \sum_{k=1}^{m} f_{ik}u_k p_{kj}$$

and hence

$$f_{ij}u_j \geq \sum_{k \in S} f_{ik}u_k p_{kj}.$$

Using the fact that $f_{ij} = 1$ for all $i, j \in S$, this simplifies to

$$u_j \geq \sum_{k \in S} u_k p_{kj}. \tag{7.1.6}$$

By Eq. (7.1.5) $\sum_{j \in S} u_j$ is finite. Summing Eq. (7.1.6) on j gives

$$\sum_{j \in S} u_j \geq \sum_{k \in S} u_k \sum_{j \in S} p_{kj} = \sum_{k \in S} u_k.$$

A strict inequality is impossible and hence equality in Eq. (7.1.6) holds, i.e.,

$$u_j = \sum_{k \in S} u_k p_{kj}.$$

Recall that $u_j > 0$. Let $v_j = u_j / \sum_{k \in S} u_k$. Then $\{v_j\}$ is a stationary distribution. But then it has been proved that $v_j = u_j = 1/\mu_j$. Hence $\sum_{j \in S} u_j = 1$. $\square$

A few remarks concerning this theorem are called for. If we work through Theorem 7.1.9 when the MC is finite considerable simplification occurs. In fact, the complications that arose over the interchange of limits and sums are no longer a problem and all the inequalities are automatically equalities.

Theorem 7.1.9 includes both the aperiodic and periodic cases. The proof is based upon a proof given by Levinson (1965). Alternative proofs can be given for the aperiodic case (see, e.g., Prabhu, 1965; Feller, 1968), which are slightly simpler since in this case the n-step transition probabilities all have limits as well as Cesaró limits. (Our proof is based upon the existence of the Cesaró limit via Corollary 7.1.7A.) The alternative proofs are basically a generalization of Theorem 7.1.2.

In the aperiodic case a limit distribution exists when the states are persistent nonnull (i.e., ergodic) and is given by the stationary distribution. We summarize this in following easily deduced corollary.

COROLLARY 7.1.9A: In every *aperiodic irreducible* MC, $\lim_{n \to \infty} p_j^{(n)} = u_j$ ($j \in S$) exists, independent of the initial probability distribution $\{p_k^{(0)}\}$, where $u_j = 0$ for all $j \in S$ if the MC is transient or persistent null or $u_j = 1/\mu_j$ if the MC is persistent nonnull (ergodic). The $\{u_j\}$ sequence is the unique probability distribution satisfying the stationary equations $u_j = \sum_{k \in S} u_k p_{kj}$. $\square$

The term *equilibrium distribution* appears in some texts when limiting and stationary distributions are being considered. The interpretation given is usually based upon a "physical" analog by regarding the states of the MC as small reservoirs connected by pipes through which liquid can flow, with valves to ensure that the flow goes only in the direction of the arrows. The probability p_{ij} associated with any arc is thought of as the fraction of liquid in reservoir i that will pass to reservoir j in one transition time unit. One unit of liquid is poured into the system initially and assigned to the states according to the initial distribution $\{p_j^{(0)}\}$. After a while a "steady state" or "equilibrium" situation is reached where the liquid continues to flow but the amount in each reservoir remains constant. When this situation is reached the flow into any reservoir must equal the flow out. If this condition is met for all reservoirs and π_k is the proportion of liquid at state k $(k \in S)$, then for all $j \in S$

$$\text{Flow in} = \sum_{i \in S} \pi_i p_{ij},$$

$$\text{Flow out} = \sum_{k \in S} \pi_j p_{jk} = \pi_j \sum_{k \in S} p_{jk} = \pi_j.$$

Equating these flows implies that $\pi_j = \sum_{i \in S} \pi_i p_{ij}$ subject to $\sum_{j \in S} \pi_j = 1$. Thus for an irreducible MC the term "equilibrium distribution" refers to the "stationary distribution," which coincides with the "limiting distribution" in the case of an ergodic MC.

We conclude this section with a variety of theorems, due in the main to Foster (1953), that are often very useful in aiding us in determining whether an infinite irreducible MC is transient, persistent null or persistent nonnull. Some theorems have already been deduced to aid us in this quest, in particular Theorems 6.3.4 and 7.1.9(c) and Corollary 7.1.9A. Without loss of generality we shall assume that the state space of the MC is $S = \{0, 1, 2, 3, \ldots\}$ and that the transition matrix is $P = [p_{ij}]$.

*THEOREM 7.1.10: *Foster's Theorem.* An irreducible aperiodic MC is persistent nonnull (ergodic) if there exists a nonnull solution of the equations

$$\sum_{i=0}^{\infty} x_i p_{ij} = x_j, \qquad j = 0, 1, 2, \ldots, \tag{7.1.7}$$

such that $\sum_{i=0}^{\infty} |x_i| < \infty$; and only if this property is possessed by any nonnegative solution of the inequalities

$$\sum_{i=0}^{\infty} x_i p_{ij} \leq x_j, \qquad j = 0, 1, 2, \ldots. \tag{7.1.8}$$

Proof: From Theorem 7.1.9, $\lim_{n \to \infty} p_{ij}^{(n)} = \pi_j$ always exists and is independent of i, and further either $\pi_j > 0$ for all j or $\pi_j = 0$. The system is ergodic if and only if $\pi_j > 0$. For any nonnull absolutely convergent solution $\{x_i\}$ of Eq. (7.1.7)

$$\sum_{i=0}^{\infty} x_i p_{ij}^{(n)} = x_j, \qquad j = 0, 1, 2, \ldots \tag{7.1.9}$$

for all n and so

$$\sum_{i=0}^{\infty} x_i \pi_j = x_j, \qquad j = 0, 1, \ldots. \tag{7.1.10}$$

Therefore $\pi_j > 0$ (for otherwise the solution would be null), and so the system is ergodic.

Conversely, suppose the system is ergodic and thus $\pi_j > 0$. Let $\{x_j\}$ be any nonnegative solution of Eq. (7.1.8). Then we have also Eq. (7.1.9) and (7.1.10) with inequalities. Therefore $\sum x_j < \infty$. $\quad\square$

*THEOREM 7.1.11: An irreducible MC is transient if and only if there exists a bounded nonconstant solution of the equations

$$\sum_{j=0}^{\infty} p_{ij} y_j = y_i, \qquad i = 1, 2, \dots . \tag{7.1.11}$$

*Proof: The system will be persistent if and only if there is probability unity that the zero state is eventually reached from any state $i \neq 0$. Consider therefore the modified system, as in the proof of Theorem 6.3.4, in which the zero is made absorbing. Denote the modified transition matrix by $[p'_{ij}]$ and thus $p'_{00} = 1$, $p'_{ij} = p_{ij}$ for $i \neq 0$. Now $\lim_{n \to \infty} p'^{(n)}_{ij} = \pi'_{ij}$ where $\pi'_{ij} = 0$ for $j \neq 0$ [Theorem 7.1.9(b)], and π'_{i0} is the probability that the zero state is eventually attained from the ith state [Theorem 5.2.8(b)], i.e., $\pi'_{i0} = f'_{i0}$. Now $\sum_{j=0}^{\infty} p'_{ij} \pi'_{j0} = \pi'_{i0}$ for all i (by Theorem 5.2.9 with $f'_{00} = 1$). If the original system is transient, $\pi'_{i0} < 1$ for some i and in all cases $\pi'_{00} = 1$. Therefore, defining $y_j = \pi'_{j0}$ we have a bounded nonconstant solution for Eq. (7.1.11).

Conversely, now assume that we have a nonconstant, bounded solution $\{y_i\}$ of Eq. (7.1.11). Since the constant vector is also a solution of Eq. (7.1.11) then $z_i = a y_i + b_i$ is a solution, which for a suitable choice of a, b will satisfy $z_0 = 1$, $0 \leq z_i \leq 2$. We may therefore assume that $y_0 = 1$ and $0 \leq y_i \leq 2$. Then $\sum_{j=0}^{\infty} p'_{ij} y_j = y_i$ for all $i \geq 0$, and iterating we have for all $i \geq 0$ and $n \geq 1$, $\sum_{j=0}^{\infty} p'^{(n)}_{ij} y_j = y_i$.

Now, since the original system is irreducible, $p^{(n)}_{i0} > 0$ for each i and some n. Therefore, each of the states $j = 1, 2, \dots$ must be transient in the modified system so that $p'^{(n)}_{ij} \to 0$ for $j \neq 0$ and $p'^{(n)}_{i0} \to f'_{i0}$. Therefore, since for all $i \geq 0$,

$$p'^{(n)}_{i0} = p'^{(n)}_{i0} y_0 \leq \sum_{j=0}^{\infty} p'^{(n)}_{i0} y_j = y_i,$$

letting $n \to \infty$ we have $f'_{i0} \leq y_i$. Two possible cases arise. Either there exists a y_k ($k \neq 0$), such that $y_k < 1$ or there exists a y_k ($k \neq 0$), such that $y_k > 1$. In the first case, $f'_{k0} \leq y_k < 1$, which implies that the original process is transient, since by hypothesis k can be reached from state 0 and the probability of returning is less than 1. In the second case we apply the preceding argument to the solution $z_i = 2 - y_i$ of Eq. (7.1.11) and obtain $f_{k0} \leq z_k < 1$, implying again that the process is transient. $\square$

*THEOREM 7.1.12: An irreducible MC is persistent if there exists a solution $\{y_i\}$ of the inequalities

$$\sum_{j=0}^{\infty} p_{ij} y_j \leq y_i, \qquad i = 1, 2, \dots , \tag{7.1.12}$$

such that $y_i \to \infty$ as $i \to \infty$.

Proof: Using the same "matrix modification" technique as in the previous theorem, we have

$$\sum_{j=0}^{\infty} p'_{ij} y_j \leq y_i \qquad \text{for all } i. \tag{7.1.13}$$

Since $z_i = y_i + b$ satisfies Eq. (7.1.12), we may assume that $y_i > 0$ for all $i \geq 0$. Iterating inequality (7.1.13) we have, for all $m \geq 1$,

$$\sum_{j=0}^{\infty} p'^{(m)}_{ij} y_j \leq y_i.$$

Given $\varepsilon > 0$ we choose $M(\varepsilon)$ such that $1/y_i \leq \varepsilon$ for $i \geq M(\varepsilon)$. Now

$$\sum_{j=0}^{M-1} p'^{(m)}_{ij} y_j + \sum_{j=M}^{\infty} p'^{(m)}_{ij} y_j \leq y_i,$$

and so

$$\sum_{j=0}^{M-1} p'^{(m)}_{ij} y_j + y \sum_{j=M}^{\infty} p'^{(m)}_{ij} \leq y_i,$$

where $y = \min_{r \geq M} y_r$. Since $\sum_{j=0}^{\infty} p'^{(m)}_{ij} = 1$ we have

$$\sum_{j=0}^{M-1} p'^{(m)}_{ij} y_j + y \left(1 - \sum_{j=0}^{M-1} p'^{(m)}_{ij} \right) \leq y_i.$$

As observed in the proof of Theorem 7.1.11, $p'^{(n)}_{ij} \to 0$ for $j \neq 0$, and thus, passing to the limit as $m \to \infty$, we obtain for each fixed i

$$f'_{i0} y_0 + y(1 - f'_{i0}) \leq y_i$$

or

$$1 - f'_{i0} \leq (1/y)(y_i - f'_{i0} y_0) \leq \varepsilon K$$

where $K = y_i - f'_{i0} y_0$. Since ε was arbitrary and $f'_{i0} \leq 1$, we have that $f'_{i0} = 1$ for each i proving that the original MC is persistent. $\square$

Foster (1953) also examines some additional variants to these theorems and if the reader is interested in pursuing the matter further he should consult this reference.

Exercises 7.1

1. Derive the result of Theorem 7.1.2(a) by using the results of Corollary 4.5.3A.

2. Let $\{X_n\}$ $(n = 0, 1, 2, \ldots)$ be a Markov chain with state space $S = \{1, 2, 3\}$ and transition matrix given by

$$P = \begin{bmatrix} r_0 & p_0 & 0 \\ q_1 & r_1 & p_1 \\ 0 & q_2 & r_2 \end{bmatrix} \qquad (p_0 + r_0 = p_1 + q_1 + r_1 = q_2 + r_2 = 1).$$

(a)   Deduce a set of conditions (in terms of p_0, p_1, q_1, and q_2) that will ensure that the MC is irreducible.

(b)   Are the conditions in (a) sufficient to ensure the MC is regular?

(c)   Find a stationary distribution for this MC.

(d)   Under what conditions is the stationary distribution unique?

3. The Ehrenfest diffusion model can be represented by a Markov chain. The model consists of N balls numbers respectively $1, 2, \ldots, N$, being initially distributed at random between two containers. At each trial one of the integers $1, 2, \ldots, N$ is selected at random and the ball with that number is transferred from the container it is in to the other container. Let $S = \{0, 1, \ldots, N\}$ denote the state space for the number of balls in the first container. Find the transition probabilities for this MC and justify the existence of a unique stationary distribution. Verify that the stationary distribution is given by

$$\pi_j = \binom{N}{j} \bigg/ 2^N, \qquad j = 0, 1, \ldots, N.$$

4. N black balls and N white balls are placed in two urns so that each urn contains N balls. The number of black balls in the first urn is the state of the system. At each step one ball is selected at random from each urn and the two balls interchanged. Find the transition probabilities of this MC, p_{ij} $(i, j = 0, 1, \ldots, N)$. Show that a stationary distribution $\{\pi_j\}$ exists and establish that

$$\pi_j = \binom{N}{j}^2 \bigg/ \binom{2N}{N}, \qquad j = 0, 1, \ldots, N.$$

5. Let P be the transition matrix of a finite irreducible MC and let π' be the unique stationary probability vector of the MC. Show that for any $0 < \delta < 1$

$$\lim_{n \to \infty} \sum_{i=0}^{n} \binom{n}{i} \delta^{n-i} (1 - \delta)^i P^i = \Pi = e\pi'.$$

[*Hint*: Take $P_\delta = \delta I + (1 - \delta)P$ and show that P_δ is the transition matrix of a regular MC.] (Such a convergence implies that P^n is Euler-summable to Π.) [Kemeny and Snell (1960).]

6. Let $P = [p_{ij}]$ be the transition matrix of a finite irreducible Markov chain with state space $S = \{1, 2, \ldots, m\}$ and let π' be its unique stationary probability vector. Show that a necessary and sufficient condition for $\pi_j = 1/m, j = 1, 2, \ldots, m$, is that $\sum_{i=1}^{m} p_{ij} = 1, j = 1, 2, \ldots, m$, i.e., a finite irreducible MC has equal stationary probabilities iff the transition matrix is doubly stochastic.

7. Let $\{X_n\}$ be a MC with transition matrix $P = [p_{ij}]$ and for each state i define the *holding time r.v.* τ_i as the number of occurrences of state i before the next different state is entered.
   (a)   Show that

   $$P[\tau_i = n] = (1 - p_{ii})p_{ii}^{n-1} \qquad (n = 1, 2, \ldots),$$

   with

   $$\mathsf{E}\tau_i = 1/(1 - p_{ii})$$

   and

   $$\mathrm{var}\,\tau_i = p_{ii}/(1 - p_{ii})^2.$$

   (b)   It is possible to define a new MC $\{Y_n\}$ that consists only of those $\{X_n\}$ for which a change of state is observed, i.e., $\{Y_n\}$ records only the "real" transitions. Let the transition probabilities of the $\{Y_n\}$ chain be p_{ij}^r. Show

   $$p_{ij}^r = \begin{cases} \dfrac{p_{ij}}{1 - p_{ii}}, & j \neq i, \\[2mm] 0, & j = i \end{cases}$$

   and then if the $\{Y_n\}$ chain has a stationary probability vector $\pi^{r\prime} = (\pi_1^r, \pi_2^r, \ldots, \pi_m^r)$ show that the stationary probability vector of the $\{X_n\}$ chain is given by $\pi' = (\pi_1, \pi_2, \ldots, \pi_m)$, where

   $$\pi_i = \pi_i^r \mathsf{E}\tau_i \bigg/ \sum_{i=1}^{m} \pi_i^r \mathsf{E}\tau_i.$$

*8. Deduce the necessary and sufficient conditions for the transience of the infinite irreducible birth and death chain, as obtained in Example 6.3.7, using the technique of Theorem 7.1.11. [*Hint:* Show $y_0 = 0$, $y_n = \sum_{i=0}^{n-1} \rho_i$ $(n \geq 1)$ is a solution of Eq. (7.1.11) as well as $y_n = 1$ $(n \geq 0)$.]

7.2   Techniques for Obtaining Stationary Distributions

In this section we derive a variety of techniques for obtaining stationary distributions.

In the first instance we restrict attention to finite irreducible MC's. For such chains the stationary probability vector π' can be determined by solving the constrained system of linear equations

$$\pi'(I - P) = \mathbf{0}' \qquad \text{subject to} \quad \pi'\mathbf{e} = 1, \qquad (7.2.1)$$

where $I - P$ is a singular matrix.

If the MC has a small number of states, the above system of equations can be easily solved, as in Examples 7.1.2 and 7.1.4. Note that since in this case $\lambda = 1$ is a simple eigenvalue of P (Theorems 6.1.5 and 6.1.6), $I - P$ has a zero eigenvalue [Theorem 4.4.5(e)] that is also simple. This implies that $\det(I - P) = 0$ and that $r(I - P) = m - 1$ when the MC has m states. Thus we can omit one of the equations from the system $\pi' = \pi'P$ and using the condition $\pi'\mathbf{e} = 1$ obtain m independent equations in m unknowns, which can then be solved.

In Section 4.3 we showed that a system of linear equations can be solved by using generalized inverses. However, in order that such techniques can be applied to Eqs. (7.2.1), we require a generalized inverse of the singular matrix $I - P$. To this end we can make use of the following key theorem due, independently, to Kemeny (1981) and Hunter (1982).

THEOREM 7.2.1: Let P be the transition matrix of a finite irreducible MC with stationary probability vector π'. Let $\mathbf{u}'$ and $\mathbf{t}$ be any vectors such that $\mathbf{u}'\mathbf{e} \neq 0$ and $\pi'\mathbf{t} \neq 0$. Then

(a) $I - P + \mathbf{tu}'$ is nonsingular.
(b) $[I - P + \mathbf{tu}']^{-1}$ is a generalized inverse of $I - P$.

Proof: (a) From Corollary 4.2.4A,

$$\det(I - P + \mathbf{tu}') = \det(I - P) + \mathbf{u}'[\operatorname{adj}(I - P)]\mathbf{t}. \qquad (7.2.2)$$

Firstly $\det (I - P) = 0$ and secondly Theorem 4.2.2(a) implies that

$$[\operatorname{adj}(I - P)](I - P) = (I - P)[\operatorname{adj}(I - P)] = \det(I - P)I = 0.$$

Thus from Theorems 7.1.2(e) and 7.1.6(c) if $A = \operatorname{adj}(I - P)$, $A(I - P) = (I - P)A = 0$ so that $A = PA = AP$ and hence $A = k\Pi$ for some k. Furthermore, k is nonzero. In fact, if the MC has m states, since P is irreducible, the eigenvalues of $I - P$ are $1 - \lambda_1, 1 - \lambda_2, \ldots, 1 - \lambda_m$ where $1 - \lambda_1 = 0$ is the only zero eigenvalue. Corollary 4.5.5A shows that $\operatorname{tr}(\operatorname{adj}((I - P)) = \prod_{j=2}^{m}(1 - \lambda_j) \neq 0$. But $\operatorname{tr}(\operatorname{adj}(I - P)) = \operatorname{tr}(k\Pi) = k\sum_{j=1}^{m}\pi_j = k$ and thus $k \neq 0$.

From Eq. (7.2.2),

$$\det(I - P + \mathbf{tu}') = \mathbf{u}'[k\Pi]\mathbf{t} = k\mathbf{u}'\mathbf{e}\pi'\mathbf{t} = k(\mathbf{u}'\mathbf{e})(\pi'\mathbf{t}) \neq 0,$$

establishing the required nonsingularity.

(b) First observe that

$$(I - P + \mathbf{tu}')(I - P + \mathbf{tu}')^{-1} = I,$$

so that

$$(I - P)(I - P + \mathbf{tu'})^{-1} = I - \mathbf{tu'}(I - P + \mathbf{tu'})^{-1}. \qquad (7.2.3)$$

Now note, using Eq. (7.2.1), that

$$\pi'(I - P + \mathbf{tu'}) = \pi'(I - P) + \pi'\mathbf{tu'} = (\pi'\mathbf{t})\mathbf{u'}$$

and thus

$$\pi' = (\pi'\mathbf{t})\mathbf{u'}(I - P + \mathbf{tu'})^{-1}$$

or that

$$\pi'/\pi'\mathbf{t} = \mathbf{u'}(I - P + \mathbf{tu'})^{-1}. \qquad (7.2.4)$$

Substitution of the result of Eq. (7.2.4) into Eq. (7.2.3) gives

$$(I - P)(I - P + \mathbf{tu'})^{-1} = I - \mathbf{t}\pi'/(\pi'\mathbf{t}) \qquad (7.2.5)$$

and hence

$$(I - P)(I - P + \mathbf{tu'})^{-1}(I - P) = I - P - (\mathbf{t}\pi'/\pi'\mathbf{t})(I - P) = I - P,$$

showing that $(I - P + \mathbf{tu'})^{-1}$ is a generalized inverse of $I - P$ by virtue of Definition 4.2.7. $\square$

Although generalized inverses are not unique we can use Theorem 7.2.1(b) in conjunction with Theorem 4.3.4 to obtain a characterization of all possible generalized inverses of $I - P$.

COROLLARY 7.2.1A: Under the conditions of the theorem any generalized inverse of $I - P$ can be expressed by the following equivalent forms:

(a) $$G = [I - P + \mathbf{tu'}]^{-1} + \frac{\mathbf{eu'}H}{\mathbf{u'e}} + \frac{H\mathbf{t}\pi'}{\pi'\mathbf{t}} - \frac{\mathbf{eu'}H\mathbf{t}\pi'}{(\mathbf{u'e})(\pi'\mathbf{t})}, \qquad (7.2.6)$$

where H is an arbitrary matrix;

(b) $$G = [I - P + \mathbf{tu'}]^{-1} + \frac{\mathbf{eu'}F}{\mathbf{u'e}} + \frac{G\mathbf{t}\pi'}{\pi'\mathbf{t}}, \qquad (7.2.7)$$

where F and G are arbitrary matrices;

(c) $$G = [I - P + \mathbf{tu'}]^{-1} + \mathbf{ef'} + \mathbf{g}\pi', \qquad (7.2.8)$$

where $\mathbf{f}$ and $\mathbf{g}$ are arbitrary vectors.

Proof: Using a similar approach as above note that

$$(I - P + \mathbf{tu'})^{-1}(I - P) = I - (I - P + \mathbf{tu'})^{-1}\mathbf{tu'}$$

and that

$$(I - P + \mathbf{tu'})\mathbf{e} = (I - P)\mathbf{e} + \mathbf{tu'e} = (\mathbf{u'e})\mathbf{t},$$

implying that

$$e/u'e = (I - P + tu')^{-1}t. \qquad (7.2.9)$$

Consequently,

$$(I - P + tu')^{-1}(I - P) = I - eu'/u'e. \qquad (7.2.10)$$

(a) If we use the characterization given by Eq. (4.3.10) with $A = I - P$ and $A^- = (I - P + tu')^{-1}$, then the representation given by Eq. (7.2.6) follows by using Eq. (7.2.5) and (7.2.10).

(b) The proof of part (b) is similar to (a) above but using the characterisation given by Eq. (4.3.11)

(c) Equations (7.2.7) and (7.2.8) are equivalent by taking $F = ef'$ and $G = g\pi'$ or conversely by taking $f = u'F/u'e$ and $g = Gt/\pi't$. Alternatively, Eqs. (7.2.6) and (7.2.8) are equivalent by taking $H = ef' + g\pi'$ or conversely by taking, for example,

$$f' = \frac{u'H}{u'e} \quad \text{and} \quad g = \left[I - \frac{eu'}{u'e}\right]\frac{Ht}{\pi't}. \quad \square$$

A variety of generalized inverses of $I - P$ have appeared in the literature when their derivation was often obtained by ad hoc techniques. In the following corollary we exhibit their representation in terms of the characterization given by Eq. (7.2.8). The list is not exhaustive.

COROLLARY 7.2.1B: Let P be the transition matrix of a finite, m-state, irreducible MC with stationary probability vector π' and $\Pi = e\pi'$. The following are all generalized inverses of $I - P$.

(a) $G_1 = [I - P + c\Pi]^{-1}$ for all $c \neq 0$.

(b) $G_2 = [I - P + eu']^{-1}$ provided $u'e \neq 0$.

(c) $G_3 = [I - P + \Pi]^{-1} - \Pi$.

(d) $G_4 = [I - P + \alpha\pi e']^{-1} - \alpha\Pi$ where $\alpha = (m\pi'\pi)^{-1/2}$.

(e) $G_5 = [I - P]^- + \beta \operatorname{adj}(I - P)$ for any $(I - P)^-$ and $\beta \neq 0$.

(f) If

$$P = \begin{bmatrix} P_{11} & \alpha \\ \beta' & p_{mm} \end{bmatrix},$$

then $I - P$ has a generalized inverse of the form

$$G_6 = \begin{bmatrix} (I - P_{11})^{-1} & 0 \\ 0' & 0 \end{bmatrix} = [I - P + tu']^{-1} + ef',$$

where $u' = (0',1)$, $t' = (0',1)$, and $f' = -(\beta'(I - P_{11})^{-1},1)$.

Proof: (a) G_1 follows by taking $\mathbf{t} = c\mathbf{e}, \mathbf{u} = \boldsymbol{\pi}, \mathbf{f} = \mathbf{0}$, and $\mathbf{g} = \mathbf{0}$. G_1 also follows directly from Theorem 4.4.11(b) (ii) since in that theorem $\mathbf{x}_m = \mathbf{e}$ and $\mathbf{y}'_m = \boldsymbol{\pi}'$. A special case of G_1 namely $Z = [I - P + \Pi]^{-1}$ was first recognized as a generalized inverse of $I - P$ by Hunter (1969). It will reappear in the next section as the "fundamental matrix" of the irreducible transition matrix as originally introduced by Kemeny and Snell (1960).

(b) G_2 was identified as a generalized inverse of $I - P$ by Paige *et al.* (1975). Their derivation was along the lines of the proof of Theorem 7.2.1 but with $\mathbf{t} = \mathbf{e}$.

(c) G_3 follows by taking $\mathbf{t} = \mathbf{e}, \mathbf{u} = \boldsymbol{\pi}, \mathbf{f} = -\boldsymbol{\pi}$, and $\mathbf{g} = \mathbf{0}$, and is, in fact, the group inverse $(I - P)^{\#}$ (see Definition 4.2.9) of $I - P$ as first shown by Meyer (1975).

(d) G_4 follows by taking $\mathbf{t} = \boldsymbol{\pi}, \mathbf{u} = \alpha\mathbf{e}, \mathbf{f} = -\alpha\boldsymbol{\pi}, \mathbf{g} = \mathbf{0}$ and was shown by Paige *et al.* (1975) to be the Moore-Penrose generalized inverse of $I - P$, $(I - P)^{+}$ (see Definition 4.2.8).

(e) G_5 follows from the proof of Theorem 7.2.1(a) by noting that

$$\text{adj}(I - P) = k\mathbf{e}\boldsymbol{\pi}' \qquad \text{where} \quad k = \prod_{j=2}^{m} (1 - \lambda_j) \neq 0.$$

(f) This result was derived in Section 4.2 by direct verification using Definition 4.2.7. However, it is instructive to see that this form is, in fact, a special case of Eq. (7.2.8). If we take $\mathbf{t}' = \mathbf{u}' = (\mathbf{0}', 1)$, so that $\mathbf{u}'\mathbf{e} = 1 \neq 0$ and $\boldsymbol{\pi}'\mathbf{t} = \pi_m \neq 0$, then

$$I - P + \mathbf{tu}' = \begin{bmatrix} I - P_{11} & -\alpha \\ -\beta' & 2 - p_{mm} \end{bmatrix} \equiv \begin{bmatrix} A_{11} & A_{12} \\ A_{21} & A_{22} \end{bmatrix}.$$

Since $(I - P)\mathbf{e} = \mathbf{0}$ it can be seen that $1 - p_{mm} = \beta'\mathbf{e}$ and $(I - P_{11})\mathbf{e} = \alpha$ so that $1 - p_{mm} = \beta'(I - P_{11})^{-1}\alpha$. Furthermore, from Theorem 7.2.1, $[I - P + \mathbf{tu}']^{-1}$ exists and thus from Theorem 4.2.4 we have that

$$[I - P + \mathbf{tu}']^{-1} = \begin{bmatrix} B_{11} & B_{12} \\ B_{21} & B_{22} \end{bmatrix},$$

where

$$B_{22} = (A_{22} - A_{21}A_{11}^{-1}A_{12})^{-1} = 1,$$
$$B_{21} = -B_{22}A_{21}A_{11}^{-1} = \beta'(I - P_{11})^{-1},$$
$$B_{12} = -A_{11}^{-1}A_{12}B_{22} = (I - P_{11})^{-1}\alpha,$$
$$B_{11} = A_{11}^{-1} - A_{11}^{-1}A_{12}B_{21} = (I - P_{11})^{-1} + (I - P_{11})^{-1}\alpha\beta'(I - P_{11})^{-1}.$$

If we take $\mathbf{f}'$ as stated in the theorem and $\mathbf{h} = \mathbf{0}$, then the expression for G_6 follows. This result, without the particular representation in the form of Eq.

(7.2.8), was reported by Rohde (1968) as being provided from J. Hearon (personal communication). [An expression for the group inverse of $I - P$ with P so partitioned has been given by Campbell and Meyer (1979).]   □

Using the theory of generalized inverses, we can now obtain a general expression for the stationary probability vector π', first presented in Hunter (1982).

THEOREM 7.2.2:   Let P be the transition matrix of a finite irreducible MC. If $(I - P)^-$ is *any* generalized inverse of $I - P$, and if we define $A \equiv I - (I - P)(I - P)^-$, then

$$\pi' = v'A/v'Ae, \qquad (7.2.11)$$

where v' is any vector such that $v'Ae \neq 0$.

Proof: Corollary 4.3.3B gives the consistency condition for the equation $XB = C$ (i.e., $CB^-B = C$) and the general solution, namely $X = CB^- + W(I - BB^-)$ where W is arbitrary.

Observe that Eq. (7.2.1), $\pi'(I - P) = 0'$, has such a form with $X = \pi'$, $B = I - P$, and $C = 0'$. This equation is obviously consistent and the general solution is given by

$$\pi' = w'[I - (I - P)(I - P)^-] \equiv w'A,$$

where w' must be chosen so that $\pi'e = 1 = w'Ae$.

Let v' be any vector such that $v'Ae \neq 0$ and define $w' = v'/v'Ae$ $(\neq 0')$. Then, for such a choice, $\pi' = v'A/v'Ae$ and $\pi'e = v'Ae/v'Ae = 1$.   □

There are a variety of ways we can use Theorem 7.2.2. Firstly, suppose we are given a computer subroutine for generating generalized inverses. Can we use this package? In order that we can use Eq. (7.2.11) we have to be sure that we can in fact find a suitable v'. The following corollary establishes the required verification providing us with an affirmative answer to the query.

COROLLARY 7.2.2A: Under the conditions of the theorem, if $(I - P)^-$ is *any* generalized inverse of $I - P$, then $Ae \neq 0$ and thus we can always find a v' such that $v'Ae \neq 0$.

Proof: Since any generalized inverse of $I - P$ can be characterized by the form given by Eq. (7.2.8) we see that

$$A = I - (I - P)[(I - P + tu')^{-1} + ef' + g\pi']$$

$$= \left[\frac{t}{\pi't} - (I - P)g\right]\pi',$$

using Eq. (7.2.5). Thus, since $\pi'\mathbf{e} = 1$, $A\mathbf{e} = \mathbf{t}/\pi'\mathbf{t} - (I - P)\mathbf{g}$. Now suppose that $A\mathbf{e} = \mathbf{0}$. This then implies that

$$\mathbf{t}/\pi'\mathbf{t} = (I - P)\mathbf{g},$$

so that $1 = \pi'\mathbf{t}/\pi'\mathbf{t} = \pi'(I - P)\mathbf{g} = 0$, a contradiction, and thus $A\mathbf{e} \neq \mathbf{0}$.

Now let $A\mathbf{e} = \boldsymbol{\delta} = (\delta_1, \delta_2, \ldots, \delta_m)$, and since $\boldsymbol{\delta} \neq \mathbf{0}$ there exists at least one $\delta_i \neq 0$, say δ_j. Let $\mathbf{v} = \mathbf{e}_j$. Then $\mathbf{v}'A\mathbf{e} = \delta_j \neq 0$. (Note also that if $\mathbf{v}' = \pi'$, then $\mathbf{v}'A\mathbf{e} = 1 \neq 0$.) □

Another way to use Theorem 7.2.2 is take a particular form of generalized inverse of $I - P$ that does not directly involve prior knowledge of π and use such a generalized inverse in the procedure described by Eq. (7.2.11). (Observe that there is plenty of flexibility in choosing $\mathbf{t}$ in Theorem 7.2.2 without knowing π, e.g., take any nonzero, nonnegative vector $\mathbf{t}$.)

COROLLARY 7.2.2B: If we take

$$(I - P)^- = (I - P + \mathbf{t}\mathbf{u}')^{-1} + \mathbf{e}\mathbf{f}',$$

where $\mathbf{t}, \mathbf{u}'$ are such that $\mathbf{u}'\mathbf{e} \neq 0$, and $\pi'\mathbf{t} \neq 0$, and $\mathbf{f}'$ is arbitrarily chosen, then

$$\pi' = \frac{\mathbf{u}'[I - P + \mathbf{t}\mathbf{u}']^{-1}}{\mathbf{u}'[I - P + \mathbf{t}\mathbf{u}']^{-1}\mathbf{e}}. \tag{7.2.12}$$

Proof: With $(I - P)^-$ so chosen

$$A = I - (I - P)[(I - P + \mathbf{t}\mathbf{u}')^{-1} + \mathbf{e}\mathbf{f}'],$$
$$= \mathbf{t}\mathbf{u}'(I - P + \mathbf{t}\mathbf{u}')^{-1},$$

using Eq. (7.2.3). Substitution in Eq. (7.2.11) yields

$$\pi' = \frac{\mathbf{v}'\mathbf{t}\mathbf{u}'(I - P + \mathbf{t}\mathbf{u}')^{-1}}{\mathbf{v}'\mathbf{t}\mathbf{u}'(I - P + \mathbf{t}\mathbf{u}')^{-1}\mathbf{e}}.$$

Equation (7.2.12) now follows by dividing the numerator and denominator by $\mathbf{v}'\mathbf{t}$ (since we can obviously find a $\mathbf{v}'$ so that $\mathbf{v}'\mathbf{t} \neq 0$). Note also that by Eq. (7.2.4) the denominator $\mathbf{u}'(I - P + \mathbf{t}\mathbf{u}')^{-1}\mathbf{e}$ is nonzero. □

The above corollary gives a very general procedure for finding π' in that there is still considerable flexibility in the choice of $\mathbf{u}'$ and $\mathbf{t}$. Studies have yet to be carried out to determine efficient choices to simplify the computation.

The generality of the procedure outlined in Theorem 7.2.2 is such that all known explicit methods for finding stationary distributions of finite irreducible MC's can be expressed as special cases. We show how some of the more well-known techniques can be deduced from this theorem.

COROLLARY 7.2.2C: Let P be the transition matrix of a finite m-state irreducible MC.

(a) For any $\mathbf{u}$ such that $\mathbf{u}'\mathbf{e} \neq 0$,

$$\boldsymbol{\pi}' = \mathbf{u}'[I - P + \mathbf{e}\mathbf{u}']^{-1}. \tag{7.2.13}$$

(b) Let $(I - P)^-$ be a generalized inverse of $I - P$ with the property that $\mathbf{e}'[I - (I - P)(I - P)^-]\mathbf{e} \neq 0$. Then

$$\boldsymbol{\pi}' = \frac{\mathbf{e}'[I - (I - P)(I - P)^-]}{\mathbf{e}'[I - (I - P)(I - P)^-]\mathbf{e}}. \tag{7.2.14}$$

(c) If D_j is the determinant of the matrix formed by removing the jth row and jth column from $I - P$, then

$$\boldsymbol{\pi}' = \frac{\mathbf{e}'[\mathrm{adj}(I - P)]_d}{\mathrm{tr}(\mathrm{adj}(I - P))} = \frac{1}{(\sum_{j=1}^m D_j)}(D_1, D_2, \ldots, D_m). \tag{7.2.15}$$

(d) The matrix $(I - P)_j = I - P + \mathbf{t}_j\mathbf{e}_j$, where $\mathbf{e}_j$ is the jth elementary vector and $\mathbf{t}_j = \mathbf{e} - (I - P)\mathbf{e}_j$, and which can be formed from $I - P$ by replacing its jth column by $\mathbf{e}$, is nonsingular. Its inverse is a generalized inverse of $I - P$ and

$$\boldsymbol{\pi}' = \mathbf{e}_j'[(I - P)_j]^{-1} \qquad (j = 1, 2, \ldots, m). \tag{7.2.16}$$

(e) If

$$P = \begin{bmatrix} P_{11} & \boldsymbol{\alpha} \\ \boldsymbol{\beta}' & p_{mm} \end{bmatrix},$$

then

$$\boldsymbol{\pi}' = \frac{(\boldsymbol{\beta}'(I - P_{11})^{-1}, 1)}{(\boldsymbol{\beta}'(I - P_{11})^{-1}, 1)\mathbf{e}}. \tag{7.2.17}$$

Proof: (a) In Eq. (7.2.12) take $\mathbf{t} = \mathbf{e}$, and Eq. (7.2.13) follows by observing that $\mathbf{u}'(I - P + \mathbf{e}\mathbf{u}')^{-1}\mathbf{e} = 1$, which can be established from Eq. (7.2.4) by postmultiplying by $\mathbf{t} = \mathbf{e}$. This particular result, with a direct proof, was given earlier by Paige *et al.* (1975); see also Kemeny (1981).

(b) Equation (7.2.14) follows immediately from Eq. (7.2.11) with $\mathbf{v} = \mathbf{e}$. [Decell and Odell (1967) derived this result under the additional assumption that $(I - P)(I - P)^-$ should be symmetric. This is a rather severe restriction although it certainly establishes the nonzero nature of $k = \mathbf{e}'[I - (I - P)(I - P)^-]\mathbf{e}$, since if $\boldsymbol{\alpha} = [I - (I - P)(I - P)^-]\mathbf{e}$, then $\boldsymbol{\alpha}' = \mathbf{e}'[I - (I - P)(I - P)^-]$ and it follows, upon simplification, that $\boldsymbol{\alpha}'\boldsymbol{\alpha} = k$. Now $\boldsymbol{\alpha}'\boldsymbol{\alpha} > 0$ iff $\boldsymbol{\alpha} \neq 0$. Thus $k = 0$ only if $\boldsymbol{\alpha} = \mathbf{0}$; however, this would imply that $\mathbf{e} = (I - P)(I - P)^-\mathbf{e}$ and hence $\boldsymbol{\pi}'\mathbf{e} = 1 = \boldsymbol{\pi}'(I - P)(I - P)^-\mathbf{e} = 0$, a contradiction.]

(c)   From Eq. (7.2.12)

$$\pi' = \frac{\mathbf{u}' \operatorname{adj}(I - P + \mathbf{t}\mathbf{u}')}{\mathbf{u}' \operatorname{adj}(I - P + \mathbf{t}\mathbf{u}')\mathbf{e}}.$$

By continuity arguments we would expect, as $\mathbf{t}$ approaches $\mathbf{0}$,

$$\pi' = \frac{\mathbf{u}' \operatorname{adj}(I - P)}{\mathbf{u}' \operatorname{adj}(I - P)\mathbf{e}}.$$

A formal proof of this result can be established from Theorem 7.2.2 by taking $\mathbf{v}' = \mathbf{u}' \operatorname{adj}(I - P)$, since, for any $(I - P)^-$, $\mathbf{v}'A = \mathbf{v}'$. Although Eq. (7.2.15) will follow by taking a general $\mathbf{u}$ [in which case the numerator and denominator of Eq. (7.2.15) will be multiplied by $\mathbf{u}'\mathbf{e}$] for convenience we take $\mathbf{u}' = \mathbf{e}_1'$.

Now, from the proof of Theorem 7.2.1, we see that

$$\operatorname{adj}(I - P) = k\Pi = [k\pi_j] \qquad \text{where} \quad k = \operatorname{tr}(\operatorname{adj}(I - P)) \neq 0.$$

Also, $\operatorname{adj}(I - P) = [\alpha_{ji}]$ where α_{ij} is the cofactor of the (i, j)th element of $I - P$. Thus

$$\alpha_{ji} = k\pi_j = \alpha_{jj} = D_j.$$

Consequently,

$$\mathbf{e}_1'[\operatorname{adj}(I - P)] = (\alpha_{11}, \alpha_{21}, \ldots, \alpha_{m1}) = (\alpha_{11}, \alpha_{22}, \ldots, \alpha_{mm})$$
$$= \mathbf{e}'[\operatorname{adj}(I - P)]_d = (D_1, D_2, \ldots, D_m)$$

and

$$\mathbf{e}'[\operatorname{adj}(I - P)]\mathbf{e} = \sum_{j=1}^{m} \alpha_{jj} = \operatorname{tr}(\operatorname{adj}(I - P))$$
$$= \sum_{j=1}^{m} D_j \quad \left(= k \sum_{j=1}^{m} \pi_j = k \neq 0 \right),$$

and Eq. (7.2.15) follows.

This representation of π' in terms of the D_j was originally due to Mihoc (see Frechet, 1950; Hunter, 1969).

(d)   The elements of the jth column of a matrix can be reduced to zeros by post multiplication by $I - \mathbf{e}_j\mathbf{e}_j'$, and thus

$$(I - P)_j = (I - P)(I - \mathbf{e}_j\mathbf{e}_j') + \mathbf{e}\mathbf{e}_j'$$
$$= I - P + (P\mathbf{e}_j - \mathbf{e}_j + \mathbf{e})\mathbf{e}_j' = I - P + \mathbf{t}_j\mathbf{e}_j'.$$

Observe that $\pi'\mathbf{t}_j = 1$ and $\mathbf{e}_j'\mathbf{e} = 1$, which, from Theorem 7.2.1, establishes the nonsingularity and generalized inverse properties of $(I - P)_j$. Expression

(7.2.16) follows from Eq. (7.2.12) upon noting that Eq. (7.2.4) implies

$$\mathbf{e}'_j[I - P + (P\mathbf{e}_j - \mathbf{e}_j + \mathbf{e})\mathbf{e}'_j]^{-1}\mathbf{e} = \pi'\mathbf{e}/\pi'(P\mathbf{e}_j - \mathbf{e}_j + \mathbf{e}) = 1.$$

This particular procedure was also suggested by Paige *et al.* (1975).

(e) With $(I - P)^-$ as given by Corollary 7.2.1B(f),

$$A = I - (I - P)(I - P)^- = \begin{bmatrix} 0 & 0 \\ \beta'(I - P_{11})^{-1} & 1 \end{bmatrix}.$$

Thus using Eq. (7.2.11) with $\mathbf{v} = \mathbf{e}$ we obtain

$$\pi' = \frac{\mathbf{e}'A}{\mathbf{e}'A\mathbf{e}} = \frac{(\beta'(I - P_{11})^{-1}, 1)}{(\beta'(I - P_{11})^{-1}, 1)\mathbf{e}}.$$

This result was reported by Rohde (1968). An alternative proof of this result via the group inverse $(I - P)^\#$ is given by Meyer (1975).

Snell (1975) and Meyer (1978) also obtained this result but used Eq. (7.2.16) with $j = m$ as the starting point for the derivation. (See Exercise 7.2.1) □

If expression (7.2.12) is used as the basis for a procedure to determine π', then there is still plenty of choice for $\mathbf{u}'$ and $\mathbf{t}$. Paige *et al.* (1975) carried out an error analysis and computational comparison between a variety of algorithms that included the techniques specified by Eqs. (7.2.13), (7.2.14), and (7.2.16) together with rank reduction, limits of matrix powers, and least-squares procedures. Their study concluded with a recommendation for the method given by Eq. (7.2.13) with $\mathbf{u}' = \mathbf{e}'_j P$ ($j = m$ for convenience) using Gaussian elimination with pivoting to solve the equation $\pi'(I - P + \mathbf{e}\mathbf{u}') = \mathbf{u}'$. This procedure gave the fastest computing times and the smallest average residual errors.

The techniques already presented in this section have been developed without taking into consideration any structure, such as symmetry, that may be present in the transition matrix. In such situations we are often able to use the standard techniques of difference equations or generating functions.

EXAMPLE 7.2.1: *Finite Birth and Death Chain.* Consider the MC with state space $S = \{0, 1, \ldots, a\}$ and transition matrix as displayed in Example 6.3.6, i.e., with transition probabilities

$$p_{i,\,i-1} = q_i > 0 \qquad (i = 1, 2, \ldots, a),$$

$$p_{i,\,i} = r_i \geq 0 \qquad (i = 0, 1, \ldots, a),$$

$$p_{i,\,i+1} = p_i > 0 \qquad (i = 0, 1, \ldots, a - 1).$$

This MC is irreducible since every state can be reached from every other state; thus the MC has a unique stationary distribution $\{\pi_i,\ i = 0, 1, \ldots, a\}$

where the π_i satisfy the stationary equations

$$\pi_j = \sum_i \pi_i p_{ij}.$$

Thus

$$\pi_0 = \pi_0 r_0 + \pi_1 q_1$$

$$\pi_i = \pi_{i-1} p_{i-1} + \pi_i r_i + \pi_{i+1} q_{i+1}, \quad i = 1, 2, \ldots, a-1,$$

$$\pi_a = \pi_{a-1} p_{a-1} + \pi_a r_a$$

Since $r_i = 1 - p_i - q_i$ $(i = 1, 2, \ldots, a-1)$ with $r_0 = 1 - p_0, r_a = 1 - q_a$ these equations reduce to

$$\pi_1 q_1 - \pi_0 p_0 = 0,$$

$$\pi_{i+1} q_{i+1} - \pi_i p_i = \pi_i q_i - \pi_{i-1} p_{i-1}, \quad i = 1, 2, \ldots, a-1,$$

$$\pi_a q_a - \pi_{a-1} p_{a-1} = 0.$$

Define $y_i = \pi_{i+1} q_{i+1} - \pi_i p_i$ $(i = 0, 1, \ldots, a-1)$. Then it is easy to see that $y_i = 0$ $(i = 0, 1, \ldots, a-1)$, from which we can deduce that if $\delta_0 = 1$,

$$\delta_i = \frac{p_0 \cdots p_{i-1}}{q_1 \cdots q_i} \quad (1 \le i \le a),$$

$$\pi_i = \pi_0 \delta_i \quad (i = 0, 1, \ldots, a).$$

Using the fact that $\sum_{i=0}^a \pi_i = 1$, it is easily seen that

$$\pi_0 = 1 \Big/ \left(\sum_{i=0}^a \delta_i \right),$$

and thus $\pi_i = \delta_i / (\sum_{i=0}^a \delta_i)$ $(i = 0, 1, \ldots, a)$ and the stationary distribution is uniquely determined.

The random walk with reflecting barriers at 0 and a is a special case of this model with $q_i = q$ $(i = 1, \ldots, a)$; $p_i = p$ $(i = 0, 1, \ldots, a)$, $r_i = 0$ $(i = 1, \ldots, a-1)$ with $r_0 = q, r_a = p$.

Substitution into the expression obtained for the stationary distribution gives

$$\pi_i = \frac{\left(\dfrac{p}{q}\right)^i \left[1 - \left(\dfrac{p}{q}\right)\right]}{1 - \left(\dfrac{p}{q}\right)^{a+1}}, \quad i = 0, 1, \ldots, a. \quad \square$$

When the MC is infinite, then obviously matrix techniques are not going to be suitable for solving the infinite number of stationary equations with an infinite number of unknowns. In such cases we can often find the stationary distribution (when it exists) by (i) taking limits of the analogous results for a related finite MC (see Example 7.2.2), (ii) using difference equation techniques (see Example 7.2.3), or (iii) using generating functions (see later in Chapter 9 when we consider queueing chains).

*EXAMPLE 7.2.2: *Infinite Birth and Death Chain.* Let us consider an irreducible birth and death chain with state space $S = \{0, 1, \ldots\}$ with transition probabilities $p_{i, i+1} = p_i > 0$ $(i \geq 0)$; $p_{ii} = r_i \geq 0$ $(i \geq 0)$; and $p_{i, i-1} = q_i > 0$ $(i \geq 1)$ $(p_i + q_i + r_i = 1)$.

In Example 6.3.7 we showed that such a chain is persistent iff

$$\sum_{i=1}^{\infty} \frac{q_1 \cdots q_i}{p_1 \cdots p_i} = \infty.$$

From Theorem 7.1.9 this chain has a stationary distribution $\{\pi_j, j = 0, 1, \ldots\}$ iff the MC is persistent nonnull. Using the results of Example 7.2.1, if

$$\delta_0 = 1,$$

$$\delta_i = \frac{p_0 \cdots p_{i-1}}{q_1 \cdots q_i} \quad (i \geq 1),$$

the stationary probabilities must satisfy $\pi_i = \delta_i \pi_0$ $(i \geq 0)$.

If $\sum_{j=0}^{\infty} \delta_j < \infty$ the MC has a unique stationary distribution given by

$$\pi_i = \delta_i \Big/ \left(\sum_{j=0}^{\infty} \delta_j \right) \quad (i \geq 0).$$

If $\sum_{j=0}^{\infty} \delta_j = \infty$, then any solution to the stationary equations is either identically zero or has infinite sum and hence there is no stationary distribution.

A consequence of this example and Example 6.3.7 leads to a classification of such chains.

If

$$\rho = \sum_{i=1}^{\infty} \frac{q_1 \cdots q_i}{p_1 \cdots p_i} \qquad \delta = \sum_{i=1}^{\infty} \frac{p_0 \cdots p_{i-1}}{q_1 \cdots q_i},$$

then the irreducible birth and death chain is

(i)   transient if $\rho < \infty$,
(ii) persistent nonnull if $\rho = \infty$, $\delta < \infty$, and
(iii) persistent null if $\rho = \infty$, $\delta = \infty$. $\square$

EXAMPLE 7.2.3: *Unrestricted Random Walk.* In this model $S = \{0, \pm 1, \pm 2, \ldots\}$ and the one-step transition probabilities are given by $p_{i, i+1} = p$

and $p_{i,\,i-1} = q\ (p + q = 1)$. In Example 6.1.1 we showed that this MC is irreducible and transient if $p \neq \frac{1}{2}$ (in which case no stationary distribution exists by Theorem 7.1.9) and persistent if $p = \frac{1}{2}$. We shall show that in this later case no stationary distribution exists and hence deduce from Theorem 7.1.9 that in the persistent case all the states are null.

Suppose a stationary distribution $\{\pi_j, j = 0, \pm1, \pm2, \ldots\}$ exists. Then since $\pi_j = \sum_i \pi_i p_{ij}$ we must have

$$\pi_j = \tfrac{1}{2}(\pi_{j-1} + \pi_{j+1}), \qquad j = 0, \pm1, \pm2, \ldots,$$

i.e.,

$$\pi_{j+1} - 2\pi_j + \pi_{j-1} = 0, \qquad j = 0, \pm1, \pm2, \ldots.$$

If we solve this difference equation using the methods of Section 2.9, we see that the auxiliary equation is

$$\lambda^2 - 2\lambda + 1 = 0,$$

which has two roots $\lambda = 1$ (twice) and hence the general solution would be

$$\pi_j = A + Bj.$$

In order that our solution be a probability distribution we require $\sum_j \pi_j = 1$. However, since the range of the index runs from $-\infty$ to $+\infty$ the series $\sum_j (A + Bj)$ will diverge unless $A = B = 0$, in which case $\pi_j = 0$ for every j. Thus no solution to the stationary equations exists that forms a probability distribution and thus the MC is not persistent nonnull.   □

We conclude this section with a further technique that combines matrix and generating function methods. If matrix generating functions have been used in an earlier analysis in finding say the n-step transition probabilities, then a simple extension leads to a determination of the stationary distribution.

THEOREM 7.2.3:   If P is the transition matrix of a finite irreducible Markov chain, then

$$\Pi = \lim_{s \uparrow 1}(1 - s)[I - sP]^{-1}.$$

Proof: From Theorem 7.1.7 and its corollaries it is easy to see that

$$\lim_{s \uparrow 1}(1 - s)P_{ij}(s) = \frac{1}{\mu_j} = \pi_j \qquad \text{for all} \quad i, j \in S.$$

Thus

$$\lim_{s \uparrow 1}(1 - s)[P_{ij}(s)] = [\pi_j] = \Pi.$$

The result now follows from Theorem 6.1.9, since $\mathbf{P}(s) = [P_{ij}(s)] = [I - sP]^{-1}$. $\square$

EXAMPLE 7.2.4: In Example 6.1.7 we saw that for the irreducible MC with transition matrix

$$P = \begin{bmatrix} 0 & 1 & 0 \\ q & 0 & p \\ 0 & 1 & 0 \end{bmatrix}$$

$$[I - sP]^{-1} = \frac{1}{1 - s^2} \begin{bmatrix} 1 - ps^2 & s & ps^2 \\ qs & 1 & ps \\ qs^2 & s & 1 - qs^2 \end{bmatrix},$$

so that

$$\Pi = \lim_{s \uparrow 1} \frac{1}{1 + s} \begin{bmatrix} 1 - ps^2 & s & ps^2 \\ qs & 1 & ps \\ qs^2 & s & 1 - qs^2 \end{bmatrix} = \frac{1}{2} \begin{bmatrix} q & 1 & p \\ q & 1 & p \\ q & 1 & p \end{bmatrix}. \quad \square$$

COROLLARY 7.2.3A: If

$$P = \begin{bmatrix} I & 0 \\ R & Q \end{bmatrix}$$

is the transition matrix of an absorbing MC, then

$$\lim_{s \uparrow 1} (1 - s)[I - sP]^{-1} = \Pi = \begin{bmatrix} I & 0 \\ NR & 0 \end{bmatrix},$$

where $N = (I - Q)^{-1}$.

Proof: From the results of Exercise 6.1.9

$$(1 - s)(I - sP)^{-1} = \begin{bmatrix} I & 0 \\ s(I - sQ)^{-1}R & (1 - s)(I - sQ)^{-1} \end{bmatrix}$$

and the corollary follows by taking the required limit. $\square$

Observe that the matrix Π in Theorem 7.2.3 and its corollary both have the interpretation as the $\lim_{n \to \infty} P^{(n)}$ when P is the transition matrix of a regular MC or an absorbing MC by virtue of Theorems 7.1.2(a) and 7.1.5(a). If we write

$$P^{(n)} = \Pi + T^{(n)},$$

then as n increases the effect of the matrix $T^{(n)}$ [called by Howard (1971) the *transient* matrix] will die away. In particular, when P is the transition matrix of a regular MC we see from Eq. (7.1.2) that if $\lambda_1 = 1, \lambda_2, \ldots, \lambda_n$ are the eigenvalues of P, then Π is the constant term arising from $\lambda_1 = 1$ and

$$T^{(n)} = \sum_{l=2}^{m} \lambda_l^n A_l \qquad \text{where} \quad |\lambda_l| < 1 \quad \text{for} \quad l = 2, \ldots, m,$$

and thus $\lim_{n \to \infty} T^{(n)} = 0$.

Of course, a significant amount of information concerning the behavior of the MC is contained in the sequence $\{T^{(n)}\}$ and in fact we shall find the following results of use in Section 7.3.

*THEOREM 7.2.4:   Let P be the transition matrix of a regular MC and let $T^{(n)} = P^{(n)} - \Pi$ $(n = 0, 1, 2, \ldots)$.

For $|s| < 1$ define $\mathbf{T}(s) = \sum_{n=0}^{\infty} T^{(n)} s^n$. Then $\mathbf{T}(s)$ has the following alternative representations:

(a) $\mathbf{T}(s) = [I - sP + s\Pi]^{-1} - \Pi.$

(b) $\mathbf{T}(s) = [I - sP]^{-1} - \dfrac{1}{1-s} \Pi.$

(c) $\mathbf{T}(s) = [I - s\Pi][I - sP]^{-1} - \Pi.$

(d) $\mathbf{T}(s) = [I - \Pi][I - sP]^{-1}.$

Furthermore, if

$$T \equiv \lim_{s \uparrow 1} \mathbf{T}(s) = \sum_{n=0}^{\infty} T^{(n)},$$

then

$$T = [I - P + \Pi]^{-1} - \Pi.$$

*Proof: By definition

$$\mathbf{T}(s) = I - \Pi + \sum_{n=1}^{\infty} (P^n - \Pi) s^n.$$

We now show that $(P - \Pi)^n = P^n - \Pi$. From Theorem 7.1.2(e) we have that $P\Pi = \Pi P = \Pi$ and thus P and Π commute. Furthermore, $\Pi^2 = \mathbf{e}\pi' \mathbf{e}\pi' = \mathbf{e}\pi' = \Pi$ and in general, by induction, $\Pi^n = \Pi$. Thus, using a binomial expansion,

$$(P - \Pi)^n = P^n + \sum_{i=0}^{n-1} \binom{n}{i} (-1)^{n-i} P^i \Pi^{n-i}$$

$$= P^n + \left[\sum_{i=0}^{n-1} \binom{n}{i} (-1)^{n-i} \right] \Pi$$

$$= P^n + [(1-1)^n - 1] \Pi = P^n - \Pi.$$

Thus

$$\mathbf{T}(s) = I - \Pi + \sum_{n=1}^{\infty} (P - \Pi)^n s^n.$$

For regular MC's, $P^n \to \Pi$ as $n \to \infty$ so that for $|s| < 1$ $(P - \Pi)^n s^n = (P^n - \Pi)s^n \to 0$ as $n \to \infty$. This means that condition (a) of Theorem 4.5.4 is satisfied implying that the expression for $\mathbf{T}(s)$ converges at least for $|s| < 1$ and that

$$\mathbf{T}(s) = \sum_{n=0}^{\infty} (P - \Pi)^n s^n - \Pi = [I - sP + s\Pi]^{-1} - \Pi.$$

This establishes (a). To derive the expression in (b) observe that

$$\mathbf{T}(s) = \sum_{n=0}^{\infty} P^{(n)} s^n - \sum_{n=0}^{\infty} \Pi s^n = [I - sP]^{-1} - \frac{1}{1-s} \Pi$$

using the result of Theorem 6.1.9.

Concerning the result contained in (c), it is easy to verify that for $s \neq 1$,

$$I - sP + s\Pi = [I - sP]\left[I - \frac{s}{1-s} \Pi \right] = \left[I - \frac{s}{1-s} \Pi \right][I - sP].$$

Now from Theorem 6.1.9 $[I - sP]^{-1}$ exists for $|s| < 1$ and, since it is easily seen that $\det[I - t\Pi] = 1 - t$ for all t, $[I - t\Pi]^{-1}$ exists for all $t \neq 1$. In fact, for such t, $[I - t\Pi]^{-1} = I + [t/(1-t)]\Pi$. Consequently, for $|s| < 1$,

$$[I - sP + s\Pi]^{-1} = \left[I - \frac{s}{1-s} \Pi \right]^{-1} [I - sP]^{-1} = [I - s\Pi][I - sP]^{-1}$$

and the result of (c) follows from (a).

Furthermore, by using the series expansion of $[I - sP]^{-1}$ (cf. Theorem 4.5.4) it is easily seen that $\Pi[I - sP]^{-1} = [1/(1-s)]\Pi$ and thus substitution for $[1/(1-s)]\Pi$ in (b) leads to the representation given by (d).

These results are of course all interconnected. An alternative derivation of (b) from (c) follows by observing that

$$\mathbf{T}(s) = [I - sP]^{-1} - s\Pi[I - sP]^{-1} - \Pi$$

$$= [I - sP]^{-1} - \frac{s}{1-s} \Pi - \Pi = [I - sP]^{-1} - \frac{1}{1-s} \Pi.$$

The limit as $s \uparrow 1$ is best effected by using expression (a) since from Corollary 7.2.1B we know that $[I - P + \Pi]^{-1}$ exists. $\square$

A few comments concerning this theorem are in order. Firstly, since $[I - Ps]^{-1}$ exists for $|s| < 1$ for all MC's with transition matrix P, $\mathbf{T}(s)$ exists for both irreducible periodic MC's and absorbing MC's as well with the interpretation of Π as given by Theorems 7.1.6(c) and 7.1.5(a), respectively. [The proof of part (a) of the theorem follows for the absorbing MC case by virtue of Theorem 7.1.5(e). However, for the irreducible periodic MC case we cannot invoke Theorem 4.5.4(a).]

The matrix T is also well defined for all finite irreducible MC's (by Corollary 7.2.1B) and also for absorbing MC's with the required form of Π. Howard (1971) calls T the *transient sum matrix* and thus it has a useful interpretation for all MC's.

Note also that T appeared earlier in Corollary 7.2.1B(d) as a generalized inverse of $I - P$ for finite irreducible MC's. In addition, Meyer (1975) showed that T is the group inverse of $I - P$ and claimed that T is the "correct" generalized inverse to use in connection with finite MC's.

We shall have further use of T or equivalently of $Z = T + \Pi$ in the next section.

Exercises 7.2

1. Derive the result of Corollary 7.2.2C(e) by using Eq. (7.2.16) with $j = m$.
2. In Exercise 7.1.3 it was required to verify that

$$\pi_j = \binom{N}{j} \Big/ 2^N, \qquad j = 0, 1, \ldots, N$$

is the stationary distribution for the Ehrenfest diffusion model. Without assuming prior knowledge of the expression above, solve the stationary equations directly as follows:
   (a) Put $y_j = (j + 1)\pi_{j+1} - (N - j)\pi_j$ in the stationary equations. Show that $y_j = 0$ for $j = 0, 1, \ldots, N - 1$, and hence solve the stationary equations.
   (b) Let $\Pi(s) = \sum_{j=0}^{N} \pi_j s^j$. Show that $N\Pi(s) = (1 + s)(d/ds)\Pi(s)$. Solve this differential equation obtaining $\Pi(s) = (1 + s)^N/2^N$ from which the π_j can be found.
3. In Exercise 7.1.4 an urn problem was discussed in which it was required to verify that

$$\pi_j = \binom{N}{j}^2 \Big/ \binom{2N}{N}, \qquad j = 0, 1, \ldots, N.$$

Solve the stationary equations directly to obtain this result. [A technique for solving the stationary equations is to put $y_j = (N - j + 1)^2\pi_{j-1} - j^2\pi_j$. Thence show $y_j = 0$ for $j = 0, 1, \ldots, N - 1$ and thus solve for the π_j.]

4. Consider an irreducible two-state MC with transition matrix

$$P = \begin{bmatrix} p_{11} & p_{12} \\ p_{21} & p_{22} \end{bmatrix}.$$

Let $\mathbf{t}' = (t_1, t_2)$ and $\mathbf{u}' = (u_1, u_2)$.
(a) Find an expression for $[I - P + \mathbf{t}\mathbf{u}']^{-1}$.
(b) Use Corollary 7.2.2B to find $\boldsymbol{\pi}'$.
(c) Show that *any* generalized inverse of $I - P$ is of the form

$$G = \begin{bmatrix} \dfrac{p_{21} + t_2 u_2}{\Delta} + f_1 + \dfrac{g_1 p_{21}}{p_{21} + p_{12}} & \dfrac{p_{12} - t_1 u_2}{\Delta} + f_2 + \dfrac{g_1 p_{12}}{p_{21} + p_{12}} \\[2.5ex] \dfrac{p_{21} - t_2 u_1}{\Delta} + f_1 + \dfrac{g_2 p_{21}}{p_{21} + p_{12}} & \dfrac{p_{12} + t_1 u_1}{\Delta} + f_2 + \dfrac{g_2 p_{12}}{p_{21} + p_{12}} \end{bmatrix},$$

where $\Delta \equiv (p_{12} t_2 + p_{21} t_1)(u_1 + u_2)$; and f_1, f_2, g_1, and g_2 are arbitrary.

*7.3 Moments of the First-Passage Time Distributions

Let $\{X_n\}$ $(n \geq 0)$ be a MC with space S. In Section 5.2 we introduced the random variable $T_{ij} = \min\{n : X_n = j \mid X_0 = i\}$, the number of trials for a first passage from state i to state j ("first return" when $i = j$ or "first entrance" when $i \neq j$).

Since $P\{T_{ij} = n\} = f_{ij}^{(n)}$ $(n \geq 1)$, we have that

$$P\{T_{ij} < \infty\} = \sum_{n=1}^{\infty} f_{ij}^{(n)} = f_{ij},$$

and hence T_{ij} is a well-defined proper r.v. when $f_{ij} = 1$. This certainly occurs for all $i, j \in S$ when the MC is irreducible and all the states are persistent (by Theorem 5.3.6).

When T_{ij} is a proper r.v. we can define its kth moment about the origin, $\mu_{ij}^{[k]} \equiv E(T_{ij}^k)$ $(k = 1, 2, \ldots)$. In particular, $\mu_{ij} \equiv \mu_{ij}^{[1]}$ is called the *mean first passage time of state j from state i* and $\mu_j \equiv \mu_{jj}$ is the mean recurrence time of state j.

When the MC is irreducible and composed of persistent states we define $M^{[k]} \equiv [\mu_{ij}^{[k]}]$ $(i, j \in S)$. In particular, $M \equiv M^{[1]}$ is called the *mean first passage time matrix*.

In this section we examine techniques for determining these expectations. As in the previous sections there are a variety of methods that we can use, ranging from direct methods based upon the definition, recursive arguments, generating functions, and more generally matrix techniques.

From the definition, $\mu_{ij}^{[k]} = \sum_{n=1}^{\infty} n^k f_{ij}^{(n)}$ so that once the $\{f_{ij}^{(n)}\}$ have been obtained for any particular i, j and provided this sequence forms a proper probability distribution these moments can be computed.

EXAMPLE 7.3.1:   Consider the MC with transition matrix given by

$$P = \begin{bmatrix} 0 & \frac{1}{2} & \frac{1}{2} \\ \frac{1}{2} & 0 & \frac{1}{2} \\ \frac{1}{2} & \frac{1}{2} & 0 \end{bmatrix}.$$

This is a special case of the transition matrix given in Example 6.2.1, and using the results contained therein it is easily seen that

$$f_{ii}^{(n)} = (\tfrac{1}{2})^{n-1}, \qquad n = 2, 3, \ldots \quad (i = 1, 2, 3),$$

and

$$f_{ij}^{(n)} = (\tfrac{1}{2})^n, \qquad n = 1, 2, \ldots \quad (i \neq j).$$

This MC is finite and irreducible and thus $f_{ij} = 1$ $(i, j = 1, 2, 3)$, which can, of course, be easily verified. Consequently the T_{ij} are proper r.v.'s with expectations given by

$$\mu_{ii} = \sum_{n=2}^{\infty} n(\tfrac{1}{2})^{n-1} = 3 \quad \text{and} \quad \mu_{ij} = \sum_{n=1}^{\infty} n(\tfrac{1}{2})^n = 2 \quad \text{for } i \neq j. \quad \square$$

Generating functions are also of use especially when the $F_{ij}(s) = \sum_{n=1}^{\infty} f_{ij}^{(n)} s^n$ have already been derived since, provided $F_{ij}(1) = 1$, $\mu_{ij} = \lim_{s \uparrow 1} (d/ds) F_{ij}(s)$.

EXAMPLE 7.3.2:   *Random Walk on the Nonnegative Integers with an Absorbing Barrier at the Origin.*   In Examples 6.2.5 and 6.3.3 it was shown that provided $p \leq \frac{1}{2}$ and $i > j \geq 0$, then $f_{ij} = 1$ and

$$F_{ij}(s) = \left[\frac{1 - (1 - 4pqs^2)^{1/2}}{2ps} \right]^{i-j}.$$

Since

$$\frac{dF_{ij}(s)}{ds} = (i - j)F_{10}(s) \left[\frac{2ps[4pqs(1 - 4pqs^2)^{-1/2}] - 2p[1 - (1 - 4pqs^2)^{1/2}]}{4p^2 s^2} \right]$$

we have that

$$\mu_{ij} = (i - j) \frac{[4pq(1 - 4pq)^{-1/2}] - [1 - (1 - 4pq)^{1/2}]}{2p}$$

$$= (i - j) \frac{1 - (1 - 4pq)^{1/2}}{2p(1 - 4pq)^{1/2}}$$

$$= \frac{i - j}{q - p},$$

using the result that $(1 - 4pq)^{1/2} = |p - q| = q - p$ for $p \leq \frac{1}{2}$. Observe that μ_{ij} is infinite when $p = \frac{1}{2}$ and finite when $p < \frac{1}{2}$. $\square$

In the above example the MC is neither irreducible nor finite. However, for such chains we can in fact show that the μ_{ij}, and for that matter $\mu_{ij}^{[k]}$ for any $k \geq 1$ must be finite for all i, j. This is established in the following theorem, the proof of which is based upon a presentation given by Neuts (1973a).

THEOREM 7.3.1:   In any irreducible m-state Markov chain with state space S,

(a)   there exists a constant α, $0 < \alpha < 1$, such that for all $n \geq m$,

$$\sum_{k=n+1}^{\infty} f_{ij}^{(k)} \leq \alpha^n \qquad (i, j) \in S;$$

(b)   the generating functions $F_{ij}(s)$ have radius convergence greater than one; and

(c)   all the moments of T_{ij} are finite.

Proof: (a)   We shall prove that for every pair (i, j) there exists a constant α_{ij}, $0 < \alpha_{ij} < 1$ such that

$$\sum_{k=n+1}^{\infty} f_{ij}^{(k)} \leq \alpha_{ij}^n \qquad \text{for every} \quad n \geq m.$$

To complete the proof of statement (a) it suffices to set $\alpha = \max_{i,j} \alpha_{ij}$.

Let $P = [p_{rs}]$ be the transition matrix of the MC and for fixed j let $P_j = P_j(0)$ be P with the jth column replaced by zeros. Thus if $P_j = [p'_{rs}]$, then $p'_{rs} = (1 - \delta_{sj})p_{rs}$.

If $p_{ij} = 1$ the theorem is trivial since in this case $f_{ij}^{(n)} = 0$ for $n \geq 2$. Thus $e'_i P_j \neq 0'$. Now

$$\sum_{k=n+1}^{\infty} f_{ij}^{(k)} = P\{T_{ij} \geq n + 1\} = P\{X_1 \neq j, \ldots, X_n \neq j \mid X_0 = i\}$$

$$= \sum_{i_1 \neq j} \cdots \sum_{i_n \neq j} p_{i i_1} p_{i_1 i_2} \cdots p_{i_{n-1} i_n}$$

$$= \sum_{i_1} \cdots \sum_{i_n} p'_{i i_1} p'_{i_1 i_2} \cdots p'_{i_{n-1} i_n}$$

$$= \sum_{i_n} (P_j^n)_{i, i_n} = e'_i P_j^n e.$$

Now Theorem 5.3.11 ensures that for fixed h, $p_{hj}^{(n)} > 0$ for some n with $1 \leq n \leq m - 1$, and thus for $h \neq j$

$$(P_j^m e)_h = P\{X_1 \neq j, \ldots, X_m \neq j \mid X_0 = h\} = P\left\{\bigcap_{k=1}^{m} (X_k \neq j) \mid X_0 = h\right\}$$

$$= 1 - P\left\{\bigcup_{k=1}^{m} (X_k = j) \mid X_0 = h\right\} \leq 1 - P\{X_n = j \mid X_0 = h\} < 1.$$

Also, $(P_j^m \mathbf{e})_j = 0$, and hence in accordance with Definition 4.1.2(e) we may write

$$P_j^m \mathbf{e} < \mathbf{e}.$$

Furthermore, since P_j is derived from a stochastic matrix P, it is clear that

$$P_j \mathbf{e} \le \mathbf{e}.$$

Multiplying the first equation successively on the left by $P_j, P_j^2, \ldots, P_j^{m-1}$ we obtain

$$P_j^{m+\nu} \mathbf{e} < P_j^\nu \mathbf{e} \le \mathbf{e} \qquad (\nu = 0, 1, \ldots, m-1).$$

For $\nu = 0, 1, \ldots, m-1$, define $\beta_\nu = \max_h (P_j^{m+\nu} \mathbf{e})_h$ so that $P_j^{m+\nu} \mathbf{e} \le \beta_\nu \mathbf{e}$ and $0 \le \beta_\nu < 1$. Next, let $\beta = \max\{\beta_0, \beta_1, \ldots, \beta_{m-1}\}$ and set $\alpha_{ij} = \beta^{1/(2m-1)}$ so that $0 \le \alpha_{ij} < 1$ and

$$1 > \alpha_{ij}^m > \alpha_{ij}^{m+1} > \cdots > \alpha_{ij}^{2m-1} = \beta.$$

It now follows from the definition of β_ν that

$$P_j^{m+\nu} \mathbf{e} \le \alpha_{ij}^{m+\nu} \mathbf{e} \qquad (\nu = 0, 1, \ldots, m-1).$$

Multiplying by P_j^{lm}, $l \ge 1$, we obtain the string of inequalities

$$P_j^{lm+\nu} \mathbf{e} \le \alpha_{ij}^{m+\nu} P_j^{(l-1)m} \mathbf{e} \le \alpha_{ij}^{2m+\nu} P_j^{(l-2)} \mathbf{e} \le \cdots \le \alpha_{ij}^{lm+\nu} \mathbf{e}.$$

This establishes the inequality $P_j^n \mathbf{e} \le \alpha_{ij}^n \mathbf{e}$ for all $n \ge m$ and thus

$$\sum_{k=n+1}^{\infty} f_{ij}^{(k)} = \mathbf{e}_i' P_j^n \mathbf{e} \le \alpha_{ij}^n,$$

as required.

(b)   By the comparison test for power series (cf. Theorem 2.2.1), we have

$$|F_{ij}(s)| = \left| \sum_{n=1}^{\infty} f_{ij}^{(n)} s^n \right| \le \left| \sum_{n=1}^{m-1} f_{ij}^{(n)} s^n \right| + \sum_{n=m}^{\infty} \alpha^n |s|^n.$$

The majorising series, and hence also $F_{ij}(s)$, converges for all s such that $|s| < R$ where $R = 1/\alpha > 1$.

(c)   Since $R > 1$, $F_{ij}^{(k)}(s)$ converges uniformly at least for $|s| \le 1$ [Theorem 2.3.1(c)] and hence [Theorem 2.2.3(a)] is continuous within this range. This implies that term-by-term differentiation up to any order of $F_{ij}(s)$ at $s = 1$ is permitted, so that the moments of all orders are finite. $\square$

The result expressed by Theorem 7.3.1(c) is well known when $i = j$ as far as it concerns the expectation $\mu_{jj} = \mu_j$. In fact, it is usually much easier to find expressions for the mean recurrence times μ_j than for the more general μ_{ij}. Knowledge of the n-step transition probabilities $p_{jj}^{(n)}$, the associated generating function $P_{jj}(s)$, or the unique stationary distribution lead to determination of the μ_j, as summarized below.

(a) If state j is persistent, then, from Theorem 5.2.7,

(i) if state j is aperiodic, then

$$\mu_j = \frac{1}{\lim_{n \to \infty} p_{jj}^{(n)}} \quad (\le \infty);$$

(ii) if state j is periodic with period d, then

$$\mu_j = \frac{d}{\lim_{n \to \infty} p_{jj}^{(nd)}} \quad (\le \infty).$$

(b) If state j is persistent, then, from Corollary 7.1.7A,

$$\mu_j = \frac{1}{\lim_{s \uparrow 1} (1 - s) P_{jj}(s)} \quad (\le \infty).$$

(c) From Theorem 7.1.9(c), an irreducible MC has a unique stationary distribution $\{\pi_j\}$ in the persistent nonnull case, in which case

$$\mu_j = 1/\pi_j < \infty. \tag{7.3.1}$$

It is not as easy, however, to find expressions for the μ_{ij}. An approach using Theorem 6.2.4 leads to the following result.

THEOREM 7.3.2: If P is the transition matrix of a finite irreducible MC, then for each $i, j \in S$

$$\mu_{ij} = e_i'[I - P_j(0)]^{-1}e, \tag{7.3.2}$$

where $P_j(0)$ is P with the jth column replaced by zeros.

Proof: From Eq. (6.2.18)

$$F_{ij}(s) = se_i'[I - sP_j(0)]^{-1}Pe_j = e_i'\left\{\sum_{k=0}^{\infty} s^{k+1}[P_j(0)]^k\right\}Pe_j,$$

where we have used Theorem 4.5.4, since as was shown in Theorem 6.2.4, $\rho(P_j(0)) < 1$. Now

$$\frac{dF_{ij}(s)}{ds} = e_i'\left\{\sum_{k=0}^{\infty} (k + 1)s^k[P_j(0)]^k\right\}Pe_j$$

and thus

$$\lim_{s \uparrow 1} \frac{dF_{ij}(s)}{ds} = e_i'\left\{\sum_{k=0}^{\infty} (k + 1)[P_j(0)]^k\right\}Pe_j,$$

implying that

$$\mu_{ij} = \mathbf{e}_i \{[I - P_j(0)]^{-1}\}^2 P \mathbf{e}_j,$$

where we have used the result of Exercise 4.5.1. Furthermore,

$$F_{ij}(1) = \mathbf{e}'_i[I - P_j(0)]^{-1} P \mathbf{e}_j = 1$$

when P is irreducible. [This result was stated in Exercise 6.3.1. Its proof follows easily from Theorem 7.2.1(a) and Eq. (7.2.9) by observing that $P_j(0) = P - \mathbf{p}_j \mathbf{e}'_j$ where $\mathbf{p}_j = P \mathbf{e}_j$ and setting $\mathbf{t} = \mathbf{p}_j$, $\mathbf{u}' = \mathbf{e}'_j$.]

Consequently, $[I - P_j(0)]^{-1} P \mathbf{e}_j = [I - P + \mathbf{p}_j \mathbf{e}'_j]^{-1} \mathbf{p}_j = \mathbf{e}$ and Eq. (7.3.2) follows upon substitution of this result in the above form for μ_{ij}. $\square$

THEOREM 7.3.3:   Let $P = [p_{ij}]$ be the transition matrix of an irreducible MC. Then, for all $i, j \in S$,

$$\mu_{ij} = 1 + \sum_{k \neq j} p_{ik}\mu_{kj}. \tag{7.3.3}$$

Proof: This result can be derived in a variety of ways.

(a)   From Theorems 5.1.6(a) and 5.1.6(b),

$$f_{ij}^{(n)} = \sum_{k \neq j} p_{ik} f_{kj}^{(n-1)} \qquad (n \geq 2),$$

with $f_{ij}^{(1)} = p_{ij}$ so that

$$\mu_{ij} = \sum_{n=1}^{\infty} n f_{ij}^{(n)} = p_{ij} + \sum_{n=2}^{\infty} n \sum_{k \neq j} p_{ik} f_{kj}^{(n-1)}$$

$$= p_{ij} + \sum_{k \neq j} p_{ik} \sum_{n=2}^{\infty} (n - 1 + 1) f_{kj}^{(n-1)}$$

$$= p_{ij} + \sum_{k \neq j} p_{ik}(\mu_{kj} + f_{kj})$$

$$= \sum_{k} p_{ik} + \sum_{k \neq j} p_{ik}\mu_{kj},$$

since $f_{kj} = 1$ for all k, j, and Eq. (7.3.3) follows by the stochastic nature of $P = [p_{ij}]$.

(b)   Alternatively, using Eq. (6.2.9)

$$F_{ij}(s) = sp_{ij} + s \sum_{k \neq j} p_{ik} F_{kj}(s),$$

we have that

$$\frac{dF_{ij}(s)}{ds} = p_{ij} + \sum_{k \neq j} p_{ik} F_{kj}(s) + s \sum_{k \neq j} p_{ik} \frac{dF_{kj}(s)}{ds}.$$

Taking the limit as $s \uparrow 1$, we obtain

$$\mu_{ij} = p_{ij} + \sum_{k \neq j} p_{ik} f_{kj} + \sum_{k \neq j} p_{ik} \mu_{kj},$$

and Eq. (7.3.3) follows since $f_{kj} = 1$ for all k, j.

(c) Bhat (1972) gives the following probabilistic derivation. The first passage $i \rightarrow j$ can occur in one or more steps. Accordingly, we can write

$$T_{ij} = \begin{cases} 1, & \text{with probability } p_{ij}, \\ T_{kj} + 1, & \text{with probability } p_{ik} \quad (k \neq j). \end{cases}$$

Taking expectations, we get

$$\mu_{ij} = 1 \cdot p_{ij} + \sum_{k \neq j} (\mu_{kj} + 1) p_{ik},$$

from which Eq. (7.3.3) follows. $\square$

COROLLARY 7.3.3A: If $S = \{1, 2, \ldots, m\}$ and $\boldsymbol{\mu}'_j = (\mu_{1j}, \mu_{2j}, \ldots, \mu_{mj})$, then for $j = 1, 2, \ldots, m$,

$$\boldsymbol{\mu}_j = [I - P_j(0)]^{-1} \mathbf{e}. \tag{7.3.4}$$

Proof: Equation (7.3.3) expressed in vector form for fixed j becomes

$$\boldsymbol{\mu}_j = \mathbf{e} + P_j(0) \boldsymbol{\mu}_j,$$

from which Eq. (7.3.4) follows. Note also $\mu_{ij} = \mathbf{e}'_i \boldsymbol{\mu}_j$ and thus we have an alternative derivation to Eq. (7.3.2). $\square$

COROLLARY 7.3.3B: If $M = [\mu_{ij}]$, then

$$(I - P)M = E - PM_d. \tag{7.3.5}$$

Proof: Expressing Eq. (7.3.3) in matrix form yields the result that

$$M = E + P(M - M_d),$$

from which Eq. (7.3.5) follows. $\square$

There are two complications concerning the solution of Eq. (7.3.5). Firstly the right-hand side involves M_d and secondly $I - P$ is a singular matrix. The first impediment is easily removed.

THEOREM 7.3.4: If P is the transition matrix of a finite irreducible MC with stationary probability vector $\boldsymbol{\pi}'$, then if $\Pi = \mathbf{e}\boldsymbol{\pi}'$,

$$M_d = (\Pi_d)^{-1}. \tag{7.3.6}$$

Proof: This follows directly from Eq. (7.3.1) since

$$M_d = \text{diag}(\mu_1, \mu_2, \ldots, \mu_m)$$

$$= \text{diag}\left(\frac{1}{\pi_1}, \frac{1}{\pi_2}, \ldots, \frac{1}{\pi_m}\right) = (\Pi_d)^{-1}$$

since

$$\Pi_d = \text{diag}(\pi_1, \pi_2, \ldots, \pi_m).$$

Alternatively, since $\Pi P = \Pi$ and $\Pi E = E$, premultiplication of Eq. (7.3.5) by Π yields

$$\Pi M_d = E.$$

Taking diagonal elements gives $\Pi_d M_d = I$ and the result follows.   □

The natural method to overcome the second complication is to use a generalized inverse of $I - P$. Observe that Eq. (7.3.5) is of the form $AX = C$ where $A = I - P$, $X = M$, and $C = E - P(\Pi_d)^{-1}$. The general procedure for solving this system of linear equations is given by Corollary 4.3.3A and by taking a general form for the generalized inverse of $I - P$. This line of attack is taken in Theorem 7.3.6 to follow.

Before we examine this procedure we present a lemma containing some useful results required in our derivation. The proofs are straightforward and consequently omitted.

LEMMA 7.3.5: Let Λ be any diagonal matrix, X any square matrix, and $D = (\Pi_d)^{-1}$. Then

(a) $(X\Lambda)_d = (X_d)\Lambda$,
(b) $(XE)_d = (X\Pi)_d D$,
(c) $E\Pi_d = \Pi$.   □

THEOREM 7.3.6: If G is *any* generalized inverse of $I - P$, then

$$M = [G\Pi - E(G\Pi)_d + I - G + EG_d]D, \tag{7.3.7}$$

where $D = (\Pi_d)^{-1}$.

Proof: Since Theorem 7.3.4 implies that $M_d = D$, Eq. (7.2.5) becomes

$$(I - P)M = E - PD,$$

which, by Corollary 4.3.3A, has a general solution given by

$$M = G(E - PD) + \{I - G(I - P)\}U, \tag{7.3.8}$$

where U is an arbitrary matrix, provided the consistency restriction

$$[I - (I - P)G](E - PD) = 0 \tag{7.3.9}$$

is satisfied. With G taken in the general form as given by Eq. (7.2.8), i.e., for suitable $\mathbf{t}$, $\mathbf{u}$, $\mathbf{f}$, and $\mathbf{g}$,

$$G = [I - P + \mathbf{tu'}]^{-1} + \mathbf{ef'} + \mathbf{g}\pi',$$

we have, using Eq. (7.2.5) together with the observation that $(I - P)\mathbf{e} = \mathbf{0}$, that

$$[I - (I - P)G](E - PD) = \left[\frac{\mathbf{t}}{\pi'\mathbf{t}} - (I - P)\mathbf{g}\right]\pi'[\mathbf{ee'} - PD]$$

$$= \left[\frac{\mathbf{t}}{\pi'\mathbf{t}} - (I - P)\mathbf{g}\right][\mathbf{e'} - \pi'D]$$

$$= 0$$

since $\pi'D = \mathbf{e'}$, showing that the consistency condition, Eq. (7.3.9), is satisfied. Now, using Eq. (7.2.10), we observe that, since $\pi'(I - P) = \mathbf{0'}$,

$$[I - G(I - P)]U = \mathbf{e}\left[\frac{\mathbf{u'}}{\mathbf{u'e}} - \mathbf{f'}(I - P)\right]U$$

$$\equiv \mathbf{eh'}U \equiv \mathbf{eb'},$$

say, where $\mathbf{b'} = \mathbf{h'}U$. Thus Eq. (7.3.8) can be written as

$$M = G(E - PD) + \mathbf{eb'}. \tag{7.3.10}$$

Observe that the m^2 arbitrary elements of U have been reduced to only m, the elements of $\mathbf{b'}$. These can be determined explicitly due to the restriction that $M_d = (\Pi_d)^{-1}$. Suppose $\mathbf{b'} = (b_1, b_2, \ldots, b_m)$ and $B = \text{diag}(b_1, b_2, \ldots, b_m)$ so that $\mathbf{eb'} = EB$. From Eq. (7.3.10), forming the matrices of diagonal elements and using Lemma 7.3.5, we obtain

$$D = (G\Pi)_d D - (GP)_d D + B,$$

implying that

$$B = [I - (G\Pi)_d + (GP)_d]D. \tag{7.3.11}$$

Since $\mathbf{eb'} = EB$ and $E = \Pi D$, substitution of Eq. (7.3.11) into Eq. (7.3.10) yields

$$M = [G\Pi - E(G\Pi)_d - GP + E(GP)_d + E]D. \tag{7.3.12}$$

Further simplification of Eq. (7.3.12) is possible since we have shown above that

$$I - G + GP = \mathbf{eh'},$$

and thus $E(I - G + GP)_d = \mathbf{ee}'(\mathbf{eh}')_d = \mathbf{eh}'$. Hence $E - EG_d + E(GP)_d = I - G + GP$, implying that $E + E(GP)_d - GP = I - G + EG_d$. Substitution into Eq. (7.3.12) gives the required result. Eq. (7.3.7) □

Theorem 7.3.6 first appeared in the literature in Hunter (1982) and presents a useful result. It has the same desirable property alluded to earlier, following Theorem 7.2.2, namely that any computer package that generates generalized inverses can be used to determine means of the first passage time distributions.

Because we have used an arbitrary generalized inverse of $I - P$ in developing Theorem 7.3.6 we have considerable flexibility in making our choice. Firstly the form of the solution for M given by Eq. (7.3.7) does not depend on the choice of $\mathbf{f}$ and $\mathbf{g}$ when we use the representation given by Eq. (7.2.8).

COROLLARY 7.3.6A: If $G = G_0 + \mathbf{ef}' + \mathbf{g}\boldsymbol{\pi}'$ where $G_0 = [I - P + \mathbf{tu}']^{-1}$ $(\boldsymbol{\pi}'\mathbf{t} \neq 0, \mathbf{u}'\mathbf{e} \neq 0)$, then

$$M = [G_0\Pi - E(G_0\Pi)_d + I - G_0 + E(G_0)_d]D, \qquad (7.3.13)$$

where $D = (\Pi_d)^{-1}$.

Proof: Let $G = G_0 + H$ where $H = \mathbf{ef}' + \mathbf{g}\boldsymbol{\pi}'$. Then

$$H\Pi = (\mathbf{f}'\mathbf{e})\Pi + \mathbf{g}\boldsymbol{\pi}' = (\mathbf{f}'\mathbf{e})\Pi + H - \mathbf{ef}'$$

and

$$E(H\Pi)_d = (\mathbf{f}'\mathbf{e})E\Pi_d + EH_d - E(\mathbf{ef}')_d = (\mathbf{f}'\mathbf{e})\Pi + EH_d - \mathbf{ef}'.$$

Consequently, $H\Pi - E(H\Pi)_d = H - EH_d$ and the result follows by substitution in Eq. (7.3.7). □

The advantage of Eq. (7.3.13) is that any computation with a more general generalized inverse is effectively the same as that performed by taking one of the simpler form $G_0 = [I - P + \mathbf{tu}']^{-1}$.

However, by placing some additional restrictions on the form of the generalized inverse we can simplify the form of the expression for M.

COROLLARY 7.3.6B: (a) If $G = [I - P + \mathbf{eu}']^{-1} + \mathbf{ef}'$ (with $\mathbf{f}'$ arbitrary), then

$$M = [I - G + EG_d]D. \qquad (7.3.14)$$

(b) If $G = [I - P + \mathbf{eu}']^{-1} + \mathbf{ef}' + \mathbf{g}\boldsymbol{\pi}'$ (with $\mathbf{f}'$ and $\mathbf{g}'$ arbitrary), then

$$M = [I - G_0 + E(G_0)_d]D, \qquad (7.3.15)$$

where $G_0 = [I - P + \mathbf{eu}']^{-1}$ and $D = (\Pi_d)^{-1}$.

Proof: (a) With G as specified,

$$G\Pi = [I - P + \mathbf{eu'}]^{-1}\mathbf{e\pi'} + \mathbf{ef'e\pi'}$$

$$= \frac{\mathbf{e\pi'}}{\mathbf{u'e}} + (\mathbf{f'e})\mathbf{e\pi'} \qquad [\text{using Eq. (7.2.9)}]$$

$$= \beta\Pi \qquad \text{where} \quad \beta = (1/\mathbf{u'e}) + \mathbf{f'e}.$$

Thus $E(G\Pi)_d = \beta E\Pi_d = \beta\Pi = G\Pi$ and the result follows from Eq. (7.3.7).

(b) With G as specified, part (a) with $\mathbf{f} = 0$ implies that $E(G_0\Pi)_d = G_0\Pi$ and the result follows from Eq. (7.3.13). □

Special cases of Corollary 7.3.6B have appeared in the literature.

COROLLARY 7.3.6C: (a) If $Z = (I - P + \Pi)^{-1}$ is the fundamental matrix of the irreducible transition matrix P, then

$$M = [I - Z + EZ_d]D. \tag{7.3.16}$$

(b) If T is the group inverse of $I - P$ when P is the transition matrix of an irreducible MC, then

$$M = [I - T + ET_d]D. \tag{7.3.17}$$

Proof: (a) This follows from Corollary 7.3.6B(a) with $\mathbf{f'} = \mathbf{0'}$ and $\mathbf{u'} = \boldsymbol{\pi'}$.

(b) This follows from Corollary 7.3.6B(a) with $\mathbf{f'} = -\boldsymbol{\pi'}$ and $\mathbf{u'} = \boldsymbol{\pi'}$, since $T = [I - P + \Pi]^{-1} - \Pi$. □

The result given by Eq. (7.3.16) was established by Kemeny and Snell (1960) while the result given by Eq. (7.3.17) was derived by Meyer (1975). The technique used in both cases involved stating the solution, verifying that it is in fact a solution of Eq. (7.3.5), and thence establishing the uniqueness of the solution. The result utilizing the fundamental matrix Z was also derived by Hunter (1969) using the method of Theorem 7.3.6 by taking Z as a particular generalized inverse of $I - P$.

[Howard (1971) gives an alternative proof of (b) based upon the representation of T as given in Theorem 7.2.4. Observe that, from Theorem 6.2.5,

$$F_{ij}(s) = \frac{P_{ij}(s) - \delta_{ij}}{P_{jj}(s)}.$$

Also, from Theorems 6.1.9 and 7.2.4(b), if $\mathbf{T}(s) = [T_{ij}(s)]$

$$T_{ij}(s) = P_{ij}(s) - \pi_j/(1 - s).$$

Consequently, substitution for $P_{ij}(s)$ yields

$$F_{ij}(s) = \frac{\pi_j + (1 - s)[T_{ij}(s) - \delta_{ij}]}{\pi_j + (1 - s)T_{jj}(s)}. \qquad (7.3.18)$$

$F_{ij}(1) = 1$ for all i, j and thus

$$\mu_{ij} = \lim_{s \uparrow 1} \frac{dF_{ij}(s)}{ds} = \lim_{s \uparrow 1} \frac{1 - F_{ij}(s)}{1 - s} \qquad \text{(Theorem 2.4.1)}$$

$$= \lim_{s \uparrow 1} \frac{\delta_{ij} - T_{ij}(s) + T_{jj}(s)}{\pi_j + (1 - s)T_{jj}(s)}$$

$$= [\delta_{ij} - t_{ij} + t_{jj}] \frac{1}{\pi_j}, \qquad (7.3.19)$$

where $T \equiv [t_{ij}]$. Expressing Eq. (7.3.19) in matrix form leads to the expression for M given by Eq. (7.3.17).] $\quad\square$

There are some computational considerations that need to be taken into account. Both Z and T effectively require the prior determination of π', which, of course, requires the use of some generalized inverse of $I - P$ in solving Eq. (7.2.1). Why not use the same generalized inverse of $I - P$ to solve Eq. (7.3.5)? For such a computational procedure Z or T would not be used. In the corollary that follows we make use of some of the generalized inverses used in establishing Corollaries 7.2.1B and 7.2.2C.

COROLLARY 7.3.6D: (a) If $G = [I - P + eu']^{-1}$, then

$$M = [I - G + EG_d][(eu'G)_d]^{-1}. \qquad (7.3.20)$$

(b) If $G = [(I - P)_j]^{-1} \equiv [g_{ij}]$, then

$$M = [I - G + EG_d]D + (e_j e' - ee_j'), \qquad (7.3.21)$$

where

$$D = \text{diag}\left(\frac{1}{g_{j1}}, \frac{1}{g_{j2}}, \ldots, \frac{1}{g_{jm}}\right).$$

(c) If

$$P = \begin{bmatrix} P_{11} & \alpha \\ \beta' & P_{mm} \end{bmatrix},$$

and if $\mathbf{a} = (I - P_{11})^{-1}\mathbf{e}$, $\mathbf{b}' = \beta'(I - P_{11})^{-1}$, and $\lambda = 1 + \mathbf{b}'\mathbf{e}$, then

$$M = \begin{bmatrix} [\mathbf{ab}' - E(\mathbf{ab}')_d + \lambda\{I - (I - P_{11})^{-1} + E((I - P_{11})^{-1})_d\}]((\mathbf{eb}')_d)^{-1} & \mathbf{a} \\ [\lambda\mathbf{e}'((I - P_{11})^{-1})_d - \mathbf{e}'(\mathbf{ab}')_d]((\mathbf{eb}')_d)^{-1} & \lambda \end{bmatrix}. \qquad (7.3.22)$$

Proof: (a) Equation (7.2.13) gives $\pi' = \mathbf{u}'G$ and the result follows from Eq. (7.3.15) by noting that $\Pi_d = (\mathbf{e}\pi')_d = (\mathbf{e}\mathbf{u}'G)_d$.

(b) From Corollary 7.2.2C(d) we see that $G = [I - P + \mathbf{t}_j\mathbf{e}'_j]^{-1}$, where $\mathbf{t}_j = \mathbf{e} - (I - P)\mathbf{e}_j$. Furthermore, from Eq. (7.2.16), $\pi' = \mathbf{e}'_jG = (g_{j1}, g_{j2}, \ldots, g_{jm})$ so that the expression for D follows.

Since $\pi'\mathbf{t}_j \neq 0$ we use the general expression for M as given by Eq. (7.3.7). First observe, however, that

$$G^{-1}\mathbf{e}_j = (I - P + \mathbf{t}_j\mathbf{e}'_j)\mathbf{e}_j = (I - P)\mathbf{e}_j + \mathbf{e} - (I - P)\mathbf{e}_j = \mathbf{e}$$

so that $G\mathbf{e} = \mathbf{e}_j$ and hence that $G\Pi = G\mathbf{e}\pi' = \mathbf{e}_j\pi'$.

Consequently,

$$\begin{aligned}
[G\Pi - E(G\Pi)_d]D &= [\mathbf{e}_j\pi' - \mathbf{e}\mathbf{e}'(\mathbf{e}_j\pi')_d]D \\
&= \mathbf{e}_j\pi'D - \mathbf{e}\mathbf{e}'\mathbf{e}_j\mathbf{e}'_j \\
&= \mathbf{e}_j\mathbf{e}' - \mathbf{e}\mathbf{e}'_j
\end{aligned}$$

and Eq. (7.3.21) follows from Eq. (7.3.7).

(c) From Corollary 7.2.1B,

$$G = \begin{bmatrix} (I - P_{11})^{-1} & \mathbf{0} \\ \mathbf{0}' & 0 \end{bmatrix}$$

is a generalized inverse of $I - P$. From Corollary 7.2.2C we have that $\pi' = (1/\lambda)(\mathbf{b}', 1)$, so that

$$G\Pi = G\mathbf{e}\pi'$$

$$= \begin{bmatrix} (I - P_{11})^{-1} & \mathbf{0} \\ \mathbf{0}' & 0 \end{bmatrix} \begin{bmatrix} \mathbf{e} \\ 1 \end{bmatrix} \begin{bmatrix} \frac{1}{\lambda}\mathbf{b}', \frac{1}{\lambda} \end{bmatrix} = \begin{bmatrix} \frac{1}{\lambda}\mathbf{a}\mathbf{b}' & \frac{1}{\lambda}\mathbf{a} \\ \mathbf{0}' & 0 \end{bmatrix}.$$

Furthermore,

$$E(G\Pi)_d = \begin{bmatrix} E & \mathbf{e} \\ \mathbf{e}' & 1 \end{bmatrix} \begin{bmatrix} \frac{1}{\lambda}(\mathbf{a}\mathbf{b}')_d & 0 \\ \mathbf{0}' & 0 \end{bmatrix} = \begin{bmatrix} \frac{1}{\lambda}E(\mathbf{a}\mathbf{b}')_d & 0 \\ \frac{1}{\lambda}\mathbf{e}'(\mathbf{a}\mathbf{b}')_d & 0 \end{bmatrix},$$

$$E(G_d) = \begin{bmatrix} E & \mathbf{e} \\ \mathbf{e}' & 1 \end{bmatrix} \begin{bmatrix} ((I - P_{11})^{-1})_d & 0 \\ \mathbf{0}' & 0 \end{bmatrix} = \begin{bmatrix} E((I - P_{11})^{-1})_d & 0 \\ \mathbf{e}'((I - P_{11})^{-1})_d & 0 \end{bmatrix},$$

$$\Pi = \mathbf{e}\pi' = \begin{pmatrix} \mathbf{e} \\ 1 \end{pmatrix} \begin{pmatrix} \frac{1}{\lambda}\mathbf{b}', \frac{1}{\lambda} \end{pmatrix} = \frac{1}{\lambda} \begin{bmatrix} \mathbf{e}\mathbf{b}' & \mathbf{e} \\ \mathbf{b}' & 1 \end{bmatrix},$$

$$\Pi_d = \frac{1}{\lambda} \begin{bmatrix} (\mathbf{e}\mathbf{b}')_d & \mathbf{0} \\ \mathbf{0}' & 1 \end{bmatrix}, \qquad \Pi_d^{-1} = \begin{bmatrix} \lambda((\mathbf{e}\mathbf{b}')_d)^{-1} & \mathbf{0} \\ \mathbf{0}' & \lambda \end{bmatrix}.$$

Substitution into Eq. (7.3.7) leads to the required result. $\square$

The procedure suggested by Eq. (7.3.20) has the advantage of simplicity but it does involve the inversion of an $m \times m$ matrix, whereas the procedures suggested by Eqs. (7.3.21) and (7.3.22) require the computation of an $(m - 1) \times (m - 1)$ matrix inverse. The expression given by Eq. (7.3.21) has appeared in literature for the case $j = m$ (Meyer, 1978).

Meyer (1975) claims that generalized inverses are "forced" into the theory of Markov chains because of their equation-solving properties. His claim that this leads to cumbersome expressions that do little to enhance the theory and provides no practical or computational advantage is disputed. Theorem 7.3.6 provides a simple expression for M using *any* generalized inverse of $I - P$ and Corollaries 7.3.6B and 7.3.6D form the basis for very simple procedures for finding M. Much more elaborate techniques are needed to find Z or T. However, there is a role that Z (or equivalently $T = Z - \Pi$) can play in theoretical studies. If we restrict attention to those generalized inverses of the form $[I - P + \Pi]^{-1} + \alpha\Pi$ we are able to make some interesting observations. For the sake of convenience, we shall take $\alpha = 0$ and use Z, the fundamental matrix of the irreducible MC with transition matrix P.

Some basic properties concerning Z are contained in the following theorem.

THEOREM 7.3.7: If P is the transition matrix of a finite irreducible. MC and π' is its (unique) stationary probability vector, then $Z = [I - P + \Pi]^{-1}$ where $\Pi = e\pi'$ has the following properties:

(a)
$$Z = I + \lim_{n \to \infty} \sum_{i=1}^{n} \frac{n - i}{n} (P^i - \Pi).$$

If the MC is regular, then

$$Z = I + \sum_{n=1}^{\infty} (P^n - \Pi).$$

(b) $\quad Z = \lim_{s \uparrow 1} (I - s\Pi)(I - sP)^{-1} = \lim_{s \uparrow 1} \left[(I - sP)^{-1} - \frac{s}{1 - s} \Pi \right].$

(c) $\quad Ze = e$, implying $ZE = E$ and $Z\Pi = \Pi$.
(d) $\quad \pi'Z = \pi'$, implying $\Pi Z = \Pi$.
(e) $\quad PZ = ZP$.
(f) $\quad I - Z = \Pi - PZ = \Pi - ZP$.

Proof: (a) Observe that $Z = [I - P + e\pi']^{-1}$ exists by virtue of Theorem 7.1.1(a). By Theorem 4.5.4,

$$Z = [I - (P - \Pi)]^{-1} = I + \sum_{n=1}^{\infty} (P - \Pi)^n,$$

provided $(P - \Pi)^n \to 0$ as $n \to \infty$. By Theorems 7.1.2(e) and 7.1.6(c) $P\Pi = \Pi P = \Pi$; thus, as was shown in the proof of Theorem 7.2.4, we can deduce that $(P - \Pi)^n = P^n - \Pi$. By Theorem 7.1.2(a) for regular MC's $P^n \to \Pi$ and thus $(P - \Pi)^n \to 0$ and the result follows. By Corollary 7.1.7D for irreducible MC's $(1/n) \sum_{i=1}^n P^i \to \Pi$, so that if $K = P - \Pi$ we see that $(1/n) \sum_{i=1}^n K^i \to 0$ as $n \to \infty$. Now if $S_n = \sum_{i=1}^n [(n-i)/n] K^i$, we have after simplification that

$$S_n(I - K) = K - \frac{1}{n} \sum_{i=1}^n K^i,$$

and since $Z = (I - K)^{-1}$ exists we deduce that

$$I + S_n = (I - K)^{-1} - \left(\frac{1}{n} \sum_{i=1}^n K^i \right)(I - K)^{-1}$$

$$\to Z \qquad \text{as} \quad n \to \infty, \quad \text{using Theorem 4.5.1.}$$

(b)   These results follow direct from Theorems 7.2.4(c) and 7.2.4(b) observing that $Z = \Pi + \lim_{s \uparrow 1} \mathbf{T}(s)$.

(c)   This follows immediately from Eq. (7.2.9) with $\mathbf{t} = \mathbf{e}$ and $\mathbf{u}' = \boldsymbol{\pi}'$.

(d)   This follows, as in (c), by using Eq. (7.2.4).

(e)   and (f) follow by observing that

$$Z[I - P + \Pi] = [I - P + \Pi]Z = I,$$

and using results (c) and (d) above.   □

Theorem 7.3.7(b) serves as the basis for an alternative derivation of Z, especially if $[I - sP]^{-1}$ has been obtained for finding the n-step transition probabilities and Π has been obtained from $[I - sP]^{-1}$ using the method of Theorem 7.2.3.

EXAMPLE 7.3.3:   In Examples 6.1.7 and 7.2.4 we considered the irreducible MC with transition matrix

$$P = \begin{bmatrix} 0 & 1 & 0 \\ q & 0 & p \\ 0 & 1 & 0 \end{bmatrix},$$

and showed that

$$[I - sP]^{-1} = \frac{1}{1 - s^2} \begin{bmatrix} 1 - ps^2 & s & ps^2 \\ qs & 1 & ps \\ qs^2 & s & 1 - qs^2 \end{bmatrix}$$

and

$$\Pi = \frac{1}{2} \begin{bmatrix} q & 1 & p \\ q & 1 & p \\ q & 1 & p \end{bmatrix}.$$

It is an easy matter to show that

$$[I - sP]^{-1} - \frac{s}{1 - s}\,\Pi = \frac{1}{2(1 + s)}\begin{bmatrix} 2 + (1 + p)s & s & -ps \\ qs & 2 + s & ps \\ -qs & s & 2 + (1 + q)s \end{bmatrix},$$

leading to

$$Z = \frac{1}{4}\begin{bmatrix} 3 + p & 1 & -p \\ q & 3 & p \\ -q & 1 & 3 + q \end{bmatrix}.$$

An application of Corollary 7.3.6C gives, after an easy substitution, that

$$M = [I - Z + EZ_d]D$$

$$= \begin{bmatrix} 1 & \dfrac{1}{2} & 1 \\ \dfrac{3 + p - q}{4} & 1 & \dfrac{3 + q - p}{4} \\ 1 & \dfrac{1}{2} & 1 \end{bmatrix}\begin{bmatrix} \dfrac{2}{q} & 0 & 0 \\ 0 & 2 & 0 \\ 0 & 0 & \dfrac{2}{p} \end{bmatrix}$$

$$= \begin{bmatrix} \dfrac{2}{q} & 1 & \dfrac{2}{q} \\ \dfrac{3 + p - q}{2q} & 2 & \dfrac{3 + q - p}{2p} \\ \dfrac{2}{q} & 1 & \dfrac{2}{p} \end{bmatrix}. \quad \square$$

THEOREM 7.3.8:   If $Z = [z_{ij}]$ is the fundamental matrix of an irreducible MC with mean first passage the matrix M and stationary probability vector $\pi' = (\pi_1, \pi_2, \dots, \pi_m)$, then

(a) $$\pi'M = e'Z_dD = \left(\frac{z_{11}}{\pi_1}, \frac{z_{22}}{\pi_2}, \dots, \frac{z_{mm}}{\pi_m}\right),$$

(b) $$M\pi = \mathrm{tr}(Z)e = \left(\sum_{i=1}^{m} z_{ii}\right)e.$$

Proof: From Corollary 7.3.6C $M = [I - Z + EZ_d]D$; thus

$$\pi'M = \pi'D - \pi'ZD + \pi'ee'Z_dD$$
$$= e'Z_dD \quad [\text{since } \pi' = \pi'Z \text{ by Theorem 7.3.7(d)}].$$

Also,

$$
\begin{aligned}
M\pi &= [I - Z + EZ_d]\Pi_d^{-1}\pi \\
&= [I - Z + EZ_d]\mathbf{e} \\
&= \mathbf{ee}'Z_d\mathbf{e} \quad [\text{since } Z\mathbf{e} = \mathbf{e} \text{ by Theorem 7.3.7(c)}] \\
&= \operatorname{tr}(Z)\mathbf{e}. \quad \square
\end{aligned}
$$

Although these results are subsidiary to our central theme of finding moments of the first-passage time distributions we can give an interpretation to the result of Theorem 7.3.8(a). If we assume that the MC has gone through a large number of steps before it is observed, then the stationary probability vector is a natural choice for the initial probability vector. For such an initial vector, the ith component of the vector $\pi'M$ gives the expected number of steps for state i to first occur.

An interesting observation we can make is that once M is given, or for that matter only $\bar{M} \equiv M - M_d$, which specifies the μ_{ij} for $i \neq j$, the transition matrix of the underlying MC is completely determined. This result can be deduced from the following theorem, due to Kemeny and Snell (1960).

THEOREM 7.3.9: If M is the mean first passage time matrix of an irreducible MC and if $\bar{M} = M - M_d$, then

(a) $\bar{M}$ has an inverse,
(b) $P = I + (M_d - E)\bar{M}^{-1}$,
(c) $\pi = (\operatorname{tr}(Z) - 1)\bar{M}^{-1}\mathbf{e}$.

Proof: (a) From Eq. (7.3.5) we obtain

$$
(I - P)\bar{M} = E - M_d. \tag{7.3.23}
$$

Suppose $\bar{M}$ does not have an inverse. This implies that there is a column vector $\mathbf{x} \neq \mathbf{0}$ such that $\bar{M}\mathbf{x} = \mathbf{0}$. Postmultiplication of Eq. (7.3.23) by $\mathbf{x}$ yields

$$
\mathbf{0} = E\mathbf{x} - M_d\mathbf{x}
$$

or that $\Pi_d^{-1}\mathbf{x} = E\mathbf{x}$, i.e., $\mathbf{x} = \Pi_d\mathbf{ee}'\mathbf{x} = \pi(\mathbf{e}'\mathbf{x}) = l\pi$, where $l = \mathbf{e}'\mathbf{x} \neq 0$. Thus $\mathbf{x}$ must be a multiple of π, which means that $\bar{M}\pi = \mathbf{0}$. This is obviously false since the off-diagonal elements of $\bar{M}$ are all positive as are the elements of π. This contradiction leads to the required conclusion.

(b) This is an immediate consequence of Eq. (7.3.23).

(c)
$$
\begin{aligned}
\bar{M}\pi &= (M - D)\pi \\
&= M\pi - D\pi \\
&= \operatorname{tr}(Z)\mathbf{e} - \mathbf{e} \quad [\text{using Theorem 7.3.8(b)}]
\end{aligned}
$$

and the result follows. $\square$

Observe that if $\bar{M}$ is given, π and hence D can be found using Theorem 7.3.9(c) and thus D is determined by Theorem 7.3.9(b). Thus the MC is determined by $\bar{M}$. Kemeny and Snell (1960) point out that this is quite a surprising result since, in particular, $\bar{M}$ determines whether (or not) the chain is periodic. It would be highly desirable to find necessary and sufficient conditions that $\bar{M}$ represent a regular rather than an irreducible periodic chain. Of course, we would like such conditions to be simpler than computing P from $\bar{M}$ and then checking P.

Our studies so far in this section have been directed towards the derivation of the mean first passage time matrix, M. In any discussion centered around the higher moments some techniques to determine the variances of the r.v.'s T_{ij} are of interest.

Now

$$\text{var}(T_{ij}) = E[T_{ij}^2] - (E[T_{ij}])^2 = \mu_{ij}^{[2]} - (\mu_{ij})^2.$$

Since we have already obtained expressions for $M = [\mu_{ij}]$, it is only necesssary to find $M^{[2]} = [\mu_{ij}^{[2]}]$. By using Z as the particular generalized inverse of $I - P$ we are able to obtain simple expressions for $M^{[2]}$. [Expressions of a similar form can be obtained by using generalized inverses of the form $[I - P + \Pi]^{-1} + \alpha\Pi$. We do not, however, get a "nice" simplification when a more general form for $(I - P)^-$ is used. See Exercise 7.3.11.]

THEOREM 7.3.10:   The matrix $M^{[2]}$ satisfies the equation

$$(I - P)M^{[2]} = E + 2P[M - M_d] - PM_d^{[2]}, \qquad (7.3.24)$$

where

$$M_d^{[2]} = D(2Z_d D - I). \qquad (7.3.25)$$

The unique solution to Eq. (7.3.24) is given by

$$M^{[2]} = M(2Z_d D - I) + 2(ZM - E(ZM)_d). \qquad (7.3.26)$$

Proof: Equation (7.3.24) can be derived by using any of the arguments outlined in Theorem 7.3.3. For example,

$$\mu_{ij}^{[2]} = \sum_{n=1}^{\infty} n^2 f_{ij}^{(n)}$$

$$= p_{ij} + \sum_{n=2}^{\infty} n^2 \sum_{k \neq j} p_{ik} f_{kj}^{(n-1)}$$

$$= p_{ij} + \sum_{k \neq j} p_{ik} \sum_{n=2}^{\infty} (n - 1 + 1)^2 f_{kj}^{(n-1)}$$

$$= p_{ij} + \sum_{k \neq j} p_{ik} \{1 + 2\mu_{kj} + \mu_{kj}^{[2]}\}$$

i.e.,

$$\mu_{ij}^{[2]} = 1 + 2 \sum_{k \neq j} p_{ik} \mu_{kj} + \sum_{k \neq j} p_{ik} \mu_{kj}^{[2]}.$$

Equation (7.3.24) is obtained by expressing this above equation in matrix form.

To obtain an expression for $M_d^{[2]}$, premultiplication of Eq. (7.3.24) by Π yields

$$\Pi M_d^{[2]} = E + 2\Pi(M - M_d). \tag{7.3.27}$$

Substitution for $M - M_d$ using the results of Corollary 7.3.6B now gives

$$\Pi M_d^{[2]} = E - 2\Pi(Z - EZ_d)D = -E + 2EZ_d D,$$

since $\Pi E = E$, $\Pi Z = \Pi$, and $\Pi D = E$. Using Theorem 7.3.4 and taking diagonal elements leads to

$$\Pi_d M_d^{[2]} = -I + 2Z_d D,$$

from which Eq. (7.3.25) follows upon premultiplication by $\Pi_d^{-1} = D$.

Observe that the right-hand side of Eq. (7.3.24) is now fully determined and thus we can solve for $M^{[2]}$ using the method of Corollary 4.3.3A by taking Z as our generalized inverse of $I - P$. The consistency of these equations is not in dispute and the general solution is given by

$$M^{[2]} = ZE + 2ZP(M - M_d) - ZPM_d^{[2]} + [I - Z(I - P)]R,$$

where R is an arbitrary matrix. Note from Theorem 7.3.7(f) that $I - Z(I - P) = \Pi$ and hence $\Pi R = EB$, say, where B is an arbitrary diagonal matrix. B can now be found, as in the proof of Theorem 7.3.6, by taking diagonal elements and using Eq. (7.3.25), i.e.,

$$M_d^{[2]} = I + 2(ZPM)_d - 2(ZP)_d D - (ZP)_d M_d^{[2]} + B.$$

Substitution for the resulting expression for B gives

$$M^{[2]} = [E - ZP + E(ZP)_d]M_d^{[2]} + 2[E(ZP)_d - ZP]D + 2[ZPM - E(ZPM)_d]. \tag{7.3.28}$$

To facilitate simplification observe that Theorem 7.3.7(f) in conjunction with Lemma 7.3.5 and then Corollary 7.3.6C gives

$$\begin{aligned} E - ZP + E(ZP)_d &= E - \Pi + I - Z + E\Pi_d - E + EZ_d \\ &= I - Z + EZ_d \\ &= M\Pi_d. \end{aligned}$$

Analogously,

$$[E(ZP)_d - ZP]D = [M\Pi_d - E]D = M - ED.$$

Furthermore, from Corollary 7.3.6C and Theorem 7.3.7(f)

$$ZPM = ZM + \Pi M - M$$
$$= ZM + [EZ_d - I + Z - EZ_d]D$$
$$= ZM - [I - Z]D$$

so that

$$ZPM - E(ZPM)_d = ZM - E(ZM)_d - [I - Z - E + EZ_d]D$$
$$= ZM - E(ZM)_d - M + ED.$$

Substitution of these results into Eq. (7.3.28) and using Eq. (7.3.25) for $M_d^{[2]}$ leads immediately to Eq. (7.3.26). $\square$

The original derivation of Eq. (7.3.26) (and for that matter the results of Corollary 7.3.6C) by Kemeny and Snell (1960) were by indirect methods. The solution of Eq. (7.3.24) was stated to be that given by Eq. (7.3.26). Verification that this is a solution of Eq. (7.3.24) was then carried out and the final result followed after establishing that the solution is in fact unique.

Howard (1971) using matrix generating functions obtained an equivalent expression for $M^{[2]}$. His results and their equivalence to the results of Theorem 7.3.10 are stated and established in the next corollary. To obtain Howard's expressions we require some information concerning the limit of the first derivative of the matrix generating function $T(s)$ as s increases to 1. To maintain rigor and at the same time derive results that hold for all finite irreducible MC's we modify the proof as given by Howard.

LEMMA 7.3.11:   If P is a finite irreducible MC and $T(s) = (I - \Pi)(I - sP)^{-1}$ ($|s| < 1$), then

$$\frac{d}{ds} T(s) = \frac{1}{s} T(s)[T(s) - I] \qquad (|s| < 1).$$

In particular, if T is the transient sum matrix, then

$$T^{(1)} \equiv \lim_{s \uparrow 1} \frac{d}{ds} T(s) = T(T - I). \qquad (7.3.29)$$

Proof: Let $A(s)$ be a nonsingular square matrix whose elements are functions of s. Since $A(s)\{A(s)\}^{-1} = I$,

$$\frac{d}{ds}[A(s)\{A(s)\}^{-1}] = \left[\frac{d}{ds} A(s)\right][A(s)]^{-1} + [A(s)]\left[\frac{d}{ds}\{A(s)\}^{-1}\right]$$
$$= \frac{d}{ds} I = 0.$$

Premultiplication by $[A(s)]^{-1}$ then shows that

$$\frac{d}{ds}[A(s)]^{-1} = -[A(s)]^{-1}\left[\frac{d}{ds}A(s)\right][A(s)]^{-1}.$$

Thus

$$\frac{d}{ds}T(s) = (I - \Pi)\frac{d}{ds}(I - sP)^{-1}$$

$$= (I - \Pi)\left[-(I - sP)^{-1}\left\{\frac{d}{ds}(I - sP)\right\}(I - sP)^{-1}\right]$$

$$= (I - \Pi)(I - sP)^{-1}P(I - sP)^{-1}$$

$$= T(s)P(I - sP)^{-1}.$$

But

$$P(I - sP)^{-1} = P\sum_{n=0}^{\infty} s^n P^n = \sum_{n=0}^{\infty} s^n P^{n+1}$$

$$= \frac{1}{s}\sum_{n=1}^{\infty} s^n P^n = \frac{1}{s}[(I - sP)^{-1} - I].$$

This implies that

$$\frac{d}{ds}T(s) = \frac{1}{s}T(s)[(I - sP)^{-1} - I] = \frac{1}{s}T(s)\left[T(s) + \frac{1}{1-s}\Pi - I\right]$$

by Theorem 7.2.4. Furthermore, $T(s)\Pi = 0$ since

$$T(s)\Pi = [I - sP]^{-1}\Pi - \frac{1}{1-s}\Pi^2 = \left(\sum_{n=0}^{\infty} s^n P^n\right)\Pi - \frac{1}{1-s}\Pi$$

$$= \frac{1}{1-s}\Pi - \frac{1}{1-s}\Pi = 0$$

and the first result of the lemma is established. Since $T = \lim_{s\uparrow 1} T(s)$, Eq. (7.3.29) follows. □

COROLLARY 7.3.10A: For a finite irreducible MC with transient sum matrix T and $T^{(1)}$ as given in Lemma 7.3.11,

$$\mu_{ij}^{[2]} = \mu_{ij}\left[2\frac{t_{jj}}{\pi_j} + 1\right] + 2\left[t_{jj}^{(1)} - t_{ij}^{(1)}\right]\frac{1}{\pi_j}, \tag{7.3.30}$$

where $T = [t_{ij}]$ and $T^{(1)} = [t_{ij}^{(1)}]$. Furthermore, if $M^{[2]} = [\mu_{ij}^{[2]}]$, then

$$M^{[2]} = M(2T_dD + I) + 2(ET_d^{(1)} - T^{(1)})D \tag{7.3.31}$$

$$= M(2T_dD + I) + 2(TM - E(TM)_d) \tag{7.3.32}$$

$$= M(2Z_dD - I) + 2(ZM - E(ZM)_d). \tag{7.3.33}$$

Proof: In terms of generating functions

$$\mu_{ij}^{[2]} = F_{ij}^{(2)}(1) + F_{ij}^{(1)}(1)$$

$$= 2 \lim_{s \uparrow 1} \frac{d}{ds} \left\{ \frac{1 - F_{ij}(s)}{1 - s} \right\} + \mu_{ij} \qquad \text{(using Theorem 2.4.3)}.$$

From the alternative proof of Corollary 7.3.6C, if $\mathbf{T}(s) = [T_{ij}(s)]$, we have that

$$\frac{1 - F_{ij}(s)}{1 - s} = \frac{\delta_{ij} - T_{ij}(s) + T_{jj}(s)}{\pi_j + (1 - s)T_{jj}(s)},$$

$$\frac{d}{ds} \left\{ \frac{1 - F_{ij}(s)}{1 - s} \right\} = \frac{[-T_{ij}^{(1)}(s) + T_{jj}^{(1)}(s)]}{\pi_j + (1 - s)T_{jj}(s)}$$

$$- \frac{[\delta_{ij} - T_{ij}(s) + T_{jj}(s)][- T_{ij}(s) + (1 - s)T_{jj}^{(1)}(s)]}{[\pi_j + (1 - s)T_{jj}(s)]^2},$$

and hence, upon taking the limit as s increases to 1, that

$$\mu_{ij}^{[2]} = 2\left[t_{jj}^{(1)} - t_{ij}^{(1)}\right] \frac{1}{\pi_j} + 2\left[\delta_{ij} - t_{ij} + t_{jj}\right] \frac{t_{jj}}{\pi_j^2} + \mu_{ij},$$

which reduces to Eq. (7.3.30) upon utilizing the expression for μ_{ij} given by Eq. (7.3.19).

Observe that

$$TM = T[I - T + ET_d]D \qquad \text{[using Corollary 7.3.6C(b)]}$$
$$= TET_d - T^{(1)}D \qquad \text{[using Eq. (7.3.29)]}$$
$$= -T^{(1)}D$$

since

$$TE = ZE - \Pi E = E - E = 0 \qquad \text{[using Theorem 7.3.7(c)]}$$

Thus Eq. (7.3.32) follows from Eq. (7.3.31).

Since $\Pi M = \mathbf{e}\pi' M = \mathbf{e}\mathbf{e}' Z_d D = EZ_d D$, using Theorem 7.3.8(a), we have that

$$ZM = TM + \Pi M = EZ_d D - T^{(1)}D.$$

This implies that

$$ZM - E(ZM)_d = EZ_d D - T^{(1)}D - EZ_d D + ET_d^{(1)}D = (ET_d^{(1)} - T^{(1)})D,$$

and coupled with the result that

$$2T_d D + I = 2Z_d D - 2\Pi_d D + I = 2Z_d D - I$$

the equivalence between Eqs. (7.3.31) and (7.3.33) is established. $\square$

Howard expressed his result in the form given by Eq. (7.3.31). Meyer, using T as a generalized inverse of $I - P$, deduced the expression (7.3.32).

We have shown the both of these expressions are equivalent to the form given by Theorem 7.3.10 [Eq. (7.3.33)].

We conclude this section with a presentation of the results pertaining to a two-state MC.

EXAMPLE 7.3.4: *Two-State Markov Chain.* Let

$$P = \begin{bmatrix} p_{11} & p_{12} \\ p_{21} & p_{22} \end{bmatrix} = \begin{bmatrix} 1-a & a \\ b & 1-b \end{bmatrix} \quad (0 \le a \le 1, 0 \le b \le 1)$$

and $d = 1 - p_{12} - p_{21} = 1 - a - b$ $(|d| \le 1)$. From Example 7.1.1 (continued) it follows that, provided $-1 \le d < 1$,

$$\Pi = \frac{1}{p_{12} + p_{21}} \begin{bmatrix} p_{21} & p_{12} \\ p_{21} & p_{12} \end{bmatrix} = \frac{1}{1-d} \begin{bmatrix} b & a \\ b & a \end{bmatrix}.$$

Note that $-1 \le d < 1$ ensures that the MC is irreducible, regular if $-1 < d < 1$ and irreducible periodic when $d = -1$.

The fundamental matrix $Z = [I - P + \Pi]^{-1}$ is

$$Z = \frac{1}{1-d} \begin{bmatrix} b + \dfrac{a}{1-d} & a - \dfrac{a}{1-d} \\ b - \dfrac{b}{1-d} & a + \dfrac{b}{1-d} \end{bmatrix}.$$

The transient sum matrix is

$$T = \frac{1}{1-d} \begin{bmatrix} \dfrac{a}{1-d} & -\dfrac{a}{1-d} \\ -\dfrac{b}{1-d} & \dfrac{b}{1-d} \end{bmatrix}.$$

The mean first passage matrix M is

$$M = \begin{bmatrix} \dfrac{1-d}{b} & \dfrac{1}{a} \\ \dfrac{1}{b} & \dfrac{1-d}{a} \end{bmatrix} = \begin{bmatrix} 1 + \dfrac{p_{12}}{p_{21}} & \dfrac{1}{p_{12}} \\ \dfrac{1}{p_{21}} & 1 + \dfrac{p_{21}}{p_{12}} \end{bmatrix}$$

and the variance matrix for the first passage times is

$$[\mathrm{var}(T_{ij})] = \begin{bmatrix} \dfrac{a(1+d)}{b^2} & \dfrac{1-a}{a^2} \\ \dfrac{1-b}{b^2} & \dfrac{b(1+d)}{a^2} \end{bmatrix}. \quad \square$$

Exercises 7.3

1. Verify that the expression for μ_{ij} given by Eq. (7.3.2) satisfies the recurrence relationship Eq. (7.3.3). (*Hint:* Show that $e_i'P_j(0) = \sum_{k \neq j} p_{ik}e_k'$.)

2. Use Corollary 7.3.3A to obtain the results of the mth column of M when

$$P = \begin{bmatrix} P_{11} & \alpha \\ \beta' & p_{mn} \end{bmatrix},$$

as given in Corollary 7.3.6D(c), i.e., show

$$\mu_m' = \begin{bmatrix} (I - P_{11})^{-1}e \\ 1 + \beta(I - P_{11})^{-1}e \end{bmatrix}.$$

3. In Examples 6.2.6 and 6.3.4 a simple random walk on the nonnegative integers with a reflecting barrier at the origin was considered. In particular, it was shown that for this chain with state space $S = \{0, 1, 2, \ldots\}$ and transition probabilities $p_{00} = q$, $p_{i,i+1} = p$ $(i = 0, 1, \ldots)$, $p_{i,i-1} = q$ $(i = 1, 2, \ldots)$, it is irreducible and persistent if $p \leq q$. Show that

   (a) for $j \geq 0$,

$$\mu_{jj} = \begin{cases} \infty, & p = q = \frac{1}{2}, \\ \dfrac{q}{q-p}\left(\dfrac{q}{p}\right)^j, & p < q; \end{cases}$$

   (b) for $i < j$,

$$\mu_{ij} = \begin{cases} \infty, & p = q = \frac{1}{2}, \\ \dfrac{i-j}{q-p}, & p < q. \end{cases}$$

4. For an irreducible transition matrix P
   (a) prove that

$$M = P^k M + kE - (P + P^2 + \cdots + P^k)M_d \qquad (k = 1, 2, \ldots).$$

   (b) Use this result to prove that

$$\lim_{k \to \infty} \frac{P + P^2 + \cdots + P^k}{k} = \Pi. \qquad [\text{Pearl (1973).}]$$

5. Let P be the transition matrix of an absorbing MC with canonical form

$$P = \begin{bmatrix} I & 0 \\ R & Q \end{bmatrix}.$$

In Theorem 7.1.5 we showed that

$$\lim_{n \to \infty} P^{(n)} = \Pi = \begin{bmatrix} I & 0 \\ NR & 0 \end{bmatrix},$$

where $N = (I - Q)^{-1}$.

(a) Show that $Z = [I - P + \Pi]^{-1}$ exists for an absorbing chain and is given by

$$Z = \begin{bmatrix} I & 0 \\ -N^2 QR & N \end{bmatrix}.$$

(b) For irreducible MC's it is easily seen [cf. Theorem 7.3.7(f)] that if Z is the fundamental matrix of the MC, then

$$I - Z(I - P) = \Pi.$$

Verify that the same interpretation holds for Z as given in (a) above.

(c) Does Z satisfy the remaining properties of Theorem 7.3.7? Is Z a generalized inverse of $I - P$? (*Note*: Absorbing MC's have not been considered in this section because they are not irreducible. This means that $f_{ij} \neq 1$ all $i, j \in S$ and it is meaningless to examine moments of possibly improper first passage time distributions.)

6. Let $\{X_n, n \geq 0\}$ be a sequence of independent trials. This can be represented as an irreducible MC with transition matrix $P = [p_{ij}]$ where $p_{ij} = p_j$ ($j = 1, 2, \ldots, m$). Show that

$$Z = I,$$
$$M = ED,$$
$$M^{[2]} = ED(2D - 1),$$
$$[\text{var}(T_{ij})] = E(D^2 - D),$$

where $D = [\delta_{ij}/p_j]$.

7. If $M^{[k]} = [\mu_{ij}^{[k]}]$ show that for $k \geq 1$,

$$M^{[k]} = P[M^{[k]} - M_d^{[k]}] + \sum_{i=1}^{k-1} \binom{k}{i} P[M^{[i]} - M_d^{[i]}] + E$$

so that once $M^{[i]}$ have been determined for $i = 1, \ldots, k - 1$ it is possible to solve (theoretically) for $M^{[k]}$.

8. Consider an m-state irreducible MC. For each pair of states i, j define

$$\mu_{ij,k} = \sum_{n=1}^{k} nf_{ij}^{(n)} \quad \text{and} \quad M_k \equiv [\mu_{ij,k}].$$

The aim of this exercise is to obtain an alternative proof to the result that $\mu_{ij} = \lim_{k \to \infty} \mu_{ij,k} < \infty$ for all i, j.

(a) Show that $M_{k+1} = PM_k - P(M_k)_d + \sum_{r=1}^{k+1} F^{(r)}$. (*Hint*: Consider Theorem 5.1.6.)

(b)   Using (a) above show that $\Pi(M_k)_d \leq \Pi F = E$, and consequently deduce that $\mu_{jj} \leq 1/\pi_j, j = 1, \ldots, m$.

(c)   Show that for all k and l

$$M_{k+l} = P^l M_k + \sum_{r=1}^{k+l} F^{(r)} + \sum_{r=1}^{k+l-1} PF^{(r)} + \cdots + \sum_{r=1}^{k+1} P^{l-1} F^{(r)}$$

$$- \sum_{s=0}^{l-1} P^{l-s}(M_{k+s})_d,$$

and consequently

$$(P + P^2 + \cdots + P^l)M_d + M_{k+1} \geq P^l M_k.$$

(d)   For fixed i, j there exists an l such that $p_{ji}^{(l)} > 0$. Show that

$$\mu_{ij} \leq \frac{1}{\pi_j p_{ji}^{(l)}} (1 + p_{jj} + p_{jj}^{(2)} + \cdots + p_{jj}^{(l)}). \qquad \text{[Pearl (1973).]}$$

9. Show that if Z is the fundamental matrix of a regular MC with limit matrix Π, then

$$Z^k = \Pi + \sum_{r=0}^{\infty} \binom{r+k-1}{k-1}(P^r - \Pi) \qquad (k \geq 1).$$

[One method of proof is to use generalized inverses and induction:

(a)   Let

$$S_k = \Pi + \sum_{r=0}^{\infty} \binom{r+k-1}{k-1}(P^r - \Pi)$$

and show that $(I - P)S_k = S_{k-1} - \Pi$.

(b)   Using generalized inverses for solving systems of equations, show that $S_k = ZS_k - \Pi - \Pi R$, where R is arbitrary and such that $\Pi S_k = \Pi$, implying that $\Pi R = -\Pi$.

(c)   Use induction to show $S_k = Z^k$.]

10. A random walk on a circle consists of m states labeled clockwise $1, 2, \ldots, m$ with m followed by 1. The transition probabilities are given by $p_{i, i+1} = p$, $p_{i, i-1} = q$, and $p_{ii} = 1 - p - q$ for $i = 1, 2, \ldots, m$ (with $i + 1 = 1$ if $i = m$ and $i - 1 = m$ if $i = 1$.)

(a)   Find the stationary probability vector π' and hence show that $\mu_j = m$.

(b)   Show that the mean first passage times for $i \neq j$ are

$$\mu_{ij} = \begin{cases} \dfrac{d(m-d)}{2p} & \delta = 1, \\[3mm] \dfrac{1}{p-q}\left\{d - m\dfrac{\delta^d - 1}{\delta^m - 1}\right\}, & \delta \neq 1, \end{cases}$$

where $\delta = p/q$ and d is the clockwise distance from i to j.

[*Hints*:   (i)   By symmetry μ_{ij} depends only on d and not on the particular starting state i.

(ii)   From the equation for M write down the general and boundary difference equations for μ_{im}. Write $\mu_{im} = x_{m-i}$ and hence obtain

$$-qx_{d+1} + (p+q)x_d - px_{d-1} = 1 \qquad (2 \le d \le m-2)$$

$$(p+q)x_{m-1} - px_{m-2} = 1$$

$$-qx_2 + (p+q)x_1 = 1$$

$$-px_{m-1} - qx_1 + x_0 = 1$$

(iii)   Solve these difference equations.]

11. Show that if G is any generalized inverse of $I - P$, then

$$(\Pi_d)^2 M_d^{[2]} = I + 2[(I - \Pi)G(I - \Pi)]_d.$$

12. Consider an irreducible two-state MC with transition matrix

$$P = \begin{bmatrix} p_{11} & p_{12} \\ p_{21} & p_{22} \end{bmatrix}.$$

Let $\mathbf{t}' = (t_1, t_2)$ and $\mathbf{u}' = (u_1, u_2)$.
(a)   Using the expression given in Exercise 7.2.4 for *any* generalized inverse G of $I - P$. Show that

$$G\Pi - E(G\Pi)_d + I - G + EG_d = \begin{bmatrix} 1 & \dfrac{1}{p_{21} + p_{12}} \\ \dfrac{1}{p_{12} + p_{21}} & 1 \end{bmatrix}.$$

Using Theorem 7.3.6, deduce the expression for M as given in Example 7.3.4.
(b)   Using the expression given in Exercise 7.2.4 for G, show that
(i) $G\Pi = E(G\Pi)_d$ if and only if $t_1 - t_2 = (g_1 - g_2)(p_{12}t_2 + p_{21}t_1)$;

(ii) $EG_d - G = \dfrac{1}{p_{12}t_2 + p_{21}t_1} \begin{bmatrix} 0 & t_1 \\ t_2 & 0 \end{bmatrix}$ if and only if $g_1 = g_2$.

*7.4   Occupation Times

In Section 5.2 we examined the distribution of N_{ij}, the number of times a MC $\{X_n\}$ is in state j during an unlimited number of trials $n = 1, 2, \ldots,$

given that $X_0 = i$. In particular, we showed (Theorem 5.2.10) that

$$P\{N_{ij} = k\} = \begin{cases} 1 - f_{ij}, & k = 0, \\ f_{ij}f_{jj}^{k-1}(1 - f_{jj}), & k \geq 1. \end{cases}$$

From this result we obtained a characterization of transient and persistent states (Theorem 5.2.11).

In this section we shall restrict attention, primarily, to a finite number of trials and examine the distribution and moments of the number of visits to a particular state during these trials. In order to be entirely clear as to whether we are including the initial state or not, we take care in defining the relevant r.v.'s of interest.

DEFINITION 7.4.1:   Let $\{X_n\}$ $(n = 0, 1, \ldots)$ be a MC with state space S. Then for all $i, j \in S$ let

$N_{ij}^{(n)} \equiv$ Number of k $(1 \leq k \leq n)$ such that $X_k = j$ given $X_0 = i$,

$M_{ij}^{(n)} \equiv$ Number of k $(0 \leq k \leq n)$ such that $X_k = j$ given $X_0 = i$,

$$N_{ij} \equiv \lim_{n \to \infty} N_{ij}^{(n)}, \qquad M_{ij} \equiv \lim_{n \to \infty} M_{ij}^{(n)}. \quad \square$$

Thus $N_{ij}^{(n)}$ is the number of visits the MC makes to state j in the first n trials (with the convention that $N_{ij}^{(0)} \equiv 0$). If we wish to count the occurrence of state i at the zeroth trial when $i = j$ we use $M_{ij}^{(n)} = N_{ij}^{(n)} + \delta_{ij}$. Obviously $M_{ij} = N_{ij} + \delta_{ij}$.

There is no standard terminology for distinguishing between the *occupation time* r.v.'s $N_{ij}^{(n)}$ and $M_{ij}^{(n)}$ or between the *total occupation time* r.v.'s N_{ij} and M_{ij}. Howard (1971) does, however, refer to $N_{ij}^{(n)}$ as the *modified occupancy statistic* when we are interested only in the internal transitions of the process.

We have seen (Theorem 5.2.1) that if $i = j$ the event "return to state i" is a recurrent event, while (Theorem 5.2.2) if $i \neq j$ such an event is a delayed recurrent event. Consequently, if starting from i, the time of the rth visit to j is the sum of (possibly extended-valued) independent r.v.'s T_{ij}, $T_{jj}^{(1)}, \ldots, T_{jj}^{(r-1)}$ where $T_{jj}^{(k)}$ is distributed as T_{jj}. Thus (cf. Theorem 3.4.1 and Corollary 3.4.1A for the case $i = j$)

$$\{N_{ij}^{(n)} = r\} \equiv \left\{ T_{ij} + \sum_{v=1}^{r-1} T_{jj}^{(v)} \leq n < T_{ij} + \sum_{v=1}^{r} T_{jj}^{(v)} \right\},$$

which is more conveniently expressed as

$$\{N_{ij}^{(n)} \geq r\} \equiv \left\{ T_{ij} + \sum_{v=1}^{r-1} T_{jj}^{(v)} \leq n \right\}.$$

As a consequence of these two results we can use the theory of Section 3.4 (when $i = j$) and Section 3.5 (when $i \neq j$) to find expressions for the distribution of the $N_{ij}^{(n)}$. Such expressions are conveniently formulated using generating functions.

THEOREM 7.4.1: If $q_{ij}(n, k) = P\{N_{ij}^{(n)} = k\}$ $(k = 0, 1, \ldots, n)$, $(n = 0, 1, \ldots)$, and

$$Q_{ij}(s, t) = \sum_{n=0}^{\infty} \sum_{k=0}^{\infty} q_{ij}(n, k) s^n t^k,$$

then, for $|s| < 1$, $|t| < 1$,

$$Q_{ij}(s, t) = \frac{1 - F_{ij}(s) + t\{F_{ij}(s) - F_{jj}(s)\}}{(1 - s)\{1 - tF_{jj}(s)\}}. \tag{7.4.1}$$

Proof: Since T_{ij} is sampled from the distribution $\{f_{ij}^{(n)}\}$ with p.g.f. $F_{ij}(s)$ and the $\{T_{jj}^{(k)}\}$ $(k \geq 1)$, are sampled from the distribution $\{f_{jj}^{(n)}\}$ with p.g.f. $F_{jj}(s)$ an immediate application of Theorem 3.5.4 with $B(s) = F_{ij}(s)$ and $F(s) = F_{jj}(s)$ gives the required result. $\square$

COROLLARY 7.4.1A: If the MC $\{X_n\}$ is finite and if $\mathbf{Q}(s, t) = [Q_{ij}(s, t)]$, then, for $|s| < 1$, $|t| < 1$,

$$\mathbf{Q}(s, t) = \frac{1}{(1 - s)} [E - (1 - t)\mathbf{F}(s)\{\mathbf{I} - t\mathbf{F}_d(s)\}^{-1}]. \tag{7.4.2}$$

Proof: Equation (7.4.2) follows by expressing Eq. (7.4.1) in matrix form. $\square$

COROLLARY 7.4.1B: For $|s| < 1$,

$$\sum_{n=0}^{\infty} P\{N_{ij}^{(n)} = k\} s^n = \begin{cases} \dfrac{1 - F_{ij}(s)}{1 - s}, & (k = 0), \\[2mm] \dfrac{F_{ij}(s)\{1 - F_{jj}(s)\}\{F_{jj}(s)\}^{k-1}}{1 - s} & (k \geq 1). \end{cases}$$

Proof: Extracting the coefficient of t^k from both sides of Eq. (7.4.1) yields the above expressions (cf. Corollary 3.5.4A). $\square$

A special case of interest concerns the general two-state MC.

EXAMPLE 7.4.1: *Two-State Markov Chain.* Let $\{X_n\}$ be a MC with state space $S = \{1, 2\}$ and transition matrix

$$P = \begin{bmatrix} p_{11} & p_{12} \\ p_{21} & p_{22} \end{bmatrix}.$$

Let $d = 1 - p_{12} - p_{21}$ so that $1 + d = p_{11} + p_{22}$.

In Exercise 6.2.2 we saw that

$$\mathbf{F}(s) = [F_{ij}(s)] = \begin{bmatrix} \dfrac{s(p_{11} - ds)}{1 - p_{22}s} & \dfrac{p_{12}s}{1 - p_{11}s} \\[2ex] \dfrac{p_{21}s}{1 - p_{22}s} & \dfrac{s(p_{22} - ds)}{1 - p_{11}s} \end{bmatrix}. \tag{7.4.3}$$

Now if $0 < p_{ij} < 1$, then it is easily seen that $F = \mathbf{F}(1) = E$ and that $f_{ij} = 1$ for all i, j and the MC is in fact regular.

Substitution into Eq. (7.4.1) or Eq. (7.4.2) yields after simplification

$$\mathbf{Q}(s,t) = \begin{bmatrix} \dfrac{1 - ds}{1 - p_{22}s - p_{11}ts + dts^2} & \dfrac{1 - dst}{1 - p_{11}s - p_{22}ts + dts^2} \\[2ex] \dfrac{1 - dst}{1 - p_{22}s - p_{11}ts + dts^2} & \dfrac{1 - ds}{1 - p_{11}s - p_{22}ts + dts^2} \end{bmatrix}.$$

Furthermore, from Eq. (7.4.3) and Corollary 7.4.1B we can deduce that for $|s| < 1$

$$\sum_{n=0}^{\infty} P\{N_{22}^{(n)} = k\}s^n = s^k(1 - ds)(p_{22} - ds)^k(1 - p_{11}s)^{-(k+1)}, \qquad (k \geq 0)$$

$$\sum_{n=0}^{\infty} P\{N_{12}^{(n)} = k\}s^n = \begin{cases} (1 - p_{11}s)^{-1}, & (k = 0), \\ s^k(1 - ds)p_{12}(p_{22} - ds)^{k-1}(1 - p_{11}s)^{-(k+1)} & (k \geq 1), \end{cases}$$

with similar results for $N_{12}^{(n)}$ and $N_{11}^{(n)}$ (using symmetry).

If use is made of the binomial theorem and the binomial series:

$$(1 - x)^{-(k+1)} = \sum_{m=0}^{\infty} \binom{k+m}{m}x^m,$$

then after much tedious algebra it can be shown that extraction of the coefficient of s^n in the above expressions yields

$$P\{N_{22}^{(n)} = k\} = \sum_{r=0}^{\min(n-k,k)} \frac{(n - r - 1)!}{r!(k - r)!(n - k - r)!} p_{11}^{n-k-r-1}p_{22}^{k-r}$$
$$\times (-d)^r\{(n - r)p_{21} + kd\},$$

$$P\{N_{12}^{(n)} = k\} = \begin{cases} p_{11}^n & (k = 0), \\[1ex] \displaystyle\sum_{r=0}^{\min(n-k,k-1)} \dfrac{(n - r - 1)!}{r!(k - r - 1)!(n - k - r)!} p_{12}p_{11}^{n-k-r-1}p_{22}^{k-r-1} \\[2ex] \times (-d)^r\left\{\dfrac{(n - r)p_{21} + kd}{k}\right\} & (k \geq 1). \end{cases}$$

By analogy (interchanging the roles of states 1 and 2, which leaves d unchanged) we can obtain expressions for the distributions of $N_{11}^{(n)}$ and $N_{21}^{(n)}$. In particular, for $k \geq 0$,

$$P\{N_{11}^{(n)} = k\} = \sum_{r=0}^{\min(n-k,k)} \frac{(n-r-1)!}{r!(k-r)!(n-k-r)!} p_{11}^{k-r} p_{22}^{n-k-r-1}$$
$$\times (-d)^r \{(n-r)p_{12} + kd\}.$$

Note also that the MC has to be in either state 1 or state 2 at each trial hence, for fixed $i = 1$ or 2, $N_{i1}^{(n)} + N_{i2}^{(n)} = n$. This implies that

$$P\{N_{i1}^{(n)} = k\} = P\{N_{i2}^{(n)} = n - k\} \qquad \text{for} \quad k = 0, 1, \ldots, n,$$

and thus we have an alternative method for finding the distribution of $N_{11}^{(n)}$ and $N_{21}^{(n)}$ from the above results. In particular,

$$P\{N_{11}^{(n)} = k\} = \begin{cases} \sum_{r=0}^{\min(n-k-1, k)} \dfrac{(n-r-1)!}{r!(n-k-r-1)!(k-r)!} p_{12} p_{11}^{k-r-1} p_{22}^{n-k-r-1} \\ \qquad \times (-d)^r \left\{ \dfrac{(n-r)p_{21} + (n-k)d}{(n-k)} \right\} \qquad (k \leq n-1), \\ p_{11}^n \qquad\qquad\qquad\qquad\qquad\qquad\qquad\qquad (k = n). \end{cases}$$

The equivalence of these two expressions for the distribution of $N_{11}^{(n)}$ can also be established by algebraic means, but the proof is tedious.

The distribution of $N_{ij}^{(n)}$ for a two-state MC has been considered in the literature by various authors. Gabriel (1959) first tackled the problem using combinatorial arguments but failed to obtain expressions of the form presented above. In fact the analogous expressions obtained by Gabriel involve $n(n + 1) + 2$ terms of binomial coefficients inside the summation, considerably more than the maximum of $4 \min(n - k, k)$, i.e., at most $2n$ terms in the expressions above.

Helgert (1970) used a recurrence relationship technique and generating functions to obtain results that can be shown to be equivalent to those above.

Pedler (1971) used bivariate generating functions and expressed his results in terms of confluent hypergeometric functions and showed that Gabriel's "rather unwieldly" expressions could be replaced by expressions involving standard functions.

Of course, the technique we have used depends only on the observation that "return to a state of a MC" generates a (possibly general) recurrent event process. The rest follows from recurrent event theory and algebra. □

In general, exact distributions of the $N_{ij}^{(n)}$ are difficult to obtain. However, a generalization of Theorem 3.4.5 leads immediately to an expression for the asymptotic distribution of $N_{jj}^{(n)}$.

THEOREM 7.4.2:   If state j is a persistent state whose recurrence time distribution $\{f_{jj}^{(n)}\}$ has finite mean μ_j and finite variance σ_j^2, then for each fixed α,

$$\lim_{n \to \infty} P\left\{\frac{N_{jj}^{(n)} - n/\mu_j}{(\sigma_j^2 n/\mu_j^3)^{1/2}} \leq \alpha\right\} = \Phi(\alpha) \equiv \frac{1}{\sqrt{2\pi}} \int_{-\infty}^{\alpha} e^{-x^2/2}\, dx. \quad \square$$

In other words $N_{jj}^{(n)}$ asymptotically obeys a normal probability law with mean n/μ_j and variance $\sigma_j^2 n/\mu_j^3$.

The proof of Theorem 3.4.5 and hence the above theorem is based upon the observation that $\{T_{jj}^{(n)}\}$ is a sequence of independent and identically distributed r.v.'s with finite means and variances and hence obeys the conditions for the central limit theorem. In fact the sequence $T_{ij}, T_{jj}^{(1)}, T_{jj}^{(2)}, \ldots$ also satisfies conditions for the applicability of the central limit theorem since the addition of a finite number of terms does not affect the limit value of an average. Hence we can generalize Theorem 7.4.2 to the case of the r.v. $N_{ij}^{(n)}$.

COROLLARY 7.4.2A: If state j is persistent with a recurrence time distribution having finite mean μ_j and finite variance σ_j^2, and if $f_{ij} = 1$, then for each fixed α,

$$\lim_{n \to \infty} P\left\{\frac{N_{ij}^{(n)} - n/\mu_j}{(\sigma_j^2 n/\mu_j^3)^{1/2}} \leq \alpha\right\} = \Phi(\alpha). \quad \square$$

If the MC is irreducible and all the states are persistent nonnull, then $\mu_j = 1/\pi_j (< \infty)$ [cf. Eq. (7.3.1)] and thus we can further extend our generalization.

COROLLARY 7.4.2B: Let $\{X_n\}$ be an irreducible MC with stationary distribution $\{\pi_j\}$ and $\mathrm{var}(T_{jj}) = \sigma_j^2 < \infty$ for all $j \in S$. Then for all $i, j \in S$ and each fixed α,

$$\lim_{n \to \infty} P\left\{\frac{N_{ij}^{(n)} - n\pi_j}{\sigma_j \pi_j^{3/2} n^{1/2}} \leq \alpha\right\} = \Phi(\alpha). \quad \square$$

Note also that if the state space of the irreducible MC is finite, then σ_j^2 is automatically finite, by Theorem 7.3.1.

Let us now focus attention on evaluation of the moments of the occupation time r.v.'s and in particular the expectations $EN_{ij}^{(n)}$.

Occupation times may be usefully represented as sums of random variables. Let $\{X_n\}$ $(n = 0, 1, \ldots)$ be a MC and for all states i, j and $n = 1, 2, \ldots$ define

$$Y_{ij}^{(n)} = \begin{cases} 1 & \text{if } X_n = j, \quad \text{given } X_0 = i, \\ 0 & \text{if } X_n \neq j, \quad \text{given } X_0 = i. \end{cases}$$

In words, given that the MC starts initially in state i, $Y_{ij}^{(n)}$ is equal to 1 or 0 depending on whether or not at time n the chain is in state j. Thus we may write

$$N_{ij}^{(n)} = \sum_{k=1}^{n} Y_{ij}^{(k)} \quad \text{and with} \quad Y_{ij}^{(0)} \equiv \delta_{ij}, \quad M_{ij}^{(n)} = \sum_{k=0}^{n} Y_{ij}^{(k)}.$$

We examine techniques for evaluating $EN_{ij}^{(n)}$.

THEOREM 7.4.3: For all $i, j \in S$, for all $n \geq 1$,

$$EN_{ij}^{(n)} = \sum_{k=1}^{n} p_{ij}^{(k)}. \tag{7.4.4}$$

Thus, using matrix theory,

$$[EN_{ij}^{(n)}] = \sum_{k=1}^{n} P^k, \tag{7.4.5}$$

using generating functions, for $|s| < 1$,

$$\sum_{n=0}^{\infty} EN_{ij}^{(n)} s^n = \frac{P_{ij}(s) - \delta_{ij}}{(1 - s)} \tag{7.4.6}$$

$$= \frac{F_{ij}(s)}{(1 - s)\{1 - F_{jj}(s)\}}, \tag{7.4.7}$$

and in terms of matrix generating functions

$$\left[\sum_{n=0}^{\infty} EN_{ij}^{(n)} s^n \right] = \frac{1}{(1 - s)} \{(I - sP)^{-1} - I\}. \tag{7.4.8}$$

Proof: Equation (7.4.4) comes directly from the representation of $N_{ij}^{(n)}$ in terms of indicator r.v.'s using $EY_{ij}^{(k)} = p_{ij}^{(k)}$.

Equation (7.4.5) follows directly from Eq. (7.4.4) and Theorem 5.1.2.

Equation (7.4.6) comes from Eq. (7.4.4) using generating functions.

Equation (7.4.7) can be derived either from Theorem 3.5.7 or from Eq. (7.4.6) and Theorem 6.2.5.

Equation (7.4.8) follows immediately from Eq. (7.4.6) and Theorem 6.1.9. $\square$

Sometimes it is more convenient to consider $EM_{ij}^{(n)}$ rather than $EN_{ij}^{(n)}$. With the obvious extensions we deduce immediately from Theorem 7.4.3:

COROLLARY 7.4.3A: For all $i, j \in S$, for all $n \geq 0$,

$$EM_{ij}^{(n)} = \sum_{k=0}^{n} p_{ij}^{(k)}, \tag{7.4.9}$$

$$[EM_{ij}^{(n)}] = \sum_{k=0}^{n} P^k, \tag{7.4.10}$$

$$\sum_{n=0}^{\infty} EM_{ij}^{(n)} s^n = \frac{P_{ij}(s)}{1 - s} \qquad (|s| < 1), \tag{7.4.11}$$

and

$$\left[\sum_{n=0}^{\infty} \mathsf{E}M_{ij}^{(n)}s^n\right] = \frac{(I - sP)^{-1}}{1 - s} \qquad (|s| < 1). \quad \square \qquad (7.4.12)$$

EXAMPLE 7.4.2: *Two-State Markov Chain.* Let $\{X_n\}$ be the MC as considered in Example 7.4.1. We have a variety of techniques at our disposal. Using generating functions we have from Eqs. (7.4.3) and (7.4.7)

$$\sum_{n=0}^{\infty} \mathsf{E}N_{12}^{(n)}s^n = \frac{p_{12}s}{(1 - s)^2(1 - ds)},$$

$$\sum_{n=0}^{\infty} \mathsf{E}N_{22}^{(n)}s^n = \frac{s(p_{22} - ds)}{(1 - s)^2(1 - ds)}.$$

Extraction of the coefficient of s^n from both of these expressions yields, for $n \geq 0$,

$$\mathsf{E}N_{12}^{(n)} = p_{12}\{n(1 - d) - d + d^{n+1}\}/(1 - d)^2,$$
$$\mathsf{E}N_{22}^{(n)} = \{np_{12}(1 - d) + p_{21} - p_{21}d^{n+1}\}/(1 - d)^2.$$

It can be shown that these expressions (and the analogous ones for $N_{11}^{(n)}$ and $N_{21}^{(n)}$) are equivalent to those obtained in different form by Gabriel (1959) and Helgert (1970).

Using the form given by Eq. (7.4.12), Howard (1971) showed (using our notation) that

$$\left[\sum_{n=0}^{\infty} \mathsf{E}M_{ij}^{(n)}s^n\right] = \frac{1}{(1 - s)^2(1 - d)}\begin{bmatrix} p_{21} & p_{12} \\ p_{21} & p_{12} \end{bmatrix}$$
$$+ \frac{1}{(1 - s)(1 - d)^2}\begin{bmatrix} p_{21} & -p_{12} \\ -p_{21} & p_{12} \end{bmatrix}$$
$$+ \frac{d}{(1 - ds)(1 - d)^2}\begin{bmatrix} -p_{12} & p_{12} \\ p_{21} & -p_{21} \end{bmatrix},$$

which yields upon extraction of s^n,

$$[\mathsf{E}M_{ij}^{(n)}] = \frac{(n + 1)}{1 - d}\begin{bmatrix} p_{21} & p_{12} \\ p_{21} & p_{12} \end{bmatrix} + \frac{1 - d^{n+1}}{(1 - d)^2}\begin{bmatrix} p_{12} & -p_{12} \\ -p_{21} & p_{21} \end{bmatrix}.$$

These results are equivalent to those obtained above. $\square$

EXAMPLE 7.4.3: *Absorbing Markov Chain.* Let P be the transition matrix of an absorbing MC with structure given by

$$P = \begin{bmatrix} I & 0 \\ R & Q \end{bmatrix},$$

as in Theorem 5.4.3.

If we are interested in $\mathsf{E}M_{ij}^{(n)}$ for $i, j \in T$, the transient states, then from Eq. (7.4.9) and Corollary 6.1.1A,

$$[\mathsf{E}M_{ij}^{(n)}] = \sum_{k=0}^{n} [p_{ij}^{(k)}]_{i, j \in T} = \sum_{k=0}^{n} Q^k,$$

so that

$$[I - Q][\mathsf{E}M_{ij}^{(n)}] = I - Q^{n+1}.$$

Premultiplication by $N = (I - Q)^{-1}$ yields

$$[\mathsf{E}M_{ij}^{(n)}] = N(I - Q^{n+1}). \quad \square$$

As is the case with the distribution of $N_{ij}^{(n)}$, explicit determination of the expected value of $N_{ij}^{(n)}$ is in general a difficult task. However, in Sections 3.4 and 3.5 we were able to obtain information concerning the asymptotic behavior of N_n for recurrent events. The relevant theorems for $N_{ij}^{(n)}$ will follow with suitable identification of the appropriate parameters.

THEOREM 7.4.4: For any MC and $i, j \in S$,

$$\lim_{n \to \infty} \frac{\mathsf{E}N_{ij}^{(n)}}{n} = \begin{cases} \dfrac{1}{\mu_j} & \text{if } f_{ij} = 1, f_{jj} = 1, \mu_j < \infty, \\[2mm] \dfrac{f_{ij}}{\mu_j} & \text{if } f_{ij} < 1, f_{jj} = 1, \mu_j < \infty, \\[2mm] 0 & \text{otherwise.} \end{cases}$$

Proof: These results follow from Theorem 3.5.6 with the assignments: $b = f_{ij}$, $f = f_{jj}$, and $\mu = \mu_j$.

Alternatively we have seen that for any MC

$$\lim_{n \to \infty} \frac{1}{n} \sum_{k=1}^{n} p_{ij}^{(k)}$$

exists with the limits given as stated in the above theorem (Theorem 7.1.7 and its corollaries). The theorem then follows using Eq. (7.4.4). $\quad \square$

COROLLARY 7.4.4A: If the MC is irreducible with a stationary distribution $\{\pi_j\}$, then for all $i, j \in S$

$$\lim_{n \to \infty} \frac{1}{n} \mathsf{E}N_{ij}^{(n)} = \pi_j.$$

Proof: Under the stated conditions $f_{ij} = 1$ for all $i, j \in S$ and $1/\mu_j = \pi_j$ by Theorem 7.1.9. $\quad \square$

If we are willing to place more stringent restrictions on the MC, then we can, in fact, obtain more information concerning $N_{ij}^{(n)}$ for large n. Firstly, if we continue with our approach involving identification of the appropriate embedded recurrent events in the MC, then the following result is effectively a restatement of relevant theorems as given in Sections 3.4 and 3.5.

THEOREM 7.4.5: If state j is ergodic, $f_{ij} = 1$, μ_{ij} finite, and the recurrence time distribution of state j has finite variance σ_j^2, then

$$\mathsf{E}N_{ij}^{(n)} = \frac{n}{\mu_j} + \left(\frac{\sigma_j^2 + \mu_j + \mu_j^2}{2\mu_j^2} - \frac{\mu_{ij}}{\mu_j} \right) + o(1). \tag{7.4.13}$$

Proof: The ergodicity of state j implies that it is persistent ($f_{jj} = 1$), nonnull ($\mu_j < \infty$), and aperiodic. In the case $i \neq j$ Eq. (7.4.13) comes direct from Eq. (3.5.6) with $v = \mu_{ij}$, $\mu = \mu_j$ and $\sigma^2 = \sigma_j^2$. The case $i = j$ comes direct from Eq. (3.4.11) with $\mu = \mu_j$ and $\sigma^2 = \sigma_j^2$ or from above by taking $\mu_{ij} = \mu_j$. $\square$

To obtain matrix expressions for $[\mathsf{E}N_{ij}^{(n)}]$ or $[\mathsf{E}M_{ij}^{(n)}]$ there are a variety of approaches we can take. We can either express Eq. (7.4.13) in matrix form by making use of the relevant results for $[\mu_{ij}]$ and $[\mu_{ij}^{[2]}]$ as derived in Section 7.3 or consider a direct attack via, say, Eqs. (7.4.5) or (7.4.10). We shall reconcile these methods later (Theorem 7.4.7) but to do so we need to develop an expression for sums of powers of transition matrices that utilises generalized inverses of $I - P$. The relevant procedure is given by the following theorem due to Hunter (1982).

THEOREM 7.4.6: If P is the transition matrix of a finite irreducible MC with stationary probability vector π', then if G is *any* generalized inverse of $I - P$, and $\Pi = e\pi'$,

$$\sum_{r=0}^{n-1} P^r = \begin{cases} n\Pi + (I - \Pi)G(I - P^n), & (7.4.14) \\ n\Pi + (I - P^n)G(I - \Pi). & (7.4.15) \end{cases}$$

Proof: Let $A_n = \sum_{r=0}^{n-1} P^r$. Then it is easy to see that

$$(I - P)A_n = I - P^n \tag{7.4.16}$$

and

$$A_n(I - P) = I - P^n. \tag{7.4.17}$$

Equations (7.4.16) and (7.4.17) are in a suitable form for applying Corollaries 4.3.3A and 4.3.3B, respectively. Both equations are in fact consistent and the arbitrary constant matrix in each solution can be eliminated using the observation that the A_n are constrained by the relationships

$$A_n\Pi = \Pi A_n = n\Pi. \tag{7.4.18}$$

With $G = [I - P + \mathbf{tu'}]^{-1} + \mathbf{ef'} + \mathbf{g\pi'}$ it is easily seen, as in the proof of Theorem 7.3.6, that

$$I - (I - P)G = \left[\frac{\mathbf{t}}{\mathbf{\pi't}} - (I - P)\mathbf{g}\right]\mathbf{\pi'} \equiv \mathbf{\alpha\pi'} \tag{7.4.19}$$

and

$$I - G(I - P) = \mathbf{e}\left[\frac{\mathbf{u'}}{\mathbf{u'e}} - \mathbf{f'}(I - P)\right] \equiv \mathbf{e\beta'}. \tag{7.4.20}$$

We shall consider only Eq. (7.4.16). The procedure follows analogously for Eq. (7.4.17). The consistency condition is verified by using Eq. (7.4.19) while Eq. (7.4.20) shows that the general solution is given by

$$A_n = G(I - P^n) + \mathbf{e\beta'}U_n = G(I - P^n) + \mathbf{eu'_n}, \tag{7.4.21}$$

say. Further, from Eq. (7.4.18)

$$n\Pi = \Pi G(I - P^n) + \mathbf{eu'_n} \tag{7.4.22}$$

(since $\Pi\mathbf{e} = \mathbf{e}$). Elimination of $\mathbf{eu'_n}$ between Eqs. (7.4.21) and (7.4.22) leads to Eq. (7.4.14). $\square$

If we restrict attention to regular Markov chains then $\lim_{n \to \infty} p_{ij}^{(n)} = \pi_j$ for all $i, j = 1, 2, \ldots, m$ and thus, in terms of matrices, $\lim_{n \to \infty} P^n = \Pi$. With this observation we obtain the following information concerning the behavior of $\sum_{r=0}^{n-1} P^r$ for large n.

COROLLARY 7.4.6A: If G is *any* generalized inverse of $I - P$ where P is the transition matrix of a regular MC, then

$$\sum_{r=0}^{n-1} P^r = n\Pi + (I - \Pi)G(I - \Pi) + o(1)E. \tag{7.4.23}$$

Proof:

$$\sum_{r=0}^{n-1} P^r = n\Pi + (I - \Pi)G(I - \Pi) + (I - \Pi)G(\Pi - P^n).$$

If $(I - \Pi)G = [b_{ij}]$, then the (i, j)th element of $(I - \Pi)G(\Pi - P^n)$ is given by

$$\sum_{k=1}^{m} b_{ik}(\pi_j - p_{kj}^{(n)}) \to 0 \qquad \text{as} \quad n \to \infty$$

by Theorem 7.1.2 and Eq. (7.4.23) follows. $\square$

This corollary leads immediately to the following important theorem.

THEOREM 7.4.7: If P is the transition matrix of a regular MC and G is *any* generalized inverse of $I - P$, then

$$[\mathbf{E}M_{ij}^{(n)}] = (n + 1)\Pi + (I - \Pi)G(I - \Pi) + o(1)E \tag{7.4.24}$$

and

$$[EN_{ij}^{(n)}] = (n + 1)\Pi - I + (I - \Pi)G(I - \Pi) + o(1)E. \qquad (7.4.25)$$

Proof: Equations (7.4.24) and (7.4.25) follow from Eq. (7.4.23) and the representations given by Eqs. (7.4.5) and (7.4.10).

Alternatively we can obtain Eq. (7.4.25) from Eq. (7.4.13) and Theorem 7.3.10. First note that for a regular MC all the conditions of Theorem 7.4.5 are automatically satisfied [by Definition 5.3.5(b) and Theorem 7.3.1]. Since $\mu_{jj}^{[2]} = ET_{jj}^2 = \sigma_j^2 + \mu_j^2$, Eq. (7.4.13) can be expressed as

$$EN_{ij}^{(n)} = \frac{n}{\mu_j} + \left(\frac{\mu_{jj}^{[2]}}{2\mu_j^2} + \frac{1}{2\mu_j} - \frac{\mu_{ij}}{\mu_j}\right) + o(1),$$

which we can write in matrix form as

$$[EN_{ij}^{(n)}] = nE(M_d)^{-1} + \tfrac{1}{2}E\{(M_d)^{-1}\}^2 M_d^{[2]} + \tfrac{1}{2}E(M_d)^{-1} - M(M_d)^{-1} + o(1)E. \qquad (7.4.26)$$

Now

$$E(M_d)^{-1} = E\Pi_d = \Pi$$

using Eq. (7.3.6) and Lemma 7.3.5(c) and

$$E\{(M_d)^{-1}\}^2 M_d^{[2]} = \Pi(M_d)^{-1} M_d^{[2]},$$

using Lemma 7.3.5(c), where

$$\Pi M_d^{[2]} = E + 2\Pi(M - M_d),$$

from Eq. (7.3.27), and

$$M(M_d)^{-1} = G\Pi - E(G\Pi)_d + I - G + EG_d,$$

from Eq. (7.3.7). Substitution of these results in Eq. (7.4.26) gives after simplification

$$\begin{aligned}[EN_{ij}^{(n)}] &= n\Pi - (I - \Pi)M(M_d)^{-1} + o(1)E \\ &= (n + 1)\Pi - I + (I - \Pi)G(I - \Pi) + o(1)E,\end{aligned}$$

as required.

Note also that Eqs. (7.4.24) and (7.4.25) are connected by observing that

$$[EM_{ij}^{(n)}] = [\delta_{ij}] + [EN_{ij}^{(n)}] = I + [EN_{ij}^{(n)}]. \quad \square$$

The expressions given in Theorem 7.4.7 are due to Hunter (1982). If G is replaced by Z, the fundamental matrix, we obtain the results of Kemeny and Snell (1960).

COROLLARY 7.4.7A: For any regular MC with fundamental matrix Z and limit matrix Π, as $n \to \infty$

$$[EM_{ij}^{(n)}] = n\Pi + Z + o(1)E, \qquad (7.4.27)$$

$$[EN_{ij}^{(n)}] = n\Pi + Z - I + o(1)E. \qquad (7.4.28)$$

Proof: Simplification of Eqs. (7.4.24) and (7.4.25) give the required forms after observing that $(I - \Pi)Z(I - \Pi) = Z - \Pi$. □

Equation (7.4.27) gives a nice interpretation to the elements of Z, since if $Z = [z_{ij}]$, then

$$EM_{ij}^{(n)} = n\pi_j + z_{ij} + o(1),$$

and thus the asymptotic behavior of the occupation time r.v.'s is easily determined from Π and Z. Alternatively, we can obtain expressions involving T, the transient sum matrix, or group inverse of $I - P$.

COROLLARY 7.4.7B: For any regular MC with transient sum matrix T and limit matrix Π, as $n \to \infty$

$$[EM_{ij}^{(n)}] = (n + 1)\Pi + T + o(1)E, \qquad (7.4.29)$$

$$[EN_{ij}^{(n)}] = (n + 1)\Pi + T - I + o(1)E. \qquad (7.4.30)$$

Proof: These results follow from Corollary 7.4.7A and the observation that $Z = T + \Pi$. These results were also derived by Meyer (1975).

Howard (1971) also derived these results using generating functions. In particular, from Eq. (7.4.11) and Theorem 7.2.4(b),

$$\sum_{n=0}^{\infty} EM_{ij}^{(n)} s^n = \frac{P_{ij}(s)}{1 - s} = \frac{1}{(1 - s)} \left[\frac{\pi_j}{1 - s} + T_{ij}(s) \right] = \frac{\pi_j}{(1 - s)^2} + \frac{T_{ij}(s)}{1 - s},$$

which implies that for large n (using Theorem 2.3.3)

$$EM_{ij}^{(n)} = (n + 1)\pi_j + t_{ij} + o(1),$$

where $t_{ij} = T_{ij}(1)$ and Eq. (7.4.29) follows. □

A continuation of our study of the r.v.'s $N_{ij}^{(n)}$ and $M_{ij}^{(n)}$ leads naturally to a discussion of their second moments or, more importantly, their variance. Since $M_{ij}^{(n)} = N_{ij}^{(n)} + \delta_{ij}$, $\text{var}(M_{ij}^{(n)}) = \text{var}(N_{ij}^{(n)})$ and thus it is immaterial as to which r.v. we consider.

THEOREM 7.4.8: For all $i, j \in S$,

$$\{E(M_{ij}^{(n)})^2\} = 2\{p_{ij}^{(n)}\} * \{EM_{jj}^{(n)}\} - \{EM_{ij}^{(n)}\}, \qquad (7.4.31)$$

$$\{E(N_{ij}^{(n)})^2\} = 2\{p_{ij}^{(n)}\} * \{EN_{jj}^{(n)}\} + (1 - 2\delta_{ij})\{EN_{ij}^{(n)}\}. \qquad (7.4.32)$$

For $|s| < 1$,

$$\sum_{n=0}^{\infty} \mathsf{E}(M_{ij}^{(n)})^2 s^n = \frac{P_{ij}(s)}{1-s}\left[2P_{jj}(s) - 1\right], \tag{7.4.33}$$

$$\sum_{n=0}^{\infty} \mathsf{E}(N_{ij}^{(n)})^2 s^n = \frac{F_{ij}(s)[1 + F_{jj}(s)]}{(1-s)[1 - F_{jj}(s)]^2}. \tag{7.4.34}$$

Proof: We establish Eqs. (7.4.31) and (7.4.33). The other results are left as an exercise.

Observe that

$$(M_{ij}^{(n)})^2 = \sum_{k=0}^{n} \sum_{r=0}^{n} Y_{ij}^{(k)} Y_{ij}^{(r)},$$

and thus

$$\mathsf{E}(M_{ij}^{(n)})^2 = \sum_{k=0}^{n} \sum_{r=0}^{n} \mathsf{E}(Y_{ij}^{(k)} Y_{ij}^{(r)}),$$

where, since $Y_{ij}^{(k)} Y_{ij}^{(r)} = 1$ iff $Y_{ij}^{(k)} = 1$ and $Y_{ij}^{(r)} = 1$,

$$\mathsf{E}(Y_{ij}^{(k)} Y_{ij}^{(r)}) = P\{X_k = j, X_r = j \mid X_0 = i\} = \begin{cases} p_{ij}^{(k)} p_{jj}^{(r-k)}, & r \geq k, \\ p_{ij}^{(r)} p_{jj}^{(k-r)}, & k \geq r. \end{cases}$$

Thus

$$\begin{aligned}
\mathsf{E}(M_{ij}^{(n)})^2 &= \sum_{k=0}^{n} \sum_{r=k}^{n} p_{ij}^{(k)} p_{jj}^{(r-k)} + \sum_{r=0}^{n} \sum_{k=r}^{n} p_{ij}^{(r)} p_{jj}^{(k-r)} - \sum_{k=0}^{n} p_{ij}^{(k)} \\
&= 2 \sum_{k=0}^{n} \sum_{r=k}^{n} p_{ij}^{(k)} p_{jj}^{(r-k)} - \mathsf{E}M_{ij}^{(n)} \\
&= 2 \sum_{k=0}^{n} p_{ij}^{(k)} \sum_{t=0}^{n-k} p_{jj}^{(t)} - \mathsf{E}M_{ij}^{(n)} \\
&= 2 \sum_{k=0}^{n} p_{ij}^{(k)} \mathsf{E}M_{jj}^{(n-k)} - \mathsf{E}M_{ij}^{(n)},
\end{aligned}$$

and Eq. (7.4.31) follows.

Equation (7.4.33) is now obtained by using Eq. (7.4.11). □

We can use Theorem 3.4.11 directly to obtain an asymptotic form for var $N_{jj}^{(n)}$ when state j is ergodic, namely, as $n \to \infty$

$$\text{var } N_{jj}^{(n)} = (n\sigma_j^2/\mu_j^3) + o(n), \tag{7.4.35}$$

where μ_j and σ_j^2 are, respectively, the mean and the variance of the recurrence time distribution of state j. A modification of Theorem 3.4.11 is required to deduce an analogous form for the var $N_{ij}^{(n)}$. We base our modification upon Eqs. (7.4.33) and (7.4.29).

THEOREM 7.4.9:   For any regular MC, with fundamental matrix Z or transient sum matrix T, and limit matrix Π,

$$[\operatorname{var} M_{ij}^{(n)}] = n\Pi[2T_d + \Pi_d - I] + o(n)\,E \qquad (7.4.36)$$
$$= n\Pi[2Z_d - \Pi_d - I] + o(n)E. \qquad (7.4.37)$$

Alternatively,

$$\operatorname{var} M_{ij}^{(n)} = (n\sigma_j^2/\mu_j^3) + o(n). \qquad (7.4.38)$$

Proof: First note that from Eqs. (7.4.33) and (7.4.11)

$$\sum_{n=0}^{\infty} EM_{ij}^{(n)}(M_{ij}^{(n)} + 1)s^n = \frac{2}{1-s}\,P_{ij}(s)P_{jj}(s).$$

Using the result that $P_{ij}(s) = T_{ij}(s) + \pi_j/(1-s)$ it can be easily verified that

$$\frac{2}{1-s}\,P_{ij}(s)P_{jj}(s) = \frac{2}{1-s}\,T_{ij}(s)T_{jj}(s) + \frac{2\pi_j}{(1-s)^2}\,P_{jj}(s)$$
$$+ \frac{2\pi_j}{(1-s)^2}\,P_{ij}(s) - \frac{2\pi_j^2}{(1-s)^3}.$$

Observe that $T_{ij}(s)$ is the g.f. of $t_{ij}^{(n)} = p_{ij}^{(n)} - \pi_j \to 0$ as $n \to \infty$ and that $[1/(1-s)]T_{ij}(s)$ is the g.f. of

$$\sum_{k=0}^{n} t_{ij}^{(k)} = \sum_{k=0}^{n} p_{ij}^{(k)} - (n+1)\pi_j = EM_{ij}^{(n)} - (n+1)\pi_j \to t_{ij} \qquad \text{as} \quad n \to \infty.$$

Thus, from Corollary 2.5.6A, the coefficient of s^n in $[2/(1-s)]\,T_{ij}(s)T_{jj}(s)$ is $o(n)$. Hence, by definition and extracting the coefficient of s^n from above,

$$\operatorname{var} M_{ij}^{(n)} = EM_{ij}^{(n)}(M_{ij}^{(n)} + 1) - EM_{ij}^{(n)} + (EM_{ij}^{(n)})^2$$
$$= 2\pi_j \sum_{k=0}^{n} EM_{jj}^{(k)} + 2\pi_j \sum_{k=0}^{n} EM_{ij}^{(k)} - \pi_j^2(n+1)(n+2)$$
$$- EM_{ij}^{(n)} + (EM_{ij}^{(n)})^2 + o(n). \qquad (7.4.39)$$

Now Lemma 2.3.6 implies that

$$\sum_{k=0}^{n} \{EM_{ij}^{(k)} - (k+1)\pi_j\} \frac{1}{n+1} \to t_{ij} \qquad \text{as} \quad n \to \infty.$$

Thus

$$\sum_{k=0}^{n} EM_{ij}^{(k)} = (n+1)t_{ij} + \frac{(n+1)(n+2)}{2}\,\pi_j + o(n).$$

Also,

$$EM_{ij}^{(n)} = (n + 1)\pi_j + t_{ij} + o(1) = (n + 1)\pi_j + o(n)$$

and

$$[EM_{ij}^{(n)}]^2 = (n + 1)^2\pi_j^2 + 2(n + 1)t_{ij}\pi_j + o(n).$$

Substitution of these expressions in Eq. (7.4.39) and simplification yields

$$\text{var } M_{ij}^{(n)} = n\pi_j[2t_{jj} + \pi_j - 1] + o(n). \tag{7.4.40}$$

Expressing Eq. (7.4.40) in matrix form yields Eq. (7.4.36) from which Eq. (7.4.37) follows since $Z = T + \Pi$. Now note that

$$\frac{n\sigma_j^2}{\mu_j^3} = \frac{n}{\mu_j^3}(\mu_{jj}^{[2]} - \mu_j^2) = n(\pi_j^3\mu_{jj}^{[2]} - \pi_j).$$

Thus in matrix form

$$n\left[\frac{\sigma_j^2}{\mu_j^3}\right] = n\{E(\Pi_d)^3 M_d^{[2]} - E\Pi_d\}$$
$$= n\{\Pi(\Pi_d)^2(\Pi_d)^{-1}(2Z_d\Pi_d - I) - \Pi\}$$
$$= n\Pi[2Z_d - \Pi_d - I],$$

establishing the equivalence between Eqs. (7.4.37) and (7.4.38). □

We have concentrated solely on an examination of the distribution and moments of the occupation time r.v.'s for finite time intervals. We conclude this section with an investigation into similar properties of the *total occupation time* r.v.'s N_{ij} and M_{ij}.

The distribution of N_{ij} was considered in Theorem 5.2.10 from which it was deduced (Theorem 5.2.11) that N_{ij} is a proper r.v. (with finite expectation) only when state j is transient. In particular, for this case,

$$EN_{ij} = f_{ij}/(1 - f_{jj}). \tag{7.4.41}$$

Note also that $N_{ij} = \lim_{n \to \infty} N_{ij}^{(n)}$ (with probability one) and thus from Eq. (7.4.4)

$$EN_{ij} = \sum_{k=1}^{\infty} p_{ij}^{(k)}. \tag{7.4.42}$$

[Equations (7.4.41) and (7.4.42) are of course connected, e.g., using the limits of g.f.'s and Eq. (6.2.20).]

Consequently, let us restrict attention to chains with transient states. It is sufficient to examine absorbing MC's and we shall assume that the transition matrix is given in canonical form by

$$P = \begin{bmatrix} I & 0 \\ R & Q \end{bmatrix}.$$

To obtain "nice" expressions we consider EM_{ij}.

THEOREM 7.4.10: For $i, j \in T$, the transient states

$$EM_{ij} = \sum_{k=0}^{\infty} p_{ij}^{(k)}, \qquad (7.4.43)$$

$$EM_{ij} = \delta_{ij} + \sum_{k \in T} p_{ik} EM_{kj}. \qquad (7.4.44)$$

In matrix form, for $i, j \in T$,

$$[EM_{ij}] = N = (I - Q)^{-1}. \qquad (7.4.45)$$

Proof: Equation (7.4.43) follows by a similar derivation to Eq. (7.4.42) using Eq. (7.4.9).

Observe that

$$M_{ij} = \begin{cases} \delta_{ij} & \text{with probability} \quad \sum_{k \in \bar{T}} p_{ik}, \\ M_{kj} + \delta_{ij} & \text{with probability} \quad p_{ik}, k \in T, \end{cases}$$

by considering the first step from state i. If at this step the MC enters an absorbing state the number of visits to j is zero unless $i = j$. On the other hand, if the MC moves to $k \in T$, then from that position onward the number of visits to j is $M_{kj} + \delta_{ij}$.

Taking expectations we get

$$EM_{ij} = \sum_{k \in \bar{T}} p_{ik} \delta_{ij} + \sum_{k \in T} p_{ik} E(M_{kj} + \delta_{ij}),$$

which leads immediately to Eq. (7.4.44).

This argument can be formalized by considering

$$EM_{ij} = E[M_{ij}|X_0 = i] = \sum_{k \in S} E[M_{ij}|X_1 = k, X_0 = i] P[X_1 = k|X_0 = i].$$

Splitting the state space into T and $\bar{T}$ we have that if $k \in T$, then given $X_1 = k$ and $X_0 = i$, $M_{ij} = M_{kj} + \delta_{ij}$; while if $k \in \bar{T}$, then the only transient state involved is $X_0 = i$ and $M_{ij} = \delta_{ij}$. Thus

$$EM_{ij} = \sum_{k \in T} p_{ik} E[M_{kj} + \delta_{ij}|X_1 = k] + \sum_{k \in \bar{T}} p_{ik} \delta_{ij},$$

from which Eq. (7.4.44) follows.

Now Eq. (7.4.43) in matrix form yields

$$[EM_{ij}] = \sum_{k=0}^{\infty} Q^k = (I - Q)^{-1};$$

using Corollary 6.1.1A and Theorem 4.5.4. Alternatively, from Eq. (7.4.44),

$$[EM_{ij}] = I + Q[EM_{ij}],$$

which can be solved for $[EM_{ij}]$, yielding Eq. (7.4.45). $\square$

Expressions for second moments and hence the variances can be obtained in a similar fashion.

THEOREM 7.4.11:   For $i, j \in T$, the transient states

$$\text{var}(M_{ij}) = n_{ij}(2n_{jj} - 1 - n_{ij}) \tag{7.4.46}$$

or, alternatively,

$$[\text{var}(M_{ij})] = 2NN_d - N - N_{sq}, \tag{7.4.47}$$

where $N = [n_{ij}]$ and $N_{sq} = [n_{ij}^2]$.

Proof: Before outlining our derivation note that from Theorem 5.2.10 N_{ij} is a modified geometric r.v. and hence has finite moments of all orders. Similar remarks follow for M_{ij}. Bhat ((1972), pp. 74–75) gives an alternative (more complicated) argument to show that EM_{ij}^2 is finite.
Since

$$M_{ij}^2 = \begin{cases} \delta_{ij}^2 & \text{with probability} \quad \sum_{k \in \bar{T}} p_{ik}, \\[2ex] (M_{kj} + \delta_{ij})^2 & \text{with probability} \quad p_{ik}, \; k \in T, \end{cases}$$

taking expectations (or using the conditioning argument as in Theorem 7.4.10) we get

$$\begin{aligned} EM_{ij}^2 &= \sum_{k \in \bar{T}} p_{ik}\delta_{ij} + \sum_{k \in T} p_{ik}E(M_{kj} + \delta_{ij})^2 \\ &= \delta_{ij} + \sum_{k \in T} p_{ik}EM_{kj}^2 + 2\delta_{ij}\sum_{k \in T} p_{ik}EM_{kj}, \end{aligned}$$

since

$$\sum_{k \in \bar{T}} p_{ik} + \sum_{k \in T} p_{ik} = 1.$$

Thus in matrix form

$$[EM_{ij}^2] = I + Q[EM_{ij}^2] + 2(QN)_d,$$

where we have used Eq. (7.4.45). Solving for $[EM_{ij}^2]$ after noting that $QN = N - I$ (as shown in the proof of Theorem 6.3.1) yields

$$[EM_{ij}^2] = N(2N_d - I).$$

Equation (7.4.47) now follows since $[\text{var}(M_{ij})] = [EM_{ij}^2] - [EM_{ij}]^2$. Equation (7.4.46) is the element form of Eq. (7.4.47).   □

One of the interesting questions we can ask is how many trials does the MC spend in the transient states prior to absorption? Such a r.v. can be expressed in terms of the total occupation time r.v.'s.

Let $N_i = \sum_{j \in T} N_{ij}$ and $M_i = \sum_{j \in T} M_{ij} = N_i + 1$. Then N_i is the total number of visits to transient states (before absorption), whereas M_i can be interpreted as either the total number of trials including the initial trial the MC is in the transient states or the total number of trials not including the initial trial to absorption.

THEOREM 7.4.12:   For each $i \in T$, the transient states,

(a) $$EM_i = 1 + \sum_{k \in T} p_{ik} EM_k.$$ (7.4.48)

(b)   The vector of mean times to absorption (EM_i) is given by

$$(\mathbf{EM}_i) = N\mathbf{e}.$$ (7.4.49)

(c)   The vector of variances of times to absorption $(\text{var } \mathbf{M}_i)$ is given by

$$(\text{var } \mathbf{M}_i) = (2N - I)N\mathbf{e} - (N\mathbf{e})_{sq}.$$ (7.4.50)

Proof: (a)   Equation (7.4.48) follows from Eq. (7.4.44) by summing over $j \in T$.

(b)   From Eq. (7.4.48), $(\mathbf{EM}_i) = \mathbf{e} + Q(\mathbf{EM}_i)$, from which Eq. (7.4.49) is obtained.

(c)   Proceeding as in Theorem 7.4.11 we have

$$EM_i^2 = \sum_{k \in T} p_{ik} \cdot 1 + \sum_{k \in T} p_{ik} E(M_k + 1)^2,$$

which reduces to

$$EM_i^2 = 1 + \sum_{k \in T} p_{ik} EM_k^2 + 2 \sum_{k \in T} p_{ik} EM_k$$

or in vector form

$$(\mathbf{EM}_i^2) = \mathbf{e} + Q(\mathbf{EM}_i^2) + 2Q(\mathbf{EM}_i).$$

One can solve for $(\mathbf{EM}_i^2)$, substitute for $(\mathbf{EM}_i)$, and proceed as in Theorem 7.4.11 to obtain the required result.   □

EXAMPLE 7.4.4: *Classical Gambler's Ruin Model.*   Consider the classical gambler's ruin model as described in Example 5.1.2 (and also Examples 6.1.4, 6.2.3, and 6.3.2) and let us examine the derivation of $m_i \equiv EM_i$, the expected duration of the game until either the gambler or his opponent is ruined.

Rather than evaluate N, it is often easier to solve Eqs. (7.4.48) using standard difference equation techniques. In this case the equations become

$$m_i = 1 + qm_{i-1} + pm_{i+1} \qquad (2 \leq i \leq a - 2),$$

$$m_1 = 1 + pm_2,$$

$$m_{a-1} = 1 + qm_{a-2}.$$

If we define $m_0 \equiv 0$ and $m_a \equiv 0$, then we may write

$$m_i = 1 + qm_{i-1} + pm_{i+1} \qquad (1 \leq i \leq a - 1).$$

This is a nonhomogeneous difference equation, the homogeneous version of which has been solved earlier in Example 6.3.2. The general solution is the complementary function (CF) plus the particular solution (PS).

If $p \neq q$, the CF is $A + B(q/p)^i$ and for the PS we try $m_i = \lambda i$. Substitution shows that $\lambda = 1/(q - p)$. Thus, in this case,

$$m_i = \frac{i}{q - p} + A + B\left(\frac{q}{p}\right)^i.$$

Using the boundary conditions ($m_0 = m_a = 0$) we see that $A + B = 0$ and $A + B(q/p)^a = -a/(q - p)$, which leads to the general solution

$$m_i = \frac{i}{q - p} - \frac{a}{q - p}\left\{ \frac{1 - \left(\dfrac{q}{p}\right)^i}{1 - \left(\dfrac{q}{p}\right)^a} \right\} \qquad (1 \leq i \leq a - 1).$$

If $p = q$, the CF is $A + Bi$ and for the PS we try $m_i = \lambda i^2$. Substitution gives $\lambda = -1$ and hence $m_i = -i^2 + A + Bi$. The boundary conditions determine A and B, leading to

$$m_i = i(a - i) \qquad (1 \leq i \leq a - 1).$$

Feller (1968) remarks that this duration is considerably longer than we would naïvely expect. Note that if two players with $500 each toss a coin until one is ruined the average duration is 250,000 trials. If a gambler has only $1 and his adversary $999, the average duration is 998 trials.

We can also take the limit as $a \to \infty$ and consider a game against an infinitely rich adversary. When $p > q$ the game may go on forever, and thus it makes no sense to talk about its expected duration. When $p < q$ we obtain for the expected duration $i/(q - p)$ but when $p = q$ the expected duration is infinite. These results were found earlier in Example 7.3.2 as moments of first passage time distributions. □

Actually, given a fixed state j in an irreducible MC, then the mean first passage times $\{\mu_{ij}, i \neq j\}$ can be regarded as mean absorption times in a new MC obtained from the original MC by making state j absorbing. Following Parzen (1962), we define a new transition matrix $[q_{rs}]$ by setting

$$q_{rs} = \begin{cases} 1 & \text{if } r = j, s = j, \\ 0 & \text{if } r = j, s \neq j, \\ p_{rs} & \text{if } r \neq j, \text{ all } s. \end{cases}$$

This derived MC contains a single absorbing state, while all other states are transient. The behavior of the new MC before absorption is the same as the behavior of the original MC before visiting j for the first time. In particular, μ_{ij} in the original MC is the same as M_i in the new MC. This means that we have quite a large arsenal of techniques that can be used interchangeably.

We conclude this section with a general discussion concerning the solution of (7.4.48). If $m_i = \mathsf{E}M_i$, then the $\{m_i\}$ satisfy the equations

$$m_i = 1 + \sum_{k \in T} p_{ik} m_k. \tag{7.4.51}$$

When the MC is finite these equations have a unique solution that when expressed in matrix form is given by Eq. (7.4.49). However, when the MC is infinite, the solution of Eqs. (7.4.51) need not be unique. Suppose $\{m_i^{(1)}\}$ and $\{m_i^{(2)}\}$ are two solutions. Let $x_i = m_i^{(1)} - m_i^{(2)}$. Then

$$x_i = \sum_{k \in T} p_{ik} x_k \tag{7.4.52}$$

and then $\{x_i\}$ satisfies Eq. (6.3.7). The null solution $x_i = 0$ for all $i \in T$ is the only bounded solution of Eq. (6.3.7) if the probability of remaining forever in the transient states is zero (Corollary 6.3.3A). However, there is no reason why the $\{x_i\}$ should be unbounded and in general may be unbounded. Of course, when the chain is finite $x_i \equiv 0$, but in the general case it is not clear how to characterize the required solution.

EXAMPLE 7.4.5: Consider the MC with state space $S = \{0, 1, 2, \ldots\}$ and transition probabilities $p_{i, i+1} = q_i$ and $p_{i0} = p_i$ $(0 < p_i < 1, p_i + q_i = 1)$ for $i = 1, 2, \ldots$ with $p_{00} = 1$. Obviously, state 0 is absorbing and states $1, 2, 3, \ldots$ are transient.

Firstly, if h_i is the probability that the system remains in the transient states starting in state i it is seen, from first principles, that

$$h_i = q_i q_{i+1} \cdots = \prod_{j=1}^{\infty} (1 - p_j) \qquad (i = 1, 2, \ldots).$$

If $0 < p_j < 1$, then $\prod_{j=1}^{\infty}(1 - p_j) = 0$ for all $i \geq 1$ if and only if $\sum_{j=1}^{\infty} p_j = \infty$ (which may be proved by using the inequalities

$$1 - \sum_{j=i}^{n} p_i < \prod_{j=i}^{n}(1 - p_j) < \exp\left[-\sum_{j=i}^{n} p_j\right]\Bigg).$$

Thus, provided $\sum_{j=1}^{\infty} p_j = \infty$, $h_i = 0$ for all $i \geq 1$.

Now consider the solution of Eq. (6.3.7), or equivalently Eq. (7.4.52), i.e., $x_i = q_i x_{i+1}$ $(i = 1, 2, \ldots)$. The general solution can be found by taking $x_1 = \theta$ (arbitrary), implying

$$x_i = \frac{\theta}{q_1 q_2 \cdots q_{i-1}}, \qquad i = 2, 3, \ldots.$$

However, if $\sum p_j = \infty$, the solution with $\theta = 0$, the null solution, is the only bounded solution consistent with the earlier observation that $h_i = 0$ $(i \geq 1)$.

Now consider deriving $\mathsf{E}M_i$ from first principles. Since

$$M_i = \begin{cases} 1 & \text{with prob. } p_i, \\[2ex] n & \text{with prob. } \left(\displaystyle\prod_{j=i}^{i+n-2} q_j\right) p_{i+n-1} \qquad (n \geq 2), \\[3ex] \infty & \text{with prob. } \displaystyle\prod_{j=i}^{\infty} q_j = 0, \end{cases}$$

M_i is finite with probability one, and hence its expectation is well defined and given by

$$\mathsf{E}M_i = p_i + \sum_{n=2}^{\infty} n p_{i+n-1} \left(\prod_{j=i}^{i+n-2} q_j\right)$$

$$= 1 + \sum_{n=0}^{\infty} \left(\prod_{j=i}^{i+n} q_j\right).$$

However, this is not the unique solution of Eq. (7.4.51) since the most general solution is given by

$$m_i = \mathsf{E}M_i + x_i$$

$$= 1 + \sum_{n=0}^{\infty} \left(\prod_{j=i}^{i+n} q_j\right) + \theta \Bigg/ \prod_{j=1}^{i-1} q_j,$$

i.e., we have an infinite number of solutions depending on one arbitrary parameter.   □

We leave this chapter with an unsolved problem. How many arbitrary parameters are there involved in the general solution of Eq. (7.4.51) and what extra criteria are required to assign the true value to the arbitrary parameters to obtain $\{EM_i\}$?

Exercises 7.4

1. Show that for any MC

$$\lim_{n \to \infty} \frac{EN_{ij}^{(n)}}{1 + EN_{jj}^{(n)}} = f_{ij}.$$

(The transient and persistent cases require separate proof.)

2. Let $Y_{ij}^{(n)}$ be the indicator r.v.'s as introduced in Section 7.4.
   (a)   Verify that the entries in the accompanying table are the joint probabilities $P\{Y_{ij}^{(n)} = r,\ Y_{ij}^{(n+m)} = s\}$:

		s	
r	0		1
0	$1 - p_{ij}^{(n+m)} - p_{ij}^{(n)}(1 - p_{jj}^{(n)})$		$p_{ij}^{(n+m)} - p_{ij}^{(n)}p_{jj}^{(m)}$
1	$p_{ij}^{(n)}(1 - p_{jj}^{(m)})$		$p_{ij}^{(n)}p_{jj}^{(m)}$

   (b)   Show that

$$\mathrm{cov}(Y_{ij}^{(n)}, Y_{ij}^{(n+m)}) = p_{ij}^{(n)}(p_{jj}^{(m)} - p_{ij}^{(n+m)}),$$

   and hence deduce that $Y_{ij}^{(n)}$ and $Y_{ij}^{(n+m)}$ are not independent.
   (c)   Given that state j was visited at time n, the probability of a visit at time $n + m$ is $p_{jj}^{(m)}$, independent of n. On the other hand,

$$P\{X_{n+m} = j | X_n \neq j\} = P\{Y_{ij}^{(n+m)} = 1 | Y_{ij}^{(n)} = 0\}$$
$$= [p_{ij}^{(n+m)} - p_{ij}^{(n)}p_{jj}^{(m)}]/(1 - p_{ij}^{(n)}).$$

Extend these results by considering the joint distribution of $Y_{ij}^{(n)}$, $Y_{ij}^{(n+m)}$, $Y_{ij}^{(n+m+v)}$ and explain how the observation of a visit to j at n and $n + m$ can be used to predict whether or not a visit will occur at time $n + m + v$ (Heathcote, 1971).

3. Let $\{X_n\}$ be a regular MC with stationary probability distribution $\{\pi_j\}$. Then given any $i, j \in S$ for any $\varepsilon > 0$,

$$\lim_{n \to \infty} P\left(\left| \frac{1}{n} EN_{ij}^{(n)} - \pi_j \right| > \varepsilon \right) = 0.$$

[*Hint*: It is sufficient to show that

$$\lim_{n \to \infty} E\left[\left(\frac{1}{n} N_{ij}^{(n)} - \pi_j\right)^2\right] = 0.$$

Further details of a proof along these lines is given by Bhat (1972).]

4. Let $\{X_n\}$ be a sequence of independent trials with m possible outcomes at each trial with probability of outcome j being p_j. Thus $\{X_n\}$ is a MC with transition matrix $P = [p_{ij}]$ with $p_{ij} = p_j$. Show that

   (a) $EN_{ij}^{(n)} = np_j,$
   (b) $E[(N_{ij}^{(n)})^2] = n(n - 1)p_j + np_j^2 \ (n \geq 1),$
   (c) $\text{var}[N_{ij}^{(n)}] = np_j(1 - p_j) \ (n \geq 1).$

5. Let

$$P = \begin{bmatrix} p_{11} & p_{12} \\ p_{21} & p_{22} \end{bmatrix}$$

be the transition matrix of a two-state MC and let $d = 1 - p_{12} - p_{21}$. Let $N_i^{(n)}$ be the number of times state 2 (a "success," S) is immediately followed by state 1 (a "failure," F) during the first n steps starting from state i at the zeroth step, $(i = 1, 2,)$, i.e., $N_i^{(n)}$ is the number of times the sequence SF occurs in the first n trials starting from state i.

(a) Let X_i be the number of trials required for the first completion of the pattern SF when the initial state is i. Find, from first principles, expressions for the probability distributions of the X_i. Hence show that

$$Es^{X_1} = p_{12}p_{21}s^2/(1 - p_{11}s)(1 - p_{22}s),$$
$$Es^{X_2} = p_{21}s/(1 - p_{11}s).$$

(b) Using recurrent event theory (in particular Corollary 3.5.4A), show that

$$P\{N_1^{(n)} = r\} = \begin{cases} 1 & (r = 0, n = 0) \\[2mm] p_{11}^n + \displaystyle\sum_{a=0}^{n-1} p_{11}^a p_{22}^{n-a-1} p_{12} & (r = 0, n \geq 1), \\[4mm] p_{12}^r p_{21}^r \displaystyle\sum_{k=r}^{n-r} \binom{k}{r}\binom{n-k}{r} p_{11}^{k-r} p_{22}^{n-k-r-1}\{p_{12} + rd/(n-k)\} \\[2mm] & (r \geq 1, n \geq 2r), \\[2mm] 0 & (r \geq 1, n < 2r); \end{cases}$$

and

$$P\{N_2^{(n)} = r\} = \begin{cases} p_{22}^n & (r = 0, n \geq 0), \\ 0 & (r \geq 1, n < 2r - 1), \\ p_{12}^{r-1}p_{21}^r \sum_{k=r}^{n-r+1} \binom{k-1}{r-1}\binom{n-k+1}{r} p_{11}^{k-r}p_{22}^{n-r-k} \\ \quad \times \{p_{12} + rd/(n - k + 1)\} & (r \geq 1, n \geq 2r - 1). \end{cases}$$

(c) Using recurrent event theory (in particular Theorem 3.5.5) or otherwise show that

$$EN_1^{(n)} = p_{12}p_{21}[n(1 - d) - 1 + d^n]/(1 - d)^2 \qquad (n \geq 0),$$

$$EN_2^{(n)} = p_{21}[n(1 - d)p_{01} + p_{10}\{1 - d^n\}]/(1 - d)^2 \qquad (n \geq 0).$$

(d) Let N_n be the number of occurrences of the sequence SF in n independent Bernoulli trials with constant probability of success p and failure $q = 1 - p$. Show, using the above results, that

$$P\{N_n = r\} = \sum_{k=r}^{n-r} \binom{k}{r}\binom{n-r}{r} p^k q^{n-k},$$

$$EN_n = (n - 1)pq.$$

(For further details, see Hunter, 1973.)

6. Let $\{X_n\}$ $(n \geq 0)$ be a two-state MC with transition matrix

$$P = \begin{bmatrix} p_{11} & p_{12} \\ p_{21} & p_{22} \end{bmatrix}.$$

Let $_iN_{jk}^{(n)}$ be the number of times state j is immediately followed by state k during the first n steps starting from state i at the zeroth step $(i, j, k = 1,2)$.

(a) Show that the expectations $E[_iN_{jk}^{(n)}]$ satisfy the system of difference equations

$$E[_iN_{jk}^{(n)}] = \sum_{r=1}^{2} p_{ir}\{E[_rN_{jk}^{(n-1)}] + \delta_{ij}\delta_{rk}\} \qquad (i,j,k = 1,2),$$

where δ_{ij} is the Kronecker delta.

(b)   Let

$$\alpha_{jk}(n) = \begin{bmatrix} E_1N_{jk}^{(n)} \\ E_2N_{jk}^{(n)} \end{bmatrix}.$$

Show that for suitable β_{jk},

$$\alpha_{jk}(n) = P\alpha_{jk}(n-1) + \beta_{jk}.$$

(c)   Show that

$$\alpha_{jk}(n) = \left[\sum_{r=0}^{n-1} P^r\right]\beta_{jk}.$$

(d)   Using Theorem 7.4.6, or otherwise, show that

$$E[_iN_{jk}^{(n)}] = \left[\frac{np_{2-j,\,1+j}}{1-d} + \frac{(-1)^{i+j}p_{1+i,\,2-i}}{(1-d)^2}\{1-d^n\}\right]p_{jk},$$

where $d = 1 - p_{12} - p_{21}$. (*Note*: In Exercise 7.4.5 an alternative technique for finding $E[_iN_{21}^{(n)}]$ was examined.)

7. (a)   Deduce Eq. (7.4.32) from Eq. (7.4.31) using the fact that $M_{ij}^{(n)} = N_{ij}^{(n)} + \delta_{ij}$.
   (b)   Deduce Eq. (7.4.32) along the lines given in the proof of Theorem 7.4.8 for Eq. (7.4.31).
   (c)   Deduce Eq. (7.4.34) from Eq. (7.4.32) (cf. Corollary 3.5.5A).

8. A simplistic model for flow through a university undergraduate program is as follows. Let the state space be $\{1, 2, \dots, 6\}$ where the states represent (1) student has failed and dropped out, (2) student has graduated, (3) student is in 4th year, (4) student is in 3rd year, (5) student is in 2nd year, and (6) student is in 1st year. Assume that student progression through the program is a MC with the above state space such that each year a student has probability p of failing and dropping out, a probability q of having to repeat a year, and the probability r of passing onto the next year.
   (a)   Write down the transition matrix in the form

$$\begin{bmatrix} I & 0 \\ R & Q \end{bmatrix}.$$

   (b)   Find the fundamental matrix $N = (I - Q)^{-1}$. [Simplify by writing $t = r/(p+r)$.]
   (c)   Interpret the elements of N.
   (d)   Find NR and interpret the elements.
   (e)   Find (EM_i).
   (f)   Find $[\text{var } M_{ij}]$.
   (g)   Evaluate the above expressions when $p = 0.2$, $q = 0.1$, and $r = 0.7$. In particular, show that a student must reach the 3rd year before he has a better than even chance of graduating (Kemeny and Snell, 1960).

9. If $f(M_{ij})$ is a function of M_{ij} with finite expectation show that

$$Ef(M_{ij}) = f(\delta_{ij}) \sum_{k \in \tilde{T}} p_{ik} + \sum_{k \in T} p_{ik} E[f(M_{kj} + \delta_{ij})].$$

Use this result to show that

$$[EM_{ij}^3] = N[6N_d^2 - 6N_d + I],$$

where $N = (I - Q)^{-1}$, as in Theorem 7.4.10.

10. Show that in the terminology of Theorem 7.4.10.

$$EM_{ij}M_{ik} = n_{ij}n_{jk} + n_{ik}n_{kj} - \delta_{jk}n_{ij},$$

where $N = [n_{ij}]$.

11. Let Y_i be the total number of *different* transient states the MC visits before absorption including the initial state $i \in T$. Then

$$Y_i = \sum_{j \in T} V_{ij} \quad \text{where} \quad V_{ij} = \begin{cases} 1 & \text{if } X_n = j \text{ for some } n \text{ given } X_0 = i \\ 0 & \text{otherwise.} \end{cases}$$

Show

(a)
$$EV_{ij} = \begin{cases} 1 & \text{if } i = j, \\ f_{ij} & \text{if } i \neq j. \end{cases}$$

(b)   The vector of means (EY_i) is given by

$$(EY_i) = [F_4 + (I - F_{4d})]e = NN_d^{-1}e,$$

where F_4 is as given in Theorem 6.3.1.

12. r points, labeled $1, 2, \ldots, r$ can be visualized as lying on the circumference of a circle. A particle moves on these points in the following manner: If the particle is at point i at trial n, then at the next trial, $n + 1$, it either moves to point $i + 1$ (or point 1 if $i = r$) with probability p_i or remains at point i with probability q_i ($p_i + q_i = 1$).
(a) Write down the transition matrix for this MC.
(b)   Under what conditions is the MC an absorbing MC with just one absorbing state, say point 1. Show in this case that if M_i ($i = 2, \ldots, r$) is the number of trials to absorption at point 1 starting at point i, then

$$EM_i = \sum_{j=i}^{r} 1/p_j.$$

(c)   Can you find an easy way of establishing (b) above?

13. For the generalized gambler's ruin problem (Exercise 6.3.6) show that the expected duration of the game when the gambler's initial capital is

k units and his opponent has $a - k$ units is given by

$$\frac{1}{q - p}\left[k - a\frac{(q/p)^k - 1}{(q/p)^a - 1}\right]$$

in the case $q \neq p$.

14. Show that if G is *any* generalized inverse of a regular MC with limit matrix Π the expression

$$(I - \Pi)G(I - \Pi)$$

is invariant and is equivalent to the group inverse T (see Hunter, 1982).

Chapter 8

Applications of Discrete Time Markov Chains

8.1 Branching Chains

The concept of a branching chain was introduced earlier in Example 5.1.6. In this section we examine some of the properties of this special Markov chain (MC) in some detail, utilizing some of the techniques we have developed.

A branching chain can be visualized as follows. Consider a species of individuals (men, neutrons, electrons, bacteria, etc.) that can generate new individuals of like kind. Suppose we start with a single individual and consider it to be the original or zeroth generation. This initial individual produces new individuals that form the first generation. Members of this generation produce additional individuals that comprise the second generation. This procedure continues with the descendants of the nth generation being the members of the $(n + 1)$th generation. Specifically, we assume that each individual produces offspring independently of the other members of its generation and, furthermore, it is also assumed that each individual has the same probability p_k of creating exactly k new individuals ($k = 0, 1, 2, \ldots$) with $\sum_{k=0}^{\infty} p_k = 1$. Let X_n denote the number of individual members in the nth generation. Then $\{X_n, n \geq 0\}$ is a branching chain.

Observe that if $X_n = 0$, then no more individuals can be produced and the branching chain dies out or is said to become extinct. One of the important issues we shall examine is conditions under which such a phenomenon will occur.

If $X_n \geq 1$, let $Y_i^{(n)}$ ($i = 1, \ldots, X_n$) be the number of offspring produced by the ith individual of the nth generation. Then

$$X_{n+1} = \begin{cases} 0, & X_n = 0, \\ Y_1^{(n)} + \cdots + Y_{X_n}^{(n)}, & X_n \geq 1. \end{cases} \tag{8.1.1}$$

Under the assumptions stated above we have that the $Y_i^{(n)}$ are independent random variables, each distributed as a random variable Y. Further, observe that X_{n+1} depends solely on the value of X_n and thus we are led to the following formal definition.

DEFINITION 8.1.1: Let Y_n, $n \geq 1$, be a sequence of independent and identically distributed, nonnegative integer-valued random variables with common probability distribution (the *offspring distribution*) given by $p_k = P\{Y = k\}$ ($k \geq 0$). A Markov chain $\{X_n, n \geq 0\}$ is called a *branching chain* with state space $S = \{0, 1, 2, \ldots\}$ if its transition probabilities are given by

$$p_{ij} = P\{X_{n+1} = j \mid X_n = i\} = P\{Y_1 + \cdots + Y_i = j\} \tag{8.1.2}$$

for $i \geq 1, j \geq 0$ with $p_{00} = 1$. $\square$

What we have constructed is a model for a wide variety of situations. In many cases it models quite realistically what is occurring, for example, in the splitting activity of certain atomic particles, as in neutron chain reactions. When the model is applied to biological processes the assumption of independence may not be strictly valid and this may limit the applicability of the model.

Historically the study of branching chains began with Galton and Watson (1874) when they formulated the model to describe a possible mechanism of extinction of distinguished English families. Galton had noted that families descending from famous men seemed to die out more than he thought they should and he wanted to determine the probability of extinction of a family name. In the context described above, since the family name is inherited by males only, X_n will be the number of males in the nth generation.

Watson provided a partial solution to the problem. A complete solution for the extinction problem was provided by Steffensen (1930).

The model was also used by Fisher (1922, 1930) to study the survival of the progeny of a mutant gene and to study variations in gene frequencies. Haldane (1927) also applied the model to some questions in population genetics.

Let us assume that $X_0 = 1$, i.e., we start with a single individual and consider finding the probability distribution of X_n, the number of individuals in the nth generation. Observe that Eq. (8.1.1) implies that X_{n+1} is a compound r.v. being a random sum of a sequence of independent and identically distributed r.v.'s. We saw in Section 2.6 that generating functions are a very effective tool in the determination of the probability distribution of such r.v.'s. An application of Theorem 2.6.1 in the above setting leads to the following basic theorem.

THEOREM 8.1.1:   Let $\{X_n, n \geq 1\}$ be a branching chain with $X_0 = 1$ and let the offspring distribution $\{p_k\}$ have p.g.f. $P(s)$. Let

$$P_n(s) = \sum_{k=0}^{\infty} P\{X_n = k\}s^k$$

be the p.g.f. of X_n. Then, for $n \geq 0$ and $|s| \leq 1$,

$$P_{n+1}(s) = P_n(P(s)) = P(P_n(s)). \tag{8.1.3}$$

Proof: Using the representation given by Eq. (8.1.1), observe that $X_1 = Y_1^{(0)}$ where $Y_1^{(0)}$ has p.g.f. $P(s)$ and thus

$$P_1(s) = P(s). \tag{8.1.4}$$

Now $X_2 = Y_1^{(1)} + \cdots + Y_{X_1}^{(1)}$, where each $Y_i^{(1)}$ has p.g.f. $P(s)$ and X_1 has p.g.f. $P_1(s)$, and thus from Theorem 2.6.1

$$P_2(s) = P_1(P(s)) = P(P_1(s)), \tag{8.1.5}$$

using Eq. (8.1.4).

In general, since X_{n+1} is the sum of X_n r.v.'s, each having the same distribution, the offspring distribution, application of Theorem 2.6.1 gives, for $n \geq 1$,

$$P_{n+1}(s) = P_n(P(s)). \tag{8.1.6}$$

An inductive argument will show that for $n \geq 1$

$$P_{n+1}(s) = P(P_n(s)). \tag{8.1.7}$$

(Equation (8.1.5) shows that Eq. (8.1.7) holds for $n = 1$. Assume Eq. (8.1.7) is true for $n = k$. Then

$$
\begin{aligned}
P_{k+2}(s) &= P_{k+1}(P(s)) &&\text{[using Eq. (8.1.6)]} \\
&= P(P_k(P(s))) &&\text{[using Eq. (8.1.7) with } n = k] \\
&= P(P_{k+1}(s)) &&\text{[using Eq. (8.1.6)],}
\end{aligned}
$$

and thus Eq. (8.1.7) holds for $n = k + 1$.)

Note also that $P_0(s) = s$ and thus Eq. (8.1.3) holds not only for $n \geq 1$ but also for $n = 0$. □

The Markov property of the branching chain follows naturally from Eq. (8.1.1) and the homogeneity of the MC follows by Eq. (8.1.2) since the law of reproduction is the same from generation to generation (cf. Definition 5.1.2).

Observe that the coefficient of s^j in $P_n(s)$, namely $P\{X_n = j\}$, is really the conditional probability $P\{X_n = j | X_0 = 1\}$ so that Theorem 8.1.1 gives us a technique for finding the transition probabilities. Note that Eq. (8.1.2) implies that

$$p_{ij} = P\{Y_1 + \cdots + Y_i = j\} = \text{coefficient of } s^j \text{ in } [P(s)]^i,$$

since by Corollary 2.5.2B the p.g.f. of $Y_1 + \cdots + Y_i$, the sum of i i.i.d. r.v.'s, is given by $[P(s)]^i$. Furthermore, the n-step transition probabilities are given by

$$
\begin{aligned}
p_{ij}^{(n)} &= P\{X_n = j | X_0 = i\} \\
&= P\{X_n^{(1)} + \cdots + X_n^{(i)} = j\} \\
&= \text{coefficient of } s^j \text{ in } [P_n(s)]^i,
\end{aligned}
\tag{8.1.8}
$$

since each of the initial members of the zeroth generation acts independently and each give rise to a separate branching chain with $X_n^{(k)}$ having a p.g.f. $P_n(s)$ for $k = 1, 2, \ldots, i$.

Before we can examine the properties and behavior of this MC in any detail we need to classify the states of the chain. The nature of the state space and the classification depends critically on the offspring distribution $\{p_n\}$. From Definition 8.1.1 it is obvious that the state space will contain the state 0 that, if it is ever reached, is an absorbing state since $p_{00} = 1$.

In the special case where $p_0 + p_1 = 1$ the state space reduces to $S = \{0, 1\}$ and at most one individual will be present in the nth generation. In particular when $p_0 = 1$ and $p_1 = 0$ we have that $X_1 = 1$, $X_n = 0$ for $n > 1$ so that state 1 is transient and state 0 is reached with probability one. If $p_0 = 0$ and $p_1 = 1$, then $X_n = 1$ for all n and state 1 is persistent (absorbing) and state 0 is never reached. If $0 < p_0 < 1$ with $p_1 = 1 - p_0$, then $X_0 = 1, \ldots,$ $X_{N-1} = 1$, $X_n = 0$ for $n \geq N$ when N is a geometric r.v. In this case state 1 is transient and state 0 is reached with probability one (Exercise 8.1.4).

Observe also that if $p_0 = 0$, then the sequence $\{X_n\}$ is nondecreasing and then state 0 is never reached. In this case if $0 \leq p_1 < 1$ the state space is $S = \{0, 1, 2, \ldots\}$ and all states other than 0 are transient.

If $0 < p_0 \leq p_0 + p_1 < 1$, then it is seen that states 1, 2, 3, ... are all transient with state 0 absorbing. However, state 0 may or may not be reached and it is this question that we wish to consider in some detail. In other words, starting with $X_0 = 1$, we wish to determine conditions under which the branching chain becomes extinct or, in MC terminology, the probability that the MC is absorbed in state 0, f_{10}.

Now

$$f_{10} = \sum_{n=1}^{\infty} f_{10}^{(n)},$$

where

$$f_{10}^{(n)} = P\{X_n = 0, X_k \neq 0, k = 1, \ldots, n - 1 \,|\, X_0 = 1\}$$
$$= P\{\text{Branching chain is absorbed } at \text{ the } n\text{th generation}\}.$$

Further,

$$p_{10}^{(n)} = P\{X_n = 0 \,|\, X_0 = 1\}$$
$$= P\{\text{Branching chain is absorbed } by \text{ the } n\text{th generation}\}.$$

These three probabilities can all be expressed in terms of the p.g.f. $P_n(s)$ evaluated at $s = 0$.

THEOREM 8.1.2:   Let $\{X_n, n \geq 0\}$ be a branching chain with $X_0 = 1$. Then, if $P_n(s)$ is the p.g.f. of X_n,

(a) $\qquad\qquad\qquad p_{10}^{(n)} = P_n(0) \qquad\qquad (n \geq 1), \qquad\qquad (8.1.9)$

(b) $\qquad\qquad\qquad f_{10}^{(n)} = P_n(0) - P_{n-1}(0) \qquad (n \geq 1), \qquad\qquad (8.1.10)$

(c) $\qquad\qquad\qquad f_{10} = \lim_{n \to \infty} P_n(0). \qquad\qquad\qquad\qquad (8.1.11)$

Proof: (a)   Equation (8.1.9) follows directly from Eq. (8.1.8).

(b)   Since state 0 is absorbing we have that $f_{00}^{(k)} = 1$ for $k = 1$ and 0 otherwise. Equation (5.2.4) with $j = 0$ implies that

$$f_{10}^{(n)} = p_{10}^{(n)} - p_{10}^{(n-1)} \qquad\qquad (8.1.12)$$

(with $p_{10}^{(0)} \equiv 0$) and Eq. (8.1.10) follows from Eq. (8.1.9).

(c)   Since $f_{10}^{(n)} \geq 0$ we have, from Eq. (8.1.12) that $\{p_{10}^{(n)}\}$ is a non-decreasing, bounded sequence and hence has a limit in the interval $[0, 1]$.

Furthermore, since state 0 is absorbing, $p_{00}^{(k)} = 1$ for $k \geq 1$ and Eq. (5.2.3) leads to the result that

$$p_{10}^{(n)} = \sum_{k=1}^{n} f_{10}^{(n)} \qquad (n \geq 1). \qquad\qquad (8.1.13)$$

Thus

$$\lim_{n \to \infty} p_{10}^{(n)} = \lim_{n \to \infty} \sum_{k=1}^{n} f_{10}^{(k)} \qquad \text{[by Eq. (8.1.13)]}$$

$$= \sum_{k=1}^{\infty} f_{10}^{(k)} \qquad \text{(by definition of a limit)}$$

$$= f_{10} \qquad \text{(by definition)},$$

and the result follows by Eq. (8.1.9).

Alternatively, since state 0 is absorbing it is also aperiodic, persistent, nonnull (with mean recurrence time $\mu_0 = 1$) and hence ergodic. Theorem 5.2.8(c) now implies that $p_{10}^{(n)}$ tends to a limit, f_{10}. □

From Theorem 8.1.1 we have that $P_{n+1}(0) = P(P_n(0))$ and hence, since $P(s)$ is continuous for $0 \le s \le 1$ (at $s = 1$ by Abel's convergence theorem, Theorem 2.3.2), we deduce from Theorem 8.1.2(c) that $f_{10} = P(f_{10})$. However, the equation $s = P(s)$ does not necessarily have a unique solution in $[0, 1]$. Certainly $s = 1$ is always a solution, since $P(s)$ is a p.g.f. The following lemma details the location of the remaining roots in the interval $[0, 1)$.

LEMMA 8.1.3: Let $P(s) = \sum_{n=0}^{\infty} p_n s^n$ be the p.g.f. of a nonnegative integer-valued r.v. with mean $\mu (\le +\infty)$.

(a) If $p_1 = 1$, then every real number in $[0, 1)$ is a root of $s = P(s)$.
(b) If $p_1 < 1$ and $\mu \le 1$, then the equation $s = P(s)$ has no roots in $[0, 1)$.
(c) If $p_0 = 0$ and $\mu > 1$, then the equation $s = P((s)$ has a root at 0 and no other root in $(0, 1)$.
(d) If $p_0 > 0$ and $\mu > 1$, then the equation $s = P(s)$ does not have a root at 0 but has a unique root in $(0, 1)$.

Proof: If $p_1 = 1$, then by definition $P(s) = s$ for all $s \in [0, 1]$ and the result given by (a) follows.

For the remainder of the proof we shall assume, without loss of generality, that $p_1 < 1$ since if $p_1 = 1$ we have that $\mu = 1$.

The roots of the equation $s = P(s)$ are those points s where the two curves $y = s$ and $y = P(s)$ intersect. Furthermore, observe that $\mu = P^{(1)}(1)$ is the slope of the curve $y = P(s)$ at $s = 1$. We shall establish that graphs of $y = P(s)$ for $s \in [0, 1]$ have the forms as displayed in Fig. 8.1.1, in the three typical cases corresponding to $\mu < 1$, $\mu = 1$, and $\mu > 1$.

Now $P(s) = \sum_{n=0}^{\infty} p_n s^n$ is a monotone nondecreasing function and from Theorem 2.3.7 $P(s)$ is differentiable for $s \in [0, 1)$ with $P^{(1)}(s) = \sum_{n=1}^{\infty} n p_n s^{n-1}$ for s in this range and $P^{(1)}(1) \equiv \lim_{s \uparrow 1} P^{(1)}(s)$ $(\le \infty)$ in accordance with

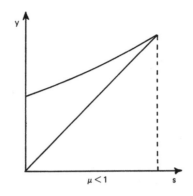

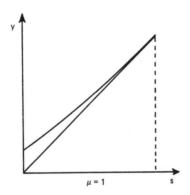

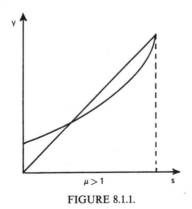

FIGURE 8.1.1.

the notation established following Theorem 2.3.8. Note also that

$$P^{(k)}(s) = \sum_{n=k}^{\infty} n(n-1) \cdots (n-k+1)p_n s^{k-n} \geq 0$$

$$\text{for} \quad s \in [0,1) \quad \text{and} \quad k = 1, 2, \ldots.$$

Suppose that $\mu < 1$. Then $\lim_{s \uparrow 1} P^{(1)}(s) < 1$ and since $P^{(1)}(s)$ is nondecreasing in s, $0 \leq s < 1$, we conclude that $P^{(1)}(s) < 1$ for $0 \leq s < 1$. Now suppose that $\mu = 1$ and $p_1 < 1$. Then $p_n > 0$ for some $n \geq 2$ (otherwise, $p_0 = 1 - p_1 > 0$, which implies $\mu < 1$, a contradiction). Therefore, $P^{(1)}(s)$ is strictly increasing in s for $s \in [0, 1)$. Since $\lim_{s \uparrow 1} P^{(1)}(s) = 1$ we again conclude that $P^{(1)}(s) < 1$ for $0 \leq s < 1$. Thus if $\mu \leq 1$ and $p_1 < 1$ we have shown that $P^{(1)}(s) < 1$ for $s \in [0, 1)$. Consequently $(d/ds)[P(s) - s] < 0$ for $s \in [0, 1)$ and hence $P(s) - s$ is strictly decreasing on $[0, 1]$. Since $P(1) - 1 = 0$ we see that $P(s) - s > 0$ for $0 \leq s < 1$ and hence $s = P(s)$ has no roots on $[0, 1)$. This proves part (b) of the theorem.

Next suppose that $\mu > 1$. Then $\lim_{s \uparrow 1} P^{(1)}(s) > 1$. Since $\lim_{s \downarrow 0} P^{(1)}(s) = p_1 < 1$ the continuity of $P^{(1)}(s)$ implies that there exists an $s_0 \in (0, 1)$ such that $P^{(1)}(s_0) = 1$. Furthermore, since $\mu > 1$ there exists a $p_n > 0$ for some $n \geq 2$ (for, if not, $p_0 + p_1 = 1$, in which case $\mu < 1$, a contradiction), which implies that $P^{(2)}(s) > 0$ for $s \in [0, 1)$, from which we deduce that $P^{(1)}(s)$ is a strictly monotone increasing function with $P^{(1)}(s) > 1$ for $s \in (s_0, 1)$. By the mean-value theorem there exists an $s_1 \in (s_0, 1)$ such that

$$P(1) = P(s_0) + (1 - s_0)P^{(1)}(s_1).$$

Since $P(1) = 1$ we have that

$$\frac{1 - P(s_0)}{1 - s_0} = P^{(1)}(s_1) > 1$$

and consequently $P(s_0) - s_0 < 0$. Further, since $P(s) - s$ is continuous in s and nonnegative at $s = 0$, the intermediate-value theorem states that there must be a $\rho \in [0, s_0)$ such that $P(\rho) - \rho = 0$.

We now show that there is only one such root ρ in $[0, 1)$. Suppose there are at least two distinct roots ρ_0, ρ_1 with $0 \leq \rho_0 < \rho_1 < 1$. Consequently, the function $f(s) = P(s) - s$ is such that $f(\rho_0) = f(\rho_1) = f(1) = 0$. Two applications of Rolle's theorem show that there exists a $c_0 \in (\rho_0, \rho_1)$ and a $c_1 \in (\rho_1, 1)$ such that $f^{(1)}(c_0) = 0$ and $f^{(1)}(c_1) = 0$. A further application of Rolle's theorem shows that there exists a $d \in (c_0, c_1)$ such that $f^{(2)}(d) = P^{(2)}(d) = 0$. But $\mu > 1$ implies, from above, the existence of a positive p_n for $n \geq 2$ and hence

$$P^{(2)}(s) = \sum_{n=2}^{\infty} n(n-1)p_n s^{n-2} > 0 \quad \text{for} \quad s \in [0, 1).$$

This contradiction shows that there can be at most one $\rho \in [0, 1)$ such that $P(\rho) = \rho$. Observe that when $p_0 = 0$ we have a root at $s = 0$ that must be the only root in $[0, 1)$. $\square$

We utilize the results of Lemma 8.1.3 to develop a procedure for finding f_{10} for a branching chain.

THEOREM 8.1.4: Let $\{X_n, n \geq 0\}$ be a branching chain with $X_0 = 1$ and having an offspring distribution $\{p_n\}$ with mean μ.

(a) If $p_0 = 0$, then $f_{10} = 0$.
(b) If $p_0 > 0$ and $\mu \leq 1$, then $f_{10} = 1$.
(c) If $p_0 > 0$ and $\mu > 1$, then f_{10} is the unique positive root less than 1 of the equation $s = P(s)$.

Proof: If $p_0 = 0$, then each individual always produces at least one offspring and $X_n > 0$ for all n. Thus $P_n(0) = 0$ for all n and extinction is impossible, implying $f_{10} = 0$ [consistent with Eq. (8.1.11)].

If $p_0 = 1$, then each individual produces no offspring and $X_n = 0$ for $n > 0$. Thus $P_n(0) = 1$ and extinction is certain with $f_{10} = 1$. This is a special case of (b) since $p_0 = 1$ implies $\mu = 0$.

Thus in what follows we may assume $0 < p_0 < 1$. From Theorem 8.1.2(c) we have that $f_{10} = \lim_{n \to \infty} P_n(0)$ and from the observation following Theorem 8.1.2, f_{10} is a root of the equation $s = P(s)$.

If $0 < p_0 < 1$, then $p_1 < 1$; thus when $\mu \leq 1$ we have from Lemma 8.1.3(b) that $s = P(s)$ has no roots in $[0, 1)$ so that the only root occurs at $s = 1$, implying that $f_{10} = 1$.

If $0 < p_0 < 1$ and $\mu > 1$, then Lemma 8.1.3(d) shows us that $s = P(s)$ has a unique positive root, ρ say, in $(0, 1)$ together with the root at 1. The desired result will follow by deducing that $f_{10} = \rho$.

From Theorem 8.1.2(b) since $f_{10}^{(n)} \geq 0$ we see that $P_{n-1}(0) \leq P_n(0)$ for $n = 1, 2, \ldots$ and thus $\{P_n(0)\}$ is a nondecreasing sequence that, since the $P_n(0)$ are probabilities, is bounded above by 1. In fact we show that $P_n(0) \leq \rho$ for $n \geq 0$.

Now $P_0(0) = 0 \leq \rho$, showing the bound holds for $n = 0$. Assume that the bound holds for a given value of n. Since $P(s)$ is increasing in s we have, using Eq. (8.1.3), that

$$P_{n+1}(0) = P(P_n(0)) \leq P(\rho) = \rho,$$

and thus the bound holds for the next value of n and thus, by induction the bound is true for all $n \geq 0$. Letting $n \to \infty$ we see, using Eq. (8.1.11), that

$$f_{10} = \lim_{n \to \infty} P_n(0) \leq \rho.$$

Since f_{10} is one of two numbers ρ or 1 it must be the number ρ. $\square$

COROLLARY 8.1.4A: Let $\{X_n, n \geq 0\}$ be a branching chain with $X_0 = 1$ and having an offspring distribution $\{p_n\}$ with mean μ. If $p_0 > 0$, then the branching chain dies out with probability one if and only if $\mu \leq 1$. $\square$

Additional information can be extracted from the graphs of $y = P(s)$ and $y = s$. By starting at $y = p_0 = P_1(0)$ we can plot successively $P_n(0) = p_{10}^{(n)}$ for $n = 1, 2, \ldots$ on either the y axis or the s axis and demonstrate graphically the iterative convergence of $p_{10}^{(n)}$ to f_{10}. As an illustration we present in Fig. 8.1.2 a typical example when $\mu < 1$. Similar diagrams can be constructed when $\mu = 1$ and $\mu > 1$.

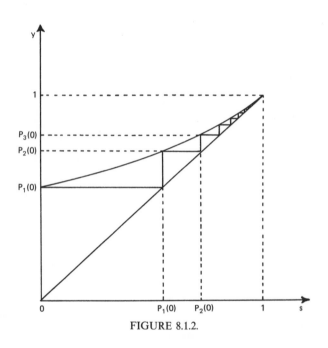

FIGURE 8.1.2.

EXAMPLE 8.1.1:   If in a branching chain $\{X_n, n \geq 0\}$ with $X_0 = 1$, each individual produces N or 0 direct descendants with probabilities p or q $(p + q = 1)$, find necessary and sufficient conditions for the probability of extinction to be unity.

Note that $P(s) = q + ps^N$ with $\mu = P^{(1)}(1) = Np$. Firstly, if $q = 0$ ($p_0 = 0$), Theorem 8.1.4(a) shows that $f_{10} = 0$. Secondly, if $q > 0$ ($p_0 > 0$), Corollary 8.1.4A shows that $f_{10} = 1$ iff $p \leq 1/N$. When $N = 1$, $q > 0$ implies $p < 1$ and thus this condition for certain extinction is automatically satisfied. Consequently the required necessary and sufficient conditions are (a) $p < 1$ if $N = 1$ and (b) $p \leq 1/N$ if $N > 1$.

Note that for general N the equation $s = P(s)$ is not easily solved and some iterative procedure is required.   □

Lotka (1931) compiled data from a sample of white males and estimated the offspring distribution $\{p_k\}$ for the number of male offspring per male. His approximation for the offspring p.g.f. was

$$P(s) = \frac{0.482 - 0.041s}{1 - 0.559s}.$$

The equation $s = P(s)$ has roots 1 and 0.86, leading to an estimate of 0.86 for the probability of extinction of the surname descended from a single male, which seems perhaps unexpectedly high.

Our presentation has thus far concentrated primarily on the case $X_0 = 1$. If, however, $X_0 = i$ we have from Eq. (8.1.8) that

$$p_{i0}^{(n)} = [P_n(0)]^i = [p_{10}^{(n)}]^i.$$

Furthermore, utilizing the fact that state 0 is absorbing, we deduce from Eq. (5.2.3) that

$$p_{i0}^{(n)} = \sum_{k=1}^{n} f_{i0}^{(k)} \qquad (n \geq 1),$$

so that

$$f_{i0} = \lim_{n \to \infty} \sum_{k=1}^{n} f_{i0}^{(k)} = \lim_{n \to \infty} p_{i0}^{(n)} = \lim_{n \to \infty} [p_{10}^{(n)}]^i = [\lim_{n \to \infty} p_{10}^{(n)}]^i,$$

implying that

$$f_{i0} = f_{10}^i \qquad (i \geq 1). \tag{8.1.14}$$

Equation (8.1.14) is basically a consequence of the observation that if $X_0 = i$, then we start with i independent branching chains and then for extinction to occur we require each of the i branching chains to die out.

A modification and extension of the analysis presented in Lemma 8.1.3 and Theorem 8.1.4 leads to a result concerning the limiting behavior of the probability generating function $P_n(s)$.

THEOREM 8.1.5:   Let $\{X_n, n \geq 0\}$ be a branching process with $X_0 = 1$ and extinction probability $f_{10} \equiv \rho$. Then if $P_n(s)$ is the p.g.f. of X_n and $p_1 \neq 1$, then

$$\lim_{n \to \infty} P_n(s) = \rho(s) = \begin{cases} \rho, & 0 \leq s < 1, \\ 1, & s = 1. \end{cases} \tag{8.1.15}$$

Proof: Firstly let us assume that $0 < \rho < 1$ and hence, from Theorem 8.1.4, that $p_0 > 0$ and $\mu > 1$. We shall first show that for $n = 1, 2, 3, \ldots$

$$s < P_n(s) < \rho, \qquad 0 < s < \rho, \tag{8.1.16}$$

$$P_n(s) = \rho, \qquad s = \rho, \tag{8.1.17}$$

$$\rho < P_n(s) < s, \qquad \rho < s < 1. \tag{8.1.18}$$

Under the stated conditions $P(s)$ is a strictly monotone increasing function over $[0, 1]$ and the shape of the graph $y = P(s)$ as established in Lemma 8.1.3 when $\mu > 1$ leads immediately to expressions (8.1.16), (8.1.17), and (8.1.18) when $n = 1$. The results for a general positive integer n are easily established by an inductive argument. For example, if expression (8.1.16) holds for $n = k$, then, for $0 < s < \rho$, $s < P_k(s) < \rho$. The monotonicity of $P(s)$ and the $n = 1$ result together with Theorem 8.1.4(c) implies that over this range

$$s < P(s) < P(P_k(s)) = P_{k+1}(s) < P(\rho) = \rho.$$

A further inductive argument will also establish the following results. For $n = 1, 2, \ldots$

$$P_n(s) < P_{n+1}(s), \qquad 0 < s < \rho, \tag{8.1.19}$$

$$P_n(s) = P_{n+1}(s), \qquad s = \rho, \tag{8.1.20}$$

$$P_n(s) > P_{n+1}(s), \qquad \rho < s < 1. \tag{8.1.21}$$

For $0 < s < \rho$ inequality (8.1.19) follows from inequality (8.1.16) when $n = 1$ since the monotonicity of $P(s)$ and $s < P(s)$ implies $P(s) < P(P(s)) = P_2(s)$. The general results now follow easily.

Since $P_n(s)$ is also a monotone increasing function in s (being a p.g.f. not having all the probability centered at $s = 0$) we have that for a fixed $s \in (0, \rho)$

$$P_n(0) < P_n(s) < P_n(\rho) = \rho. \tag{8.1.22}$$

Now $\{P_n(s)\}$ is a bounded increasing sequence for s in $(0, \rho)$ and thus has a limit. Taking limits in the expression (8.1.22) we obtain

$$\lim_{n \to \infty} P_n(0) \leq \lim_{n \to \infty} P_n(s) \leq \rho, \tag{8.1.23}$$

but by Eq. (8.1.11) $\rho = \lim_{n \to \infty} P_n(0)$ and thus for $s \in (0, \rho)$, $\lim_{n \to \infty} P_n(s) = \rho$.

For $s \in (\rho, 1)$ we have from inequality (8.1.21) that $\{P_n(s)\}$ is a decreasing sequence which by inequality (8.1.18) is also bounded and thus has a limit. Taking limits of inequality (8.1.18) we see that

$$\rho \leq \lim_{n \to \infty} P_n(s) \equiv \rho(s) \leq s < 1. \tag{8.1.24}$$

Since $P_{n+1}(s) = P(P_n(s))$, $\rho(s) = P(\rho(s))$ and so, for fixed s, $\rho(s)$ is a root of $x = P(s)$, which from Lemma 8.1.3(d) must be either ρ or 1. But since $\rho(s) < 1$ we must have $\rho(s) = \rho$.

Since $P_n(1) = 1$ for all n, $\lim_{n \to \infty} P_n(s) = 1$ when $s = 1$ and we have established Eq. (8.1.15) for $0 < \rho < 1$, the result for $s = 0$ follows by Eq. (8.1.11). Observe that in this case the convergence of $P_n(s)$ can be illustrated graphically, as in Fig. 8.1.3.

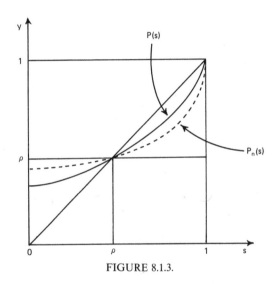

FIGURE 8.1.3.

When $\rho = 1$ we have from Theorem 8.1.4 that $p_0 > 0$ and $\mu \leq 1$ (and hence $p_1 < 1$). Similar arguments show that inequalities (8.1.16), (8.1.17), (8.1.19), and (8.1.20) still hold with ρ taken as 1. The arguments leading to inequalities (8.1.22) and (8.1.23) still apply, leading to the required result.

When $\rho = 0$ we must have $p_0 = 0$ and since $p_1 \neq 1$, by assumption, we have from Theorem 8.1.4 that $\mu > 1$. In this case it is easily seen that inequalities (8.1.17), (8.1.18), (8.1.20), and (8.1.21) still hold with ρ taken as 0. Similarly, inequalities (8.1.24) still hold and the required conclusion follows upon the relevant application of Lemma 8.1.3(c).

Note that when $p_1 = 1$, $P_n(s)$ for all n, and thus $\rho(s) = 1$ for $s \in [0, 1]$. $\square$

Observe that $\rho(s)$ is a proper p.g.f. if $f_{10} = 1$ and an improper p.g.f. if $f_{10} < 1$. In both of these cases Theorem 2.8.1 implies that

$$\lim_{n \to \infty} P\{X_n = 0\} = \lim_{n \to \infty} p_{10}^{(n)} = f_{10}$$

and

$$\lim_{n \to \infty} P\{X_n = k\} = \lim_{n \to \infty} p_{1k}^{(n)} = 0 \qquad \text{for} \quad k = 1, 2, \dots.$$

[This latter result is also a direct consequence of Theorem 5.2.8(a) since state k is transient for $k = 1, 2, \dots.$]

Now for any K,

$$P\{X_n > K\} + \sum_{k=1}^{K} P\{X_n = k\} + P\{X_n = 0\} = 1$$

and

$$\lim_{n \to \infty} P\{X_n > K\} + f_{10} = 1.$$

Hence, for any K however large,

$$P\{X_n > K\} \to 1 - f_{10} \qquad \text{as} \quad n \to \infty,$$

and thus for $f_{10} < 1$ there is a positive probability, $1 - f_{10}$, of indefinite growth. Thus when $f_{10} < 1$ the branching chain either becomes extinct (with prob. f_{10}) or else it becomes infinite (with prob. $1 - f_{10}$). Probability theory contains a lot of such "boom or bust" results, some of which are known as "zero-or-one" ("all or nothing") laws (cf. Corollaries 3.2.2A and 5.2.11A).

The following theorem examines expressions for the moments of X_n, in particular the mean, EX_n, and variance, $\text{var } X_n$.

THEOREM 8.1.6:   Let $\{X_n, n \geq 0\}$ be a branching chain with $X_0 = 1$ and having an offspring distribution with mean μ and variance σ^2. Then for $n \geq 1$,

(a) $$EX_n \equiv \mu_n = \mu^n, \tag{8.1.25}$$

(b) $$\text{var } X_n \equiv \sigma_n^2 = \begin{cases} n\sigma^2 & (\mu = 1), & (8.1.26) \\ \dfrac{\mu^{n-1}(\mu^n - 1)}{\mu - 1}\sigma^2 & (\mu \neq 1). & (8.1.27) \end{cases}$$

Proof: If the offspring distribution has p.g.f. $P(s)$, then $P(1) = 1$ and from Corollary 2.3.8A $P^{(1)}(1) = \mu$ and $P^{(2)}(1) = \sigma^2 + \mu^2 - \mu$.

By starting with the relation $P_{n+1}(s) = P_n(P(s))$ and differentiating twice, i.e.,

$$P_{n+1}^{(1)}(s) = P_n^{(1)}(P(s))P^{(1)}(s)$$

and

$$P_{n+1}^{(2)}(s) = P_n^{(2)}(P(s))[P^{(1)}(s)]^2 + P_n^{(1)}(P(s))P^{(2)}(s),$$

and taking the limit as $s \uparrow 1$, we obtain the recurrence relationships

$$\mu_{n+1} = \mu_n \mu \tag{8.1.28}$$

and

$$\sigma_{n+1}^2 + \mu_{n+1}^2 - \mu_n = (\sigma_n^2 + \mu_n^2 - \mu_n)\mu^2 + \mu_n(\sigma^2 + \mu^2 - \mu). \tag{8.1.29}$$

Since $\mu_1 = \mu$, Eq. (8.1.28) leads immediately to Eq. (8.1.25). Substitution of this result in Eq. (8.1.29) gives

$$\sigma_{n+1}^2 = \mu^2 \sigma_n^2 + \mu^n \sigma^2. \tag{8.1.30}$$

Note that if we start with the relation $P_{n+1}(s) = P(P_n(s))$ and work through the above procedure we end up with the recurrence expression

$$\sigma_{n+1}^2 = \mu \sigma_n^2 + \mu^{2n} \sigma^2. \tag{8.1.31}$$

Note also that if we use Theorem 2.6.2 by identifying S_N with X_{n+1}, X_i with $Y_i^{(n)}$ and N with X_n, in accordance with Eq. (8.1.1) we obtain Eqs. (8.1.28) and (8.1.30).

Either of Eqs. (8.1.30) or (8.1.31) leads by recursion to the result that

$$\sigma_n^2 = (\mu^{n-1} + \mu^n + \cdots + \mu^{2n-2})\sigma^2.$$

Equations (8.1.26) and (8.1.27) now follow upon summing. $\square$

When $\mu < 1$ we have seen that extinction is certain and μ_n tends to 0 as one would expect. When $\mu > 1$ it is easily seen that $0 < 1 - f_{10} \leq 1$ so that there is a positive probability that the process will not die out. In this case μ_n tends to ∞ and the crude interpretation is that the population will become infinite consistent with the convention that the mean of an improper r.v. is $+\infty$. The case $\mu = 1$ is not so easy to dispose of. In the trivial case of $p_0 = 0$ and $p_1 = 1$, $\mu = 1$, $\mu_n = 1$ and $P(X_n = 1) = 1$ for all n. However, when $p_0 \neq 0$ we have the strange situation that $\mu_n = 1$ for all n even though $P(\lim_{n \to \infty} X_n = 0) = 1$ since the chain dies out with probability one. Such a phenomenon is explained in probability courses where various modes of stochastic convergence are discussed.

We conclude this section with an investigation into the distribution of the total number of individuals that are present in a branching chain that eventually dies out. In Section 2.10 we introduced bivariate generating functions and the following theorem provides us with a suitable application.

THEOREM 8.1.7: Let $\{X_n, n \geq 0\}$ be a branching chain with $X_0 = 1$ and offspring p.g.f. $P(s)$. Let Y_n be the total number of individuals present in the branching chain up to and including the nth generation. The joint p.g.f.

$P_n(s_1, s_2)$ of (X_n, Y_n) satisfies the recurrence relationship

$$P_{n+1}(s_1, s_2) = P_n(P(s_1 s_2), s_2) \qquad (n \geq 0) \qquad (8.1.32)$$
$$= s_2 P(P_n(s_1, s_2)) \qquad (n \geq 0), \qquad (8.1.33)$$

with $P_0(s_1, s_2) = s_1 s_2$.

Proof: $P_{n+1}(s_1, s_2) = \sum\limits_{i=0}^{\infty} \sum\limits_{j=0}^{\infty} P\{X_{n+1} = i,\ Y_{n+1} = j\} s_1^i s_2^j$

$$= E(s_1^{X_{n+1}} s_2^{Y_{n+1}})$$

$$= \sum\limits_{i=0}^{\infty} \sum\limits_{j=0}^{\infty} E(s_1^{X_{n+1}} s_2^{Y_{n+1}} | X_n = i,\ Y_n = j) P(X_n = i,\ Y_n = j)$$

$$(8.1.34)$$

using Theorem 1.4.5. Since $Y_{n+1} = X_0 + \cdots + X_n + X_{n+1} = Y_n + X_{n+1}$,

$$E(s_1^{X_{n+1}} s_2^{Y_{n+1}} | X_n = i, Y_n = j) = E(s_1^{X_{n+1}} s_2^{X_{n+1} + Y_n} | X_n = i, Y_n = j)$$
$$= s_2^j E[(s_1 s_2)^{X_{n+1}} | X_n = i, Y_n = j]$$
$$= s_2^j E[(s_1 s_2)^{X_{n+1}} | X_n = i].$$

Now $X_{n+1} = \sum_{k=1}^{X_n} Z_k$ where the Z_k are independent with p.g.f. $P(s)$ so that

$$E(s_1^{X_{n+1}} s_2^{Y_{n+1}} | X_n = i, Y_n = j) = s_2^j E[(s_1 s_2)^{Z_1 + \cdots + Z_i} | X_n = i]$$
$$= s_2^j E[(s_1 s_2)^{Z_1 + \cdots + Z_i}]$$
$$= s_2^j [E(s_1 s_2)^{Z_1}]^i \qquad \text{(by Corollary 2.5.2B)}$$
$$= s_2^j [P(s_1 s_2)]^i.$$

Substitution in Eq. (8.1.34) gives Eq. (8.1.32), namely,

$$P_{n+1}(s_1, s_2) = \sum\limits_{i=0}^{\infty} \sum\limits_{j=0}^{\infty} s_2^j \{P(s_1 s_2)\}^i P(X_n = i,\ Y_n = j) = P_n(P(s_1 s_2), s_2).$$

Furthermore,

$$P_1(s_1, s_2) = P_0(P(s_1 s_2), s_2) = s_2 P(s_1 s_2)$$

so that relationship (8.1.33) holds for $n = 0$. Assume that Eq. (8.1.33) holds for $n = k$. Then

$$P_{k+2}(s_1, s_2) = P_{k+1}(P(s_1 s_2), s_2) \qquad \text{[by Eq. (8.1.32) for } n = k+1]$$
$$= s_2 P(P_k(P(s_1 s_2), s_2)) \qquad \text{[by Eq. (8.1.33) for } n = k]$$
$$= s_2 P(P_{k+1}(s_1, s_2)) \qquad \text{[by Eq. (8.1.32) for } n = k]$$

so that relationship (8.1.33) holds for $n = k + 1$ and hence generally by induction. $\square$

COROLLARY 8.1.7A: If $R_n(s)$ is the p.g.f. of Y_n, then

$$R_{n+1}(s) = sP(R_n(s)) \qquad (n \geq 0), \qquad (8.1.35)$$

with $R_0(s) = s$.

Proof: By Theorem 2.10.1 and Eq. (8.1.33)

$$R_{n+1}(s) = P_{n+1}(1, s) = sP(P_n(1, s)) = sP(R_n(s))$$

with $R_0(s) = P_0(1, s) = s$. □

Observe also that the results of Theorem 8.1.1 also follow from Theorem 8.1.7 since $P_n(s) = P_n(s, 1)$ is the (marginal) p.g.f. of X_n.

The limiting behavior of $R_n(s)$ can be examined by using a similar procedure to that used in Theorems 8.1.4 and 8.1.5.

THEOREM 8.1.8:

$$\lim_{n \to \infty} R_n(s) = R(s) \qquad \text{for} \quad 0 \leq s \leq 1,$$

where $R(s) = \sum_{k=1}^{\infty} r_k s^k$ with r_k being the probability that the total progeny consists of k individuals. Furthermore,

(a) $R(s)$ is given by the unique positive root of $t = sP(t)$.
(b) $\sum_{k=1}^{\infty} r_k = f_{10}$, the probability of extinction.
(c) $R(s)$ is a proper p.g.f. if and only if $f_{10} = 1$.

Proof: First observe that if $p_0 = 1$, $\mu = 0$, $f_{10} = 1$, implying that $P(s) = 1$ and $R_n(s) = s$ for all $n \geq 0$. The conclusions of the theorem are thus trivially derived with $R(s) = s$ and $r_1 = 1$.

Consequently we shall assume $p_0 \neq 1$ with $P(s)$ not identically 1. For $0 < s < 1$, $R_1(s) = sP(s) < s = R_0(s)$ and thus if $R_k(s) < R_{k-1}(s)$ we have that

$$R_{k+1}(s) = sP(R_k(s)) < sP(R_{k-1}(s)) = R_k(s)$$

so that the sequence $\{R_k(s)\}$ is a bounded decreasing sequence and thus tends to a limit $R(s) \equiv \sum_{k=1}^{\infty} r_k s^k$, where by Theorem 2.8.1, r_k is the probability that the total progeny consists of k individuals with $\sum r_k \leq 1$. (Note $r_0 = 0$, since total progeny must include the zeroth generation individual.)

Taking the limit in Eq. (8.1.35) we see that

$$R(s) = sP(R(s)) \qquad \text{for} \quad 0 < s < 1$$

so that for fixed $s < 1$ the value of $R(s)$ is a root of the equation

$$t = sP(t). \qquad (8.1.36)$$

We show that this root is unique. Let the smallest positive root of $x = P(x)$ be $\rho = f_{10} \leq 1$. With s held fixed $(0 < s < 1)$, $y = sP(t)$ is a monotone increasing function (a straight line if $p_0 + p_1 = 1$ and convex if $p_0 + p_1 < 1$) over $[0,1]$, as displayed in Fig. 8.1.4.

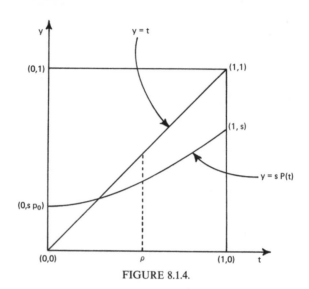

FIGURE 8.1.4.

When the left side of Eq. (8.1.36) assumes the values 0, ρ, and 1 the right side assumes, respectively, the values sp_0, $s\rho$, and s. Thus when $p_0 \neq 0$ (in which case $\rho > 0$) $0 < sp_0$, $\rho > s\rho$, and $1 > s$ so that Eq. (8.1.36) has exactly one root between 0 and ρ and no root between ρ and 1. $R(s)$ is uniquely characterized by this root and furthermore $R(s) < \rho$. [When $p_0 = 0$, $\rho = 0$ and $R(s) = 0$ for $0 \leq s < 1$ and the theorem is trivially true.]

Since $R(1)$ is obviously a root of $t = P(t)$ and since ρ is the smallest root of this equation it is clear that $R(1) = \rho$, and conclusions (b) and (c) follow.   □

If we let $A(s)$ be the p.g.f. of the total number of particles ever to have existed, not including the first, then $sA(s) = R(s)$ so that

$$A(s) = P(sA(s)). \qquad (8.1.37)$$

This result was originally developed by Good (1949).

EXAMPLE 8.1.2: Consider a model for binary fission by taking $P(s) = q + ps^2$ where p is the probability that a particle splits into two and q is the probability that the particle dies ($0 < p < 1$).

Since $\mu = 2p$, extinction is certain if and only if $p \leq \frac{1}{2}$.

If $p > \frac{1}{2}$, the probability of extinction is given by the root of $s = P(s)$ in the interval $(0, 1)$, leading to $f_{10} = q/p$.

If $p \leq \frac{1}{2}$, $f_{10} = 1$ and we investigate the determination of $R(s)$ and hence the $\{r_k\}$. Now Eq. (8.1.36) reduces to the quadratic equation

$$pst^2 - t + qs = 0,$$

from which we deduce

$$t = [1 \pm (1 - 4pqs^2)^{1/2}]/2ps.$$

$R(s)$ is given by the positive root of Eq. (8.1.36) so that one of the above two solutions must be eliminated.

When $s = 1$, t must be 1 in order that we obtain a proper p.g.f. Note now that

$$(1 - 4pq)^{1/2} = [(p + q)^2 - 4pq]^{1/2} = [(p - q)^2]^{1/2}$$

$$= \begin{cases} q - p, & p \leq q, \\ p - q, & p > q. \end{cases}$$

Since $p \leq \frac{1}{2}$ implies $p \leq q$ we must take $(1 - 4pq)^{1/2} = q - p$. Taking the limit as $s \to 1$,

$$R(1) = 1 = \frac{1 \pm (1 - 4pq)^{1/2}}{2p} = \frac{1 \pm (q - p)}{2p}.$$

It is easily seen that the positive square root is incompatible with the required conclusion and hence we are lead to

$$R(s) = \frac{1 - (1 - 4pqs^2)^{1/2}}{2ps}.$$

This p.g.f. when expanded gives non-negative coefficients with

$$r_{2n-1} = \frac{1}{n}\binom{2n - 2}{n - 1}p^{n-1}q^n \qquad (n \geq 1),$$

$$r_{2n} = 0 \qquad (n \geq 0).$$

(cf. Example 3.2.2). □

Exercises 8.1

1. Using the notation of Theorem 8.1.1 show that for any $k = 0, 1, \ldots, n$

$$P_{n+1}(s) = P_{n-k}(P_{k+1}(s)).$$

2. Let $\{X_n, n \geq 0\}$ be a branching chain with $X_0 = 1$. Suppose that each individual produces 0, 1, or 2 particles with equal probabilities of $\frac{1}{3}$. Find the p.g.f. for X_1, X_2, and X_3.

3. Let $\{X_n, \ n \geq 0\}$ be a branching chain with $X_0 = i$, a constant. If the offspring distribution has a p.g.f. $P(s)$ show that the p.g.f. of X_n, $P_n(s)$, is n for $n \geq 0$ by

$$P_{n+1}(s) = P_n(P(s)) \quad \text{with} \quad P_0(s) = s^i.$$

4. Let $\{X_n, \ n \geq 0\}$ be a branching chain wih $X_0 = 1$ and offspring distribution $\{p_n\}$ with $0 < p_0 < 1$ and $p_1 = 1 - p_0$. In this case $X_0 = 1, \ldots, X_{N-1} = 1$ and $X_n = 0$ for $n \geq N$. Find the probability distribution of N and show that N is finite with probability one. Hence deduce, from first principles, that state 0 is reached with probability one.

5. If the offspring distribution $\{p_n\}$ has a strictly monotone increasing p.g.f. $P(s)$ (which rules out the case $p_0 = 1$), show, using an inductive argument, that $\{P_n(0)\}$ is a strictly monotone increasing sequence and hence that $f_{10}^{(n)} > 0$ for $n = 1, 2, \ldots$. (See the proof of Theorem 8.1.4.)

6. Consider the branching chain $\{X_n, \ n \geq 0\}$ with $X_0 = 1$ and offspring distribution $\{p_n\}$ having p.g.f. $P(s) = p_0 + p_1 s + p_2 s^2$. Find the probability of extinction of the branching chain for all possible values of p_0, p_1, and p_2.

7. Suppose that every man in a certain society has exactly three children, which independently have probability one-half of being a boy and one-half of being a girl. *The number of males in the* nth generation forms a branching chain.

    (a)   Show that the probability that the male line of a given man eventually becomes extinct is $\sqrt{5} - 2$.

    (b)   If a given man has two boys and one girl, what is the probability that his male line will continue forever?

8. Given a population of N individuals at time $t = 0$. Suppose that an individual alive at time t units dies during the interval $[t, t+1]$ with probability p and survives during this interval with probability q. Assuming that individuals are independent of one another, show that the probability distribution of the size of the population at time n, for n a positive integer, is binomial.

9. Let $\{X_n\}$ $(n = 0, 1, 2, \ldots)$ be a branching chain where X_n denotes the number of individuals in the nth generation. Suppose $X_0 = 1$ and the number of direct descendents has a geometric distribution, i.e.,

$$P\{X_1 = i\} = p(1 - p)^i, \quad i \geq 0, \quad 0 < p < 1.$$

    (a)   If $P_n(s)$ is the p.g.f. of X_n, show that

$$P_n(s) = \frac{a_n + b_n s}{c_n + d_n s} \quad \text{for suitable} \quad a_n, b_n, c_n, d_n.$$

(b) Show that the probability that the process is extinct by the end of the nth generation is

$$p(q^n - p^n)/(q^{n+1} - p^{n+1}), \qquad \text{where} \quad q = 1 - p.$$

(c) Find x, the probability of ultimate extinction.

10. Consider the following discrete branching chain: Let

$$p_j = P[\text{an individual has } j \text{ offspring}]$$
$$= bc^{j-1} \qquad (j = 1, 2, \ldots)$$

(where $0 < c < 1, 0 < b < 1 - c$) and

$$p_0 = 1 - \sum_{j=1}^{\infty} p_j.$$

(a) Find the p.g.f. $P(s)$ of the p_j's and the mean μ.
(b) Show that the equation $s = P(s)$ has roots 1 and $s_1 = (1 - b - c)/c(1 - c)$.
(c) Show that $s_1 \geq 1$ if and only if $\mu \leq 1$ [Parzen (1962)].

11. Consider a branching chain $\{X_n, n \geq 0\}$ with $X_0 = i$ and offspring distribution having mean μ and variance σ^2. Find expressions for EX_n and var X_n.

12. Consider a branching chain $\{X_n, n \geq 0\}$ with X_0 having a distribution with mean μ_0 and variance σ_0^2. If the offspring distribution has mean μ and variance σ^2, derive expressions for EX_n and var X_n. Obtain the results of Exercise 8.1.11 as a special case.

13. Let $\{X_n, n \geq 0\}$ be a branching chain with $X_0 = 1$ and offspring distribution having a mean μ and variance σ^2. If the p.g.f. for X_n is $P_n(s)$, show that the joint p.g.f. of the pair of random variables (X_m, X_n) for $n > m$ is given by

$$P(s_1, s_2) = P_m(s_1 P_{n-m}(s_2)).$$

Furthermore, if $\mu \neq 1$ show that

$$\text{cov}(X_m, X_n) = \frac{\mu^{n-1}(\mu^m - 1)}{\mu - 1} \sigma^2.$$

14. Let $\{Y_n, n \geq 0\}$ be a process where Y_n denotes the number of individuals present at time n. Each of the Y_n individuals can become a family of r individuals at time $n + 1$ with probability g_r and in the time interval $(n, n + 1)$ s individuals can immigrate into the population with probability f_s. If $F(s)$, $G(s)$, and $P_n(s)$ are, respectively, the p.g.f.'s of the sequences $\{f_n\}$, $\{g_n\}$, and the r.v. Y_n show that

$$P_{n+1}(s) = F(s)P_n(G(s)) \qquad (n \geq 0).$$

If $Y_0 = 0$ show that

$$P_{n+1}(s) = P_n(s)F(G(G(G(\ldots G(s))))) \qquad (n \geq 0),$$

where there are n successive iterations of $G(s)$ [based on Cox and Miller (1965)].

15. Let $\{X_n, n \geq 0\}$ be a branching chain with $X_0 = 1$ and offspring distribution $\{p_n\}$. Let $\rho(s)$ be the p.g.f. of the total number of progeny that appear in the chain. If $p_n = pq^n$ $(n = 0, 1, 2, \ldots), (p + q = 1)$, show that

$$\rho(s) = [1 - (1 - 4pqs)^{1/2}]/2q.$$

Hence find the probability q_n that the total number of progeny is n.

16. Let $\{X_n, n \geq 0\}$ be a branching chain with $X_0 = 1$. Let $\{Z_n, n \geq 1\}$ be specified by $Z_n = X_1 + \cdots + X_n$, the total number of descendants in the first n generations of the branching chain.

(a)   If $P(s) = E(s^{X_1})$ and $F_n(s) = E(s^{Z_n})$ show that

$$F_{n+1}(s) = P(sF_n(s)).$$

(b)   If $\mu = EX_1$ show that

$$EZ_n = \begin{cases} \dfrac{\mu(1 - \mu^n)}{1 - \mu}, & \mu \neq 1, \\[2mm] n, & \mu = 1. \end{cases}$$

(c)   If $\mu = 1$ and var $X_1 = \sigma^2$ show that

$$\text{var}(Z_{n+1}) = \sigma^2(n + 1)^2 + \text{var}(Z_n)$$

and hence that

$$\text{var}(Z_n) = \sigma^2 n(n + 1)(2n + 1)/6.$$

Chapter 9

Discrete Time Queueing Models

9.1 Introduction

A simple description of a queueing model is that we have customers arriving at a facility demanding service of some kind. If the request can be handled at the time of arrival the service of the arriving customer commences at that time; otherwise he waits in a "queue" or "waiting line" until such time that the service center has spare capacity. The customer leaves the system once his request has been satisfied.

Such a model has three distinct components: (i) the *arrival pattern*—a description of how the customers arrive in the system, (ii) the *service mechanism*—a description of how long it takes for the customers service request to be discharged, and (iii) the *queue discipline*—a procedure for assigning waiting customers to the service facility when it is ready to handle a further task.

In order to examine the properties of any queueing system we need to have detailed specifications of the forms of these components. Before we examine each component in turn we shall see how such systems can arise. Although the literature of queueing theory deals largely with continuous time models, where the arrival of requests and their service times are allowed to assume real values, we shall formulate the problems within a discrete time framework. This is often the more natural approach to take. For example, consider a computer communication system. In such systems, the term *customer* refers to a number of data units (a *byte*, a *character*, a *packet*—a block of characters of fixed length, or a *message* consisting of one or more packets), the term *server* or *service facility* refers to a *processor, transmission line, channel,* or *terminal user,* while a *queue* is represented by *buffer storage* or *memory.* Examples of the natural elementary unit of time in a given system are the machine cycle time of a processor, the bit or byte duration of signals on a channel or transmission line, or a pulse duration of any data unit of fixed size. In such systems, all events are allowed to occur only at definite regularly spaced time points.

(There has been considerable interest in such models with the rapid growth and technological innovation in the field of computer communications. Some of the applications of the discrete time model appear in the study of time-sharing computer systems, where messages from a collection of terminals are assigned by a variety of different multiplexing methods to a central computer.)

With a little imagination it is easy to construct other examples of discrete time queueing models. For example, arrivals could be prospective customers for a shuttle service (bus, plane, helicopter) that leaves a certain station periodically, say every twenty minutes. For reasons of safety, the vehicles in question can carry only a fixed number of passengers, say M. Thus, at a departure time, the first M passengers waiting in the queue board and those remaining await a subsequent departure. One of the questions of interest concerns the number of customers waiting in the queue at any given departure time; this is a question that, for instance, would interest an architect planning a waiting room, or an engineer seeking to determine the frequency with which the shuttle should depart.

To initiate this study, let us examine methods for specifying the arrival pattern. Firstly, we assume that the time axis is segmented into a sequence of time intervals (*slots*) of unit duration, which correspond to the elementary unit of time in the system. If no more than one customer may arrive in a given slot, then, following the traditional approach used for continuous time models, we shall assume that the times between arrivals, the *interarrival times* $\{T_k\}$, are a sequence of independent and identically distributed (i.i.d.) positive integer-valued r.v.'s with a fixed probability distribution $f_n = P\{T_k = n\}$ $(n = 1, 2, \ldots)$. In other words, the interarrival times form a persistent recurrent event process. Such an arrival pattern is called a *general independent* pattern.

Special cases include the following:

(i) The *deterministic* pattern, where $f_1 = 1$, corresponding to one arrival in each slot with probability 1.

(ii) The *geometric* pattern, where $f_n = \lambda(1 - \lambda)^{n-1}$ $(n \geq 1)$. It is easily seen (Example 3.1.3) that this pattern is generated by a sequence of Bernoulli trials $\{X_k\}$, where X_k is the outcome of the kth trial, a sequence of i.i.d. r.v.'s that assumes the values 0 or 1 with $P\{X_k = 1\} = \lambda$ $(0 < \lambda < 1)$. For such a pattern, the total number of arrivals in the first n slots is a binomial r.v. with parameters (n, λ) (see Example 3.4.1). Since the expected number of arrivals per slot EX_k is λ, we say that λ is the *arrival rate*. (In some more general discrete time systems, the arrival rate may be taken to depend on the total number already present in the system.)

If the slot size is such that it is permissible for more than one customer to enter the system during a single slot, then it may be more appropriate to assume that the sequence $\{A_k\}$, where A_k is the number of arrivals that occur during the kth slot, is a sequence of i.i.d. nonnegative integer-valued r.v.'s with fixed probability distribution $a_n = P\{A_k = n\}$ $(n = 0, 1, 2, \ldots)$. An alternative way of modeling this arrival pattern is to assume that we have *compound* or *bulk* arrivals at geometric interarrival intervals. Specifically, assume that batches arrive according to a Bernoulli process at rate λ, with the number of arrivals in the batch being a r.v. with probability distribution $\{b_n\}$ $(n = 1, 2, \ldots)$. Then $a_0 = 1 - \lambda$, and $a_n = \lambda b_n$ $(n = 1, 2, \ldots)$. [Equivalently, $\lambda = 1 - a_0$, and $b_n = a_n/(1 - a_0)$ $(n = 1, 2, \ldots)$.] Such an arrival pattern can be described as the *Geometric$^{(A)}$* pattern, where the superscript A denotes the r.v. representing the number of customers that arrive during each successive time slot.

We can make similar assumptions concerning the service mechanism, although whereas the arrival generating procedure is operating continuously, the service mechanism functions only when a customer is present in the system. The general assumption we make is that the successive service times of customers handled by a server $\{S_k\}$ are assumed to be a sequence of i.i.d. nonnegative (usually positive) integer-valued r.v.'s with a fixed probability distribution $g_n = P\{S_k = n\}$ $(n = 0, 1, \ldots)$. Analogous to the earlier discussion on arrival patterns, we say that such a mechanism gives rise to a *general* service pattern. If $g_1 = 1$, we have a *deterministic* server, while if $g_n = \mu(1 - \mu)^{n-1}$ $(n = 1, 2, \ldots)$, we have a *geometric* service procedure. In an analogous manner, we may incorporate the possibility of batch servicing (either of random or fixed size) for a single server using a *Geometric$^{(B)}$* type pattern.

Regarding the queue discipline, we shall, in general, assume that customer requests are handled on a FIFO (first in, first out of the waiting line) basis. Other disciplines include LIFO (last in, first out) and SIRO (service in random order), but such disciplines are mainly of interest when we wish to examine the waiting time of a typical customer in the queueing system or queue itself. The service center may be stocked by arrivals from different sources and contain more than one server, but, unless otherwise stated, we shall examine only single-channel systems. Furthermore, unless otherwise stated, we shall assume that the system has unlimited waiting room capacity. In a finite waiting room model, any customer arriving to find the system full departs immediately and is lost to the system.

A simple way of classifying single-channel systems with FIFO queue discipline is to specify, in order, the arrival pattern, the service mechanism, the number of servers, and the system capacity (the maximum permissible number of customers in the waiting room plus the number in the service facility). We often use the following abbreviations: G for general, GI for general independent, and D for deterministic. Thus $GI/D/I/N$ refers to the system with general independent arrivals, deterministic service times, and a single server with at most N in the system. When $N = \infty$, i.e., the system has unlimited capacity, we delete any reference to the maximum system size.

In the sections that follow, we shall examine the stochastic behavior of a variety of discrete time queueing models. Since we are primarily interested in the congestion that may or may not develop, we shall concern ourselves mainly with measures of the system size, the distribution and moments of the number of customers in the system at different time points. Alternative r.v.'s of interest are the *queueing time* (the time spent in the system prior to servicing) and the *waiting time* (the total time spent in the system) of a typical customer. From the point of view of the server, the *busy period* (an interval of uninterrupted servicing) is also of importance. We give a cursory look to these latter r.v.'s only in a couple of special models. (See Sections 9.3 and 9.4.)

9.2 The Geometric/Geometric/1 Model

The simplest discrete time model we shall look at in detail is the Geometric/Geometric/1 model, i.e., customers arrive at a single server facility according to a Bernoulli process and have service times that are geometrically distributed, and hence can also be regarded as being generated by a Bernoulli process. To be specific, with our elementary unit of time taken to be unity, we shall assume that customer arrivals are independent and that the probability of an arrival during any interval is p $(0 < p < 1)$. Furthermore, the service times are independent and the probability of completion of a service during any interval, provided the server is busy, is r $(0 < r < 1)$.

Before we proceed we must make assumptions regarding the order in which our arrivals and services take place and the time that they occur, whether at the beginning of an interval or the end of the interval. In fact we shall consider three different arrangements. The first system, which we call the *early arrival system,* allows the unit in service to be ejected from the service facility just prior to the end of the time interval, with any new arrival joining the system at the commencement of the next time interval and entering the service facility at that time, if it is free. Alternatively we may consider two variants of the *late arrival system* where arrivals occur late, just prior to the end of a time interval, with services terminating at the beginning of time intervals. In the *late arrival system with immediate access* the arriving item can enter the service facility, if it is free, for an immediate initial unit of service with the possibility of it being ejected almost instantaneously, whereas in the *late arrival system with delayed access* the arriving customer is blocked from entering an empty service facility until the servicing interval terminates.

It is very important in any stochastic modeling situation to be absolutely certain what a typical sample path of the process under consideration looks like. Even if we are given the same set of arrival times and service times, the above three systems give different sample paths. To clarify this, let $X(t)$ denote the number of customers in the *system* at time t and define $X_n \equiv X(n-)$, $Y_n \equiv X(n)$, $Z_n \equiv X(n+)$, where, for convenience, we assume without any loss of generality that for the early arrival system, services can be completed only in $(n-, n)$ and arrivals can occur only in $(n, n+)$, with the reverse arrangement operating in the late arrival systems (Fig. 9.2.1).

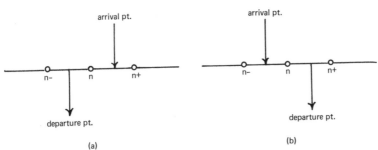

FIGURE 9.2.1. (a) Early arrival system. (b) Late arrival system.

Observe that, in the late arrival system, if an arrival occurs within $(n-, n)$ and the server is free at that time, then, with probability r, a service may be completed within $(n, n+)$ in the immediate access variant, but with probability 0 in the delayed access case.

Figure 9.2.2 details sample paths for the following realization. Arrivals occur at times 0, 1, 3, 5, 6, 7 (with interarrival times 1, 2, 2, 1, 1), with service times 2, 1, 1, 2, 1, 1. To aid in identification of the X_n, Y_n, Z_n processes we have labeled these as $X_n^{(i)}$, $Y_n^{(i)}$, $Z_n^{(i)}$, and $X_n^{(d)}$, $Y_n^{(d)}$, $Z_n^{(d)}$ for the late arrival system with immediate access and delayed access, respectively.

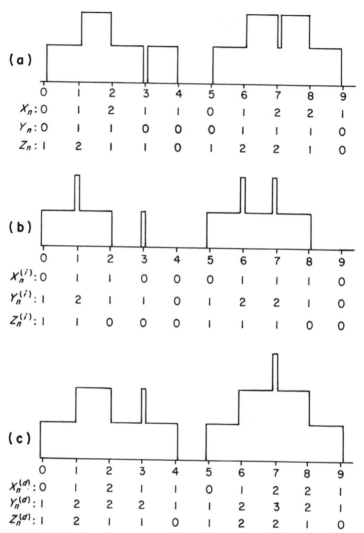

FIGURE 9.2.2. (a) Early arrival system. (b) Late arrival system with immediate access. (c) Late arrival system with delayed access.

By examining the sample paths, as in Fig. 9.2.2, or otherwise, it can be seen that $\{X_n\}$ and $\{Y_n\}$ are different processes but that $\{Z_n\} \equiv \{X_{n+1}\}$. Also, $\{X_n^{(i)}\} \equiv \{Y_n\}$, $\{Y_n^{(i)}\} \equiv \{X_{n+1}\}$, $\{Z_n^{(i)}\} \equiv \{Y_{n+1}\}$, $\{X_n^{(d)}\} \equiv \{X_n\}$, $\{Z_n^{(d)}\} \equiv \{X_{n+1}\}$ with $\{Y_n^{(d)}\}$ different to any of the above processes (see Exercise 9.2.4).

Note that Fig. 9.2.2 provides us with typical sample paths for $X(t)$, the system size r.v.. Alternatively we could have considered the *queue* size r.v., $Q(t)$, where

$$Q(t) = \begin{cases} 0 & \text{if } X(t) = 0 \text{ or } 1, \\ i - 1 & \text{if } X(t) = i \quad (i \geq 2). \end{cases} \qquad (9.2.1)$$

Note, however, that $X(t)$ provides us with more information, since when $Q(t) = 0$ we do not know whether the server is free or busy at time t. The behavior of $Q(t)$ at time points $n-$, n, and $n+$ can, of course, be found from the behavior of the r.v.'s X_n, Y_n, and Z_n, respectively, with the appropriate connection via Eq. (9.2.1).

From our observations above, the stochastic behavior of this queueing model can be examined in terms of either the $\{X_n\}$ or $\{Y_n\}$ process. We show that both of these processes are, in fact, homogeneous Markov chains. The properties of these MC's are discussed in the following two theorems.

THEOREM 9.2.1:   For the Geometric/Geometric/1 queueing model, the $\{X_n, n \geq 0\}$ process is a homogeneous MC with transition matrix given by

$$P_X = \begin{bmatrix} \bar{p} & p & 0 & 0 & \cdots \\ \bar{p}r & 1 - \bar{p}r - p\bar{r} & p\bar{r} & 0 & \cdots \\ 0 & \bar{p}r & 1 - \bar{p}r - p\bar{r} & p\bar{r} & \cdots \\ 0 & 0 & \bar{p}r & 1 - \bar{p}r - p\bar{r} & \cdots \\ \vdots & \vdots & \vdots & \vdots & \ddots \end{bmatrix}, \qquad (9.2.2)$$

where $\bar{p} = 1 - p$ and $\bar{r} = 1 - r$.

Proof: Without loss of generality, we shall consider the early arrival system. Firstly, if $X_n \geq 1$, then

$$X_{n+1} = \begin{cases} X_n + 1 & \text{if there is one arrival and no service completion,} \\ X_n & \text{if there is no arrival and no service completion or} \\ & \text{if there is one arrival and one service completion,} \\ X_n - 1 & \text{if there is no arrival and one service completion.} \end{cases}$$

If $X_n = 0$, then

$$X_{n+1} = \begin{cases} 0 & \text{if there is no arrival,} \\ 1 & \text{if there is one arrival.} \end{cases}$$

Since $\bar{p}$ is the probability of no arrival at the beginning of an interval and $\bar{r}$ is the probability of no service completion by a busy server at the end of an interval, under the assumption of independence between arrivals and departures, we have that if $X_n \geq 1$,

$$X_{n+1} = \begin{cases} X_n + 1, & \text{with probability } p\bar{r}, \\ X_n, & \text{with probability } \overline{pr} + pr, \\ X_n - 1, & \text{with probability } \bar{p}r; \end{cases} \tag{9.2.3}$$

and if $X_n = 0$,

$$X_{n+1} = \begin{cases} 1, & \text{with probability } p, \\ 0, & \text{with probability } \bar{p}. \end{cases} \tag{9.2.4}$$

Thus, by Definitions (5.1.1) and (5.1.2), $\{X_n, n \geq 0\}$ is a homogeneous MC with transition matrix given by Eq. (9.2.2). $\square$

THEOREM 9.2.2:   For the Geometric/Geometric/1 queueing model the $\{Y_n, n \geq 0\}$ process is a homogeneous MC with transition matrix given by

$$P_Y = \begin{bmatrix} 1 - p\bar{r} & p\bar{r} & 0 & 0 & \cdots \\ \bar{p}r & 1 - \bar{p}r - p\bar{r} & p\bar{r} & 0 & \cdots \\ 0 & \bar{p}r & 1 - \bar{p}r - p\bar{r} & p\bar{r} & \cdots \\ 0 & 0 & \bar{p}r & 1 - \bar{p}r - p\bar{r} & \cdots \\ \vdots & \vdots & \vdots & \vdots & \ddots \end{bmatrix}. \tag{9.2.5}$$

Proof: By analogous arguments to those used in the proof of Theorem 9.2.1, if $Y_n \geq 1$,

$$Y_{n+1} = \begin{cases} Y_n + 1, & \text{with probability } p\bar{r}, \\ Y_n, & \text{with probability } \overline{pr} + pr, \\ Y_n - 1, & \text{with probability } \bar{p}r; \end{cases} \tag{9.2.6}$$

and if $Y_n = 0$,

$$Y_{n+1} = \begin{cases} 1, & \text{with probability } p\bar{r}, \\ 0, & \text{with probability } \bar{p} + pr = 1 - p\bar{r}. \end{cases} \tag{9.2.7}$$

Equations (9.2.3) and (9.2.6) are derived by identical arguments, but there is a difference between the derivations of Eqs. (9.2.4) and (9.2.7). Note that if $Y_n = 0$, then $Y_{n+1} = 1$ only if there is an arrival in $(n, n+)$ and no service in $((n+1)-, n+1)$. Also, if $Y_n = 0$, then $Y_{n+1} = 0$ if there is no arrival (in which case the service mechanism will not function) or an arrival and a service completion.

The conclusion of the theorem now follows. $\square$

Observe that if the behavior of the $\{X_n\}$ proccess is known we can deduce the behavior of the $\{Y_n\}$ process by taking advantage of the result that if

$X_n \geq 1$, then

$$Y_n = \begin{cases} X_n, & \text{with probability } \bar{r} \text{ (no service)}, \\ X_n - 1, & \text{with probability } r \text{ (one service)}, \end{cases} \qquad (9.2.8)$$

while if $X_n = 0$, then

$$Y_n = 0. \qquad (9.2.9)$$

We now examine the properties of these MC's

COROLLARY 9.2.1A: Let $\{X_n, n \geq 0\}$ be the MC embedded in a Geometric/Geometric/1 queueing system whose transition matrix P_X is given by Eq. (9.2.2) where $0 < p < 1$ and $0 < r < 1$. Then the MC is irreducible and aperiodic. Furthermore, the MC is

(a)   transient if $\rho > 1$,
(b)   persistent null if $\rho = 1$, and
(c)   persistent nonnull if $\rho < 1$,

where $\rho = p/r$.

In the latter case the limiting distribution is given by

$$\lim_{n \to \infty} P\{X_n = j\} = \begin{cases} 1 - \rho, & j = 0, \\ \rho(1 - \alpha)\alpha^{j-1}, & j = 1, 2, \ldots, \end{cases} \qquad (9.2.10)$$

where $\alpha = p\bar{r}/\bar{p}r < 1$.

Proof: An observation of the transition matrix P_X shows that the MC is an infinite birth and death chain with $r_0 = \bar{p}$ and $p_0 = p$, and for $i = 1, 2, \ldots, q_i = \bar{p}r, r_i = 1 - \bar{p}r - p\bar{r}$, and $p_i = p\bar{r}$. The irreducibility and aperiodicity are easily established. From Example 6.3.7, since for $i = 1, 2, \ldots$

$$\rho_i \equiv \frac{q_1 \cdots q_i}{p_1 \cdots p_i} = \left(\frac{\bar{p}r}{p\bar{r}}\right)^i = \left(\frac{1}{\alpha}\right)^i, \qquad \text{with} \quad \rho_0 \equiv 1,$$

the system is transient iff $\sum_{i=0}^{\infty} \rho_i < \infty$ iff $\alpha > 1$, with the system being persistent iff $\alpha \leq 1$.

Furthermore, from Example 7.2.2,

$$\delta \equiv \sum_{i=1}^{\infty} \frac{p_0 \cdots p_{i-1}}{q_1 \cdots q_i} = \frac{p}{\bar{p}r} \sum_{i=1}^{\infty} \alpha^{i-1},$$

which is finite if $\alpha < 1$ and equal to ∞ if $\alpha \geq 1$. Thus the system is persistent nonnull if $\alpha < 1$ and persistent null if $\alpha = 1$.

Conclusions (a)–(c) follow by observing that

$$\alpha = 1 \text{ iff } p(1 - r) = (1 - p)r, \qquad \text{iff } p = r, \qquad \text{iff } \rho = 1,$$

with a similar result holding with the equality sign replaced throughout by
$<$ or $>$.

Under the conditions of nonnull persistence we can determine the
stationary distribution that, because of the aperiodicity, is also the limiting
distribution. We elect to derive this distribution using generating functions
as an illustration of the technique. The stationary distribution can also be
derived directly from the results of Example 7.2.2 (see Exercise 9.2.1).

The stationary equations, $\pi_j = \sum_{i=0}^{\infty} \pi_i p_{ij}$, reduce to

$$\pi_0 = \bar{p}\pi_0 + \bar{p}r\pi_1,$$

$$\pi_1 = p\pi_0 + (1 - \bar{p}r - p\bar{r})\pi_1 + \bar{p}r\pi_2,$$

$$\pi_j = p\bar{r}\pi_{j-1} + (1 - \bar{p}r - p\bar{r})\pi_j + \bar{p}r\pi_{j+1} \qquad (j \geq 2).$$

Multiplying the jth equation by s^j, where $|s| < 1$, and summing using
$\sum_{j=0}^{\infty} \pi_j s^j = \Pi(s)$ we get

$$\Pi(s) = p\bar{r}s \sum_{j=2}^{\infty} \pi_{j-1}s^{j-1} + (1 - \bar{p}r - p\bar{r}) \sum_{j=1}^{\infty} \pi_j s^j$$

$$+ \frac{p\bar{r}}{s} \sum_{j=0}^{\infty} \pi_{j+1}s^{j+1} + \bar{p}\pi_0 + p\pi_0 s.$$

Collecting together terms and simplifying yields

$$\Pi(s)[p\bar{r}s^2 - (\bar{p}r + p\bar{r})s + \bar{p}r] = \pi_0[\bar{p}r + (1 - \bar{p} - \bar{p}r - p\bar{r})s + (p\bar{r} - p)s^2]$$
$$= \pi_0[\bar{p}r + (pr - \bar{p}r)s - prs^2].$$

Thus

$$\Pi(s) = \frac{\pi_0 r(1 - s)(\bar{p} + ps)}{(1 - s)(\bar{p}r - p\bar{r}s)} = \frac{\pi_0 r(\bar{p} + ps)}{\bar{p}r - p\bar{r}s}.$$

For $\Pi(s)$ to be the p.g.f. of the stationary distribution we require
$\lim_{s \uparrow 1} \Pi(s) = 1$. This enables us to determine π_0 since

$$1 = \frac{\pi_0 r(\bar{p} + p)}{\bar{p}r - p\bar{r}} = \frac{\pi_0 r}{\bar{p}r - p\bar{r}}.$$

Substitution for π_0 back into the expression for $\Pi(s)$ gives us

$$\pi(s) = \frac{(\bar{p}r - p\bar{r})(\bar{p} + ps)}{\bar{p}r - p\bar{r}s} = \frac{(1 - \alpha)(\bar{p} + ps)}{1 - \alpha s},$$

where $\alpha = p\bar{r}/\bar{p}r$ (< 1). Extraction of the coefficient of s^j yields π_j. In particular, $\pi_0 = (1 - \alpha)\bar{p} = 1 - \rho$, and, for $j \geq 1$, $\pi_j = (1 - \alpha)\alpha^{j-1}(\alpha\bar{p} + p)$, which
leads to Eq. (9.2.10) by verifying that $\alpha\bar{p} + p = \rho$. $\square$

COROLLARY 9.2.2A: Let $\{Y_n, n \geq 0\}$ be the MC embedded in a Geometric /Geometric/1 queueing system whose transition matrix P_Y is given by Eq. (9.2.5) where $0 < p < 1$ and $0 < r < 1$. Then the MC is irreducible and aperiodic. Furthermore, the MC is

(a) transient if $\rho > 1$,
(b) persistent null if $\rho = 1$, and
(c) persistent nonnull if $\rho < 1$,

where $\rho = p/r$.

In this latter case the limiting distribution is given by

$$\lim_{n \to \infty} P\{Y_n = j\} = (1 - \alpha)\alpha^j, \qquad j = 0, 1, 2, \ldots, \tag{9.2.11}$$

where $\alpha = p\bar{r}/\bar{p}r < 1$.

Proof: A similar proof to that of Corollary 9.2.1A can be constructed to give the relevant results. The classification of the MC follows as before but with $\delta = \sum_{i=1}^{\infty} \alpha^i$.

Rather than proceed as in Corollary 9.2.1A, we can derive the limiting results by using Eqs. (9.2.8), (9.2.9), and (9.2.10). Firstly,

$$P[Y_n = 0] = P[X_n = 0] + P[X_n = 1]r,$$

so that

$$\lim_{n \to \infty} P[Y_n = 0] = \pi_0 + r\pi_1 = 1 - \alpha.$$

Similarly, for $j = 1, 2, \ldots,$

$$P[Y_n = j] = P[X_n = j]\bar{r} + P[X_n = j + 1]r,$$

implying that

$$\lim_{n \to \infty} P[Y_n = j] = \bar{r}\pi_j + r\pi_{j+1} = (1 - \alpha)\alpha^j. \quad \square$$

Observe that the $\{X_n\}$ MC has a modified geometric distribution for its stationary distribution, whereas the $\{Y_n\}$ chain has a geometric stationary distribution. Actually if we observe the $\{X_n\}$ process only at those time points when the system is nonempty we also obtain a geometric limiting distribution.

COROLLARY 9.2.1B: If $\{X_n, n \geq 0\}$ is the embedded MC in the Geometric/Geometric 1 model with $\rho < 1$, then

$$\lim_{n \to \infty} P\{X_n = j \mid X_n > 0\} = (1 - \alpha)\alpha^{j-1}, \qquad j = 1, 2, \ldots. \tag{9.2.12}$$

Proof: For $j \geq 1$,

$$P\{X_n = j \,|\, X_n > 0\} = \frac{P\{(X_n = j) \cap (X_n > 0)\}}{P\{X_n > 0\}} = \frac{P\{X_n = j\}}{1 - P\{X_n = 0\}}. \quad (9.2.13)$$

Since the limit of a quotient is the quotient of the limits (provided they exist), Eq. (9.2.12) follows from Eqs. (9.2.13) and (9.2.10) by noting that $\lim_{n \to \infty} [1 - P\{X_n = 0\}] = \rho$. $\square$

Geometric distributions appear frequently in many queueing problems, especially in a continuous time setting, and for this reason the $\{Y_n\}$ process appears as the main embedded process of interest. It is commonly stated that the Geometric/Geometric/1 model leads to a geometric stationary distribution, but care must always be taken as to what particular embedded process you are considering.

We can generalize the Geometric/Geometric/1 model to allow arrivals to join the queue with a probability depending on the number in the system and also permit the probability of a service completion to depend on the current number of customers in the system. Specifically, let, for $i = 0, 1, 2, \ldots$,

$p_i = P\{$arrival joins the system given that there are i
customers in the system just prior to arrival$\}$,

$r_i = P\{$service completion at the end of the time interval given
that there are i customers in the system just prior to
service completion$\}$.

Once again we need to specify the arrival system and the embedding points. For example, under the early arrival type system the $\{Y_n\}$ process has transition probabilities

$$p_{ij} = \begin{cases} p_i \bar{r}_{i+1}, & j = i + 1, & i = 0, 1, 2, \ldots, \\ p_i r_{i+1} + \bar{p}_i \bar{r}_i, & j = i, & i = 0, 1, 2, \ldots, \\ \bar{p}_i r_i, & j = i - 1, & i = 1, 2, 3, \ldots, \\ 0, & \text{otherwise}; \end{cases} \quad (9.2.14)$$

with $r_0 \equiv 0$ and $\bar{r}_0 \equiv 1$.

With the same arrival type system, the $\{X_n\}$ process has transition probabilities

$$p_{ij} = \begin{cases} p_i \bar{r}_i, & j = i + 1, & i = 0, 1, 2, \ldots, \\ p_{i-1} r_i + \bar{p}_i \bar{r}_i, & j = i, & i = 0, 1, 2, \ldots, \\ \bar{p}_{i-1} r_i, & j = i - 1, & i = 1, 2, 3, \ldots, \\ 0, & \text{otherwise}, \end{cases} \quad (9.2.15)$$

with the same definition for r_0.

The stationary distributions, when they exist, for the above two MC's can be found by using the results of Example 7.2.2.

Both of these chains are special birth and death type chains and hence models for *generalized birth and death queueing chains*. Various special cases are of interest. In particular the Geometric/Geometric/1 model with finite waiting room, say for a maximum of M customers in the system, can be modeled by taking $p_i = p$ $(0 \leq i \leq M - 1)$ and 0 otherwise, with $r_i = r$ $(i \geq 1)$. The derivation of the stationary distributions for the $\{X_n, n \geq 0\}$ and $\{Y_n, n \geq 0\}$ chains is left as an exercise (Exercise 9.2.7).

An application to buffer design of a computer communication system where the arrivals are "geometric" (and hence $p_i = p$ for all i) but where the probability of transmission during a service interval is state dependent has been discussed by Hsu and Burke (1976).

In the construction of queueing networks, where the output from one queue forms the input to another queue, "nice" results can be found if the departure process from one node has a simple structure that can be used as the arrival process to another node. The departure process of the Geometric/Geometric/1 model turns out to be a Bernoulli sequence when the initial system size distribution is appropriately chosen and thus generates events with geometric interdeparture intervals. Actually we can prove a stronger result incorporating the joint distribution of the system size at time n and the number of departures by time n.

*THEOREM 9.2.3: In a Geometric/Geometric/1 model, with early arrival system, arrivals may occur at the beginning of intervals (at times $0+$, $1+$, $2+, \ldots$) and service completions may take place just prior to the end of time intervals (at times $1-$, $2-$, $3-$, $\ldots$).

Let

$$Y_n = \text{Number of customers present in the system at time } n,$$

$$D_n = \text{Number of departures from the system by time } n.$$

If Y_0 has initial distribution, the stationary distribution for the system size, then Y_n and D_n are independent r.v.'s with Y_n distributed as a geometric $(1 - \alpha)$ r.v. and D_n distributed as a binomial (n, p) r.v.

*Proof: We wish to show that if $P\{Y_0 = j\} = (1 - \alpha)\alpha^j$ $(j \geq 0)$, then

$$a_{k,j}^{(n)} \equiv P\{D_n = k, Y_n = j\}$$

$$= \binom{n}{k} p^k \bar{p}^{n-k}(1 - \alpha)\alpha^j \quad (n = 1, 2, \ldots; j = 0, 1, 2, \ldots; k = 0, 1, \ldots, n).$$

$$(9.2.16)$$

We give a proof by induction. First observe that for $j \neq 0$

$$a_{0,j}^{(1)} = P\{D_1 = 0, Y_1 = j\} = \sum_{k=0}^{\infty} P\{D_1 = 0, Y_1 = j \mid Y_0 = k\} P\{Y_0 = k\}$$

$$= P\{\text{No arrival and no departure} \mid Y_0 = j\}(1 - \alpha)\alpha^j$$
$$\quad + P\{1 \text{ arrival and no departure} \mid Y_0 = j - 1\}(1 - \alpha)\alpha^{j-1}$$
$$= \bar{p}\bar{r}(1 - \alpha)\alpha^j + (1 - \alpha)\alpha^{j-1}$$
$$= \bar{p}(1 - \alpha)\alpha^j.$$

Similarly,

$$a_{0,0}^{(1)} = P\{\text{no arrival} \mid Y_0 = 0\}(1 - \alpha) = \bar{p}(1 - \alpha).$$

By a similar argument we are led to results for $a_{1j}^{(1)}$ and hence that

$$a_{k,j}^{(1)} = \begin{cases} \bar{p}(1 - \alpha)\alpha^j, & k = 0, \\ p(1 - \alpha)\alpha^j, & k = 1, \\ 0, & k \neq 0, 1. \end{cases}$$

These are the only probabilities of interest when $n = 1$. When $n = 2$ we need to derive $a_{0j}^{(2)}$, $a_{1j}^{(2)}$, and $a_{2j}^{(2)}$. The technique we use is to derive a general difference equation connecting together the results for general n and $n + 1$ as follows:

$$a_{k,j}^{(n+1)} = a_{k-1,j}^{(n)} P\{1 \text{ arrival and 1 departure}\}$$
$$\quad + a_{k,j-1}^{(n)} P\{1 \text{ arrival and 0 departures}\}$$
$$\quad + a_{k-1,j+1}^{(n)} P\{0 \text{ arrival and 1 departure}\}$$
$$\quad + a_{k,j}^{(n)} P\{0 \text{ arrival and 0 departures}\},$$

i.e.,

$$a_{k,j}^{(n+1)} = pra_{k-1,j}^{(n)} + p\bar{r}a_{k,j-1}^{(n)} + \bar{p}ra_{k-1,j+1}^{(n)} + \overline{pr}a_{k,j}^{(n)}, \qquad (9.2.17)$$

provided $j \neq 0$ and $k \neq 0, n + 1$. In these cases

$$a_{k,0}^{(n+1)} = pra_{k-1,0}^{(n)} + \bar{p}ra_{k-1,1}^{(n)} + \bar{p}a_{k,0}^{(n)} \qquad (k \neq 0, n + 1), \quad (9.2.18)$$

$$a_{0,j}^{(n+1)} = p\bar{r}a_{0,j-1}^{(n)} + \overline{pr}a_{0,j}^{(n)} \qquad (j > 0), \quad (9.2.19)$$

$$a_{n+1,j}^{(n+1)} = pra_{n,j}^{(n)} + \bar{p}ra_{n,j+1}^{(n)} \qquad (j \geq 0), \quad (9.2.20)$$

$$a_{0,0}^{(n+1)} = \bar{p}a_{0,0}^{(n)}. \qquad (9.2.21)$$

Using the results for $a_{kj}^{(1)}$ ($k = 0, 1$) and substitution into Eqs. (9.2.19) and (9.2.21) for $k = 0$, Eqs. (9.2.17) and (9.2.18) for $k = 1$ and Eq. (9.2.20) for $k = 2$ yields the required form for $a_{kj}^{(2)}$ ($k = 0, 1, 2$).

In like manner, if we assume the result is true for n, substitution in Eqs. (9.2.17) to (9.2.21) yields the same form as given by Eq. (9.2.16) with n replaced by $n + 1$.

For example,

$$pra_{k-1,j}^{(n)} + \bar{p}ra_{k,j-1}^{(n)} + \bar{p}ra_{k-1,j+1}^{(n)} + \overline{p}\overline{r}a_{k,j}^{(n)}$$

$$= \binom{n}{k-1} p^k \bar{p}^{n-k+1} r(1 - \alpha)\alpha^j + \binom{n}{k} p^{k+1} \bar{p}^{n-k} \bar{r}(1 - \alpha)\alpha^{j-1}$$

$$+ \binom{n}{k-1} p^{k-1} \bar{p}^{n-k+2} r(1 - \alpha)\alpha^{j+1} + \binom{n}{k} p^k \bar{p}^{n-k+1} \bar{r}(1 - \alpha)\alpha^j$$

$$= \binom{n}{k-1} p^{k-1} \bar{p}^{n-k+1} r(1 - \alpha)\alpha^j(p + \bar{p}\alpha)$$

$$+ \binom{n}{k} p^k \bar{p}^{n-k} \bar{r}(1 - \alpha)\alpha^{j-1}(p + \bar{p}\alpha).$$

Since $p + \bar{p}\alpha = p/r$ and $\bar{r}(p + \bar{p}\alpha)/\bar{p} = \alpha$ the above sum becomes

$$= \left\{ \binom{n}{k-1} + \binom{n}{k} \right\} p^k \bar{p}^{n-k+1}(1 - \alpha)\alpha^j$$

$$= \binom{n+1}{k} p^k \bar{p}^{n+1-k}(1 - \alpha)\alpha^j = a_{k,j}^{(n+1)}.$$

The inductive proof now follows. $\square$

Hsu and Burke (1976) showed, by a different method, that for a queueing chain with geometric input (parameter p) and generalized service rate (rates r_i) the output process is also geometric with parameter p.

An application of Theorem 9.2.3 to several queues or buffers in tandem shows that if each service facility has geometric service times and the input to the first buffer is geometric (parameter p) with the output from buffer i being the input to buffer $i + 1$, then the input to each buffer is geometric (parameter p). Furthermore, as a consequence of the independence between output and system size, the tandem model can be viewed as a series of independent Geometric/Geometric/1 systems (under steady state conditions).

Exercises 9.2

1. Derive the limiting distribution of $\{X_n\}$ for the Geometric/Geometric/1 model using the result of Example 7.2.2.

2. For the Geometric/Geometric/1 model with early arrival system derive the limiting distribution of $\{Y_n\}$ using the generating function technique of Corollary 9.2.1A.

3. For the Geometric/Geometric/1 model with early arrival system show that

$$X_{n+1} = \begin{cases} Y_n, & \text{with probability } \bar{p}, \\ Y_n + 1, & \text{with probability } p; \end{cases}$$

and hence derive the limiting distribution of $\{X_n\}$ from knowledge of the limiting distribution of $\{Y_n\}$.

4. Examine the $\{Y_n^{(d)}\}$ process for the Geometric/Geometric/1 model with late arrival system with delayed access. In particular,

(a)   Show that

$$Y_n^{(d)} = \begin{cases} X_n^{(d)}, & \text{with probability } \bar{p}, \\ X_n^{(d)} + 1, & \text{with probability } p. \end{cases}$$

(b)   Using the result of (a) above and the relevant results for $\{X_n^{(d)}\}$ obtain the limiting distribution of $\{Y_n^{(d)}\}$.

(c)   Show that $Y_n^{(d)}$ and $Y_{n+1}^{(d)}$ satisfy Eqs. (9.2.6) provided $Y_n^{(d)} \geq 2$. [When $Y_n^{(d)} = 1$ no information is available to ascertain whether this customer has just arrived in $(n-, n)$ or at some earlier time.]

If $Y_n^{(d)} = 1$ and this customer has just arrived, show

$$Y_{n+1}^{(d)} = \begin{cases} 2, & \text{with probability } p, \\ 1, & \text{with probability } \bar{p}. \end{cases}$$

If $Y_n^{(d)} = 1$ and this customer arrived earlier, show

$$Y_{n+1}^{(d)} = \begin{cases} 2, & \text{with probability } p\bar{r}, \\ 1, & \text{with probability } \overline{pr} + pr, \\ 0, & \text{with probability } \bar{p}r. \end{cases}$$

Is it possible to derive the transition probabilities

$$P\{Y_{n+1}^{(d)} = j \,|\, Y_n^{(d)} = 1\}?$$

If $Y_n^{(d)} = 0$ show that

$$Y_{n+1}^{(d)} = \begin{cases} 1, & \text{with probability } p, \\ 0, & \text{with probability } \bar{p}. \end{cases}$$

(d)   Does $\{Y_n^{(d)}, n \geq 0\}$ generate a Markov chain?

5. For the Geometric/Geometric/1 model with late arrival system with immediate access show, by a simple argument, that

$$Y_n^{(i)} = \begin{cases} X_n^{(i)}, & \text{with probability } \bar{p}, \\ X_n^{(i)} + 1, & \text{with probability } p. \end{cases}$$

Show also that this result may be derived from the results of Exercise 9.2.3.

6. Verify that the conclusion of Corollary 9.2.1B follows with the $\{X_n, n \geq 0\}$ process replaced by the $\{Y_n, n \geq 0\}$ process.

7. Derive the stationary distribution $\{\pi_i\}$ $(i = 0, 1, \ldots, M)$ for the Geometric/Geometric/1 model with maximum system capacity of M customers for both the $\{X_n, n \geq 0\}$ and $\{Y_n, n \geq 0\}$ chains (with early arrival system).

8. A model for a finite queue with maximum system size M is a finite MC with state space $S = \{0, 1, \ldots, M\}$ and transition matrix with first row $(1 - p, p, 0, \ldots, 0)$, last row $(0, \ldots, 0, q, 1 - q)$, and all other rows consisting of zeros except for a diagonal centered triplet $(\beta, 1 - \alpha - \beta, \alpha)$ with $0 < p < 1, 0 < q < 1, \alpha > 0, \beta > 0$, and $\alpha + \beta < 1$.

   (a) Show that the stationary distribution $\{\pi_i\}$ of this MC is given for the case $\alpha \neq \beta$ by

   $$\frac{1}{\pi_0} = 1 + r_1 r_2 r^{M-2} + r_1(1 - r^{M-1})/(1 - r),$$

   $$\pi_i = r_1 r^{i-1} \pi_0 \qquad (i = 1, 2, \ldots, M - 1),$$

   $$\pi_M = r_1 r_2 r^{M-2} \pi_0,$$

   where $r = \alpha/\beta$, $r_1 = p/\beta$, and $r_2 = \alpha/q$.

   (b) Show the limiting average system size is

   $$Mr_1 r_2 r^{M-2} \pi_0 + r_1[1 - Mr^{M-1} + (M - 1)r^M]\pi_0/(1 - r)^2.$$

   (c) Suppose that an empty queue is more attractive to customers the greater its possible length and in such a way that $p = 1 - 1/M$. In the case of $r < 1$ and M large show that the average system size is $\beta/(\beta - \alpha)(\beta - \alpha + 1)$ and that the proportion of time that the server is idle is $(\beta - \alpha)/(\beta - \alpha + 1)$.

9.3 The Geometric/G/1 Model

To introduce more flexibility into modeling discrete time queueing processes let us now consider the Geometric/G/1 model. Thus, arrivals are characterized by a Bernoulli sequence and the service times are a sequence of independent positive r.v.'s $\{S_k\}$ with probability distribution given by $g_n = P\{S_k = n\}$ $(n = 1, 2, \ldots)$.

Let us first assume that we have a late arrival system with delayed access, so that an arrival may occur at the end of the time interval and enter the system. Services are completed at the beginning of time intervals and we do not permit a departure at such a time point for a customer who has just arrived the instant previously to an empty system. In examining this model with geometric service times, i.e., the Geometric/Geometric/1 model, we saw that the $\{Z_n^{(d)}, n \geq 0\}$ process gave us a geometric distribution for the steady state system size distribution. If we look at the equivalent $\{Z_n^{(d)}, n \geq 0\}$ process for this more general model we run into some technical difficulties. The system size r.v. observed at the point $n+$ ($n = 1, 2, \ldots$), just after each point of possible departure, does not give us a MC. The reason is that if a departure does not occur at such a point we have no knowledge as to when the service will terminate, since the expended service time, which is not given as part of the state variable, is necessary (except in the case of geometric service times) for a determination of the service time yet to be expended on the same customer. To surmount this difficulty we shall observe the system only at those instants *immediately following the departure of a customer*. By examining the system at such points we see that the Markov property is preserved and as a consequence we have an *embedded Markov chain* present.

Before we examine this embedded process, consider now the other arrival systems, as described in Section 9.2, being used in conjunction with this queueing model.

Let us define

Q_n = Number of customers in the Geometric/G/1 system just
        after the service completion of the nth customer.

To signify which arrival process we are using we shall assign to Q_n, a self-explanatory superscript. Thus, if Fig. 9.2.2 gives typical sample paths for a Geometric/G/1 system, realizations of the associated Q_n process observed for each arrival type system are as follows:

Early arrival system, $\{Q_n^{(e)}\}$:	0	1	0	0	1	1	0
Late arrival system with immediate access, $\{Q_n^{(i)}\}$:	0	1	0	0	1	1	0
Late arrival system with delayed access, $\{Q_n^{(d)}\}$:	0	1	1	0	2	1	0.

Once again we see that there are two different processes with $\{Q_n^{(e)}\}$ and $\{Q_n^{(i)}\}$ being the same process and distinct from $\{Q_n^{(d)}\}$. To examine these processes in detail and to eliminate superscripts, let us define

$$Q_n \equiv Q_n^{(d)} \quad \text{and} \quad R_n \equiv Q_n^{(e)} = Q_n^{(i)}.$$

We first look at the $\{Q_n, n \geq 0\}$ process and establish the Markov property alluded to earlier.

THEOREM 9.3.1:   In the Geometric/G/1 queueing model with interarrival times generated by a geometric (parameter p) distribution (according to a late arrival system with delayed access) and service times independently distributed with distribution $\{g_n\}$ $(n \geq 1)$, let

$$Q_n \equiv \text{Number of customers in the system just after}$$
$$\text{the service completion of the } n\text{th customer.}$$

Then $\{Q_n, n \geq 0\}$ is a homogeneous MC with transition matrix given by

$$P_Q = \begin{bmatrix} k_0 & k_1 & k_2 & k_3 & \cdots \\ k_0 & k_1 & k_2 & k_3 & \cdots \\ 0 & k_0 & k_1 & k_2 & \cdots \\ 0 & 0 & k_0 & k_1 & \cdots \\ \vdots & \vdots & \vdots & \vdots & \ddots \end{bmatrix}, \tag{9.3.1}$$

where $k_j = P\{A_n = j\}, j = 0, 1, 2, \ldots$; with

$$A_n = \text{Number of customers entering the system during}$$
$$\text{the servicing of the } n\text{th customer.}$$

The p.g.f., $A(s)$, of the A_n can be found from the p.g.f. $G(s)$, of the service times $\{S_i\}$ by the relationship

$$A(s) = G(1 - p + ps). \tag{9.3.2}$$

Proof: The Markov property is established by observing that

$$Q_{n+1} = \begin{cases} Q_n - 1 + A_{n+1} & \text{if } Q_n \geq 1, \\ A_{n+1} & \text{if } Q_n = 0. \end{cases} \tag{9.3.3}$$

Relationship (9.3.3) is the crux of the whole analysis of this system, and it is important that we ensure that the r.v.'s A_n have the same distribution whether $Q_n \geq 1$ or $Q_n = 0$. With the late arrival system with delayed access, observe that if the service time of the $(n + 1)$th customer, S_{n+1}, takes k time intervals then there are k time slots within this period for arrivals to possibly occur. Figure 9.3.1 illustrates this with k taken as 3 and $A_{n+1} = 2$ for the two cases $Q_n \geq 1$ and $Q_n = 0$, respectively.

From Eqs. (9.3.3),

$$p_{ij} = P[Q_{n+1} = j \mid Q_n = i]$$
$$= \begin{cases} P[A_{n+1} = j - i + 1] = k_{j-i+1} & (i \geq 1) \\ P[A_{n+1} = j] = k_j & (i = 0), \end{cases}$$

and the expression (9.3.1) for P_Q follows.

To find the probability distribution $\{k_j\}$ of the A_n observe that

$$k_j = P[A_n = j] = \sum_{k=1}^{\infty} P[A_n = j \mid S_n = k] P[S_n = k].$$

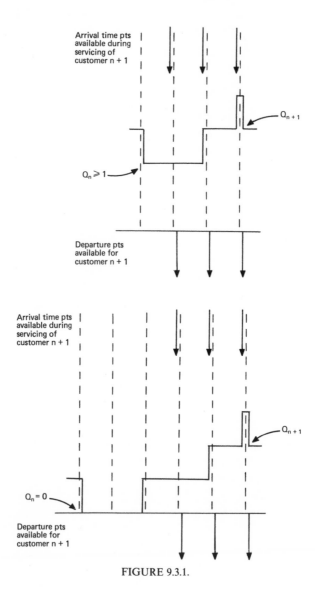

Arrival time pts
available during
servicing of
customer n + 1

$Q_{n + 1}$

$Q_n \geqslant 1$

Departure pts
available for
customer n + 1

Arrival time pts
available during
servicing of
customer n + 1

$Q_{n + 1}$

$Q_n = 0$

Departure pts
available for
customer n + 1

FIGURE 9.3.1.

By the property of geometric interarrival times, arrivals are generated by a Bernoulli sequence and thus given $S_n = k$, the probability that $A_n = j$ is equivalent to the probability of j successes in k trials with probability of success p. Thus, since $j \leq k$,

$$P[A_n = j] = \sum_{k=j}^{\infty} g_k \binom{k}{j} p^j \bar{p}^{k-j} \qquad (j \geq 0).$$

The p.g.f. of A_n is thus given by

$$\begin{aligned}
A(s) &= \sum_{j=0}^{\infty} P[A_n = j] s^j \\
&= \sum_{j=0}^{\infty} \sum_{k=j}^{\infty} g_k \binom{k}{j} p^j \bar{p}^{k-j} s^j \\
&= \sum_{k=0}^{\infty} g_k \sum_{j=0}^{k} \binom{k}{j} (sp)^j \bar{p}^{k-j} \\
&= \sum_{k=0}^{\infty} g_k (sp + \bar{p})^k \\
&= G(sp + \bar{p}) = G(1 - p + sp). \quad \square
\end{aligned}$$

Markov chains with transition matrices of the form given by Eq. (9.3.1) occur frequently in queueing models (in continuous time, as well), and before we examine the particular model presented in Theorem 9.3.1 let us determine the general structure of such MC's.

THEOREM 9.3.2:   Let $\{Q_n, n \geq 0\}$ be a homogeneous MC with transition matrix given by Eq. (9.3.1) where $\{k_j\}$ is a probability distribution.

(a) $\{Q_n, n \geq 0\}$ is an irreducible MC if and only if $0 < k_0 \leq k_0 + k_1 < 1$.
(b)   If $\{Q_n, n \geq 0\}$ is irreducible, it is aperiodic.
(c)   Under the conditions of irreducibility (and aperiodicity) the MC $\{Q_n, n \geq 0\}$ is

   (i)   transient if $\rho > 1$,
   (ii)   persistent null if $\rho = 1$, and
   (iii)   persistent nonnull if $\rho < 1$,

where $\rho = \sum_{j=1}^{\infty} j k_j$.

Proof: (a)   By drawing transition graphs it is easy to establish the following results.

If $k_0 = 0$, then $p_{ij} = 0$ for all $j < i$ and thus $i \not\leftrightarrow j$ for $j < i$ and the chain is not irreducible.

If $k_0 + k_1 = 1$, then $p_{0i} = 0$ and $p_{1i} = 0$ for $i > 1$ so that $i(i > 1)$ can never be reached from 0 or 1 and thus the MC is not irreducible.

Conversely, suppose $0 < k_0 < 1$ and $k_0 + k_1 < 1$. Then there is at least one $r > 1$ such that $k_r > 0$.

Now for any $j < i$, $p_{ij}^{(i-j)} = k_0^{i-j} > 0$ so that $i \to j$.

If $j > i > 1$, then we can find (s, l) such that $j = i + (s + 1)(r - 1) - l$ where $0 < l \le r - 1$. For such a choice $p_{ij}^{(s+1+l)} \ge k_r^{s+1} k_0^l > 0$ and thus $i \to j$. When $i = 0$ we can find (s, l) such that $j = r + s(r - 1) - l$ where $0 < l \le r - 1$, when $p_{0j}^{(s+1+l)} \ge k_r^{s+1} k_0^l > 0$ and thus $0 \to j$.

Furthermore, $i \leftrightarrow i$ since when $i = 0$, $p_{00} = k_0 > 0$, and when $i \ge 1$, $p_{ii}^{(r)} \ge k_r k_0^{r-1} > 0$.

Thus under the required condition all states communicate with each other and thus the MC is irreducible.

(b)   If the MC is irreducible, then $p_{00} = k_0 > 0$ and thus the MC is aperiodic (Exercise 5.3.2).

(c)   The conditions of transience and persistence are easily obtained by using Theorems 7.1.11 and 7.1.12.

Suppose $\rho > 1$. We show that the system of equations $\sum_{j=0}^{\infty} p_{ij} y_j = y_i$ ($i \ne 0$) admits a nonconstant bounded solution and thus by Theorem 7.1.11 the MC will be transient. Take $y_j = s^j$ for some s. Then the above system of equations takes the form, for $i \ge 1$,

$$\sum_{j=0}^{\infty} p_{ij} s^j = \sum_{j=i-1}^{\infty} k_{j-i+1} s^j = s^i$$

or

$$\sum_{j=i-1}^{\infty} k_{j-i+1} s^{j-i+1} = s$$

or

$$A(s) = \sum_{n=0}^{\infty} k_n s^n = s.$$

Now $A(0) = k_0 > 0$ and $A(1) = \sum_{n=0}^{\infty} k_n = 1$ so that if $A^{(1)}(1) = \sum_{n=1}^{\infty} n k_n = \rho > 1$, by Lemma 8.1.3(d), there exists an s_0, $0 < s_0 < 1$, such that $A(s_0) = s_0$. By taking $y_j = s_0^j$ we have found the desired bounded solution, which is clearly nonconstant.

Suppose $\rho \le 1$. By taking $y_j = j$ we show that $\sum_{j=0}^{\infty} p_{ij} y_j \le y_i$ ($i \ne 0$), and hence by Theorem 7.1.12 the MC is persistent. Provided $i \ge 1$,

$$\sum_{j=0}^{\infty} p_{ij} y_j = \sum_{j=i-1}^{\infty} k_{j-i+1} j = \sum_{j=i-1}^{\infty} k_{j-i+1} \{(j - i + 1) + i - 1\}$$

$$= \sum_{n=0}^{\infty} n k_n + i - 1 = \rho + i - 1 \le i = y_i,$$

and the result follows.

There is no easy proof for distinguishing between the persistent null and nonnull cases. Since the MC is assumed to be irreducible we need only classify a single state, say 0. To establish the required dichotomy we shall show that

$$\mu_{00} = \frac{1}{1-\rho} < \infty \qquad \text{when} \quad \rho < 1$$

and

$$\mu_{00} = \infty \qquad \text{when} \quad \rho = 1.$$

If T_{ij} is the r.v. representing the first passage from state i to state j, first observe that, for $j \geq 1$, $f_{j,j-1}^{(n)} = P\{T_{j,j-1} = n\}$ does not depend on j. In particular,

$$f_{j,j-1}^{(1)} = p_{j,j-1} = k_0,$$
$$f_{j,j-1}^{(2)} = p_{jj}p_{j,j-1} = k_1 k_0,$$

and for $n \geq 3$,

$$
\begin{aligned}
f_{j,j-1}^{(n)} &= P\{Q_1 \geq j, \ldots, Q_{n-2} \geq j, Q_{n-1} = j, Q_n = j-1 \,|\, Q_0 = j\} \\
&= \sum_{r_1 \geq j} \cdots \sum_{r_{n-2} \geq j} p_{jr_1} p_{r_1 r_2} \cdots p_{r_{n-2}j} p_{j,j-1} \\
&= \sum_{r_1 \geq j} \cdots \sum_{r_{n-2} \geq j} k_{r_1-j+1} k_{r_2-r_1+1} \cdots k_{j-r_{n-2}+1} k_0 \\
&= \sum_{s_1 \geq 0} \cdots \sum_{s_{n-2} \geq 0} k_{s_1+1} k_{s_2-s_1+1} \cdots k_{1-s_{n-2}} k_0.
\end{aligned}
\tag{9.3.4}
$$

Furthermore, $f_{00}^{(n)} = f_{10}^{(n)}$, since

$$f_{00}^{(1)} = p_{00} = k_0,$$
$$f_{00}^{(2)} = p_{01}p_{10} = k_1 k_0,$$

and for $n \geq 3$

$$
\begin{aligned}
f_{00}^{(n)} &= P\{Q_1 \geq 1, \ldots, Q_{n-2} \geq 1, Q_{n-1} = 1, Q_n = 0 \,|\, Q_0 = 0\} \\
&= \sum_{r_1 \geq 1} \cdots \sum_{r_{n-2} \geq 1} p_{0r_1} p_{r_1 r_2} \cdots p_{r_{n-2}1} p_{10} \\
&= \sum_{r_1 \geq 1} \cdots \sum_{r_{n-2} \geq 1} k_{r_1} k_{r_2-r_1+1} \cdots k_{1-r_{n-2}+1} k_0 \\
&= f_{10}^{(n)},
\end{aligned}
$$

from expression (9.3.4) with $j = 1$. Thus, if $F_{i,j}(s) = \sum_{n=0}^{\infty} f_{ij}^{(n)} s^n$ and $F(s) \equiv F_{00}(s)$, then

$$F_{j,j-1}(s) = F_{10}(s) = F_{00}(s) = F(s) \qquad (j \geq 1). \tag{9.3.5}$$

Now observe that

$$T_{j0} = T_{j,j-1} + T_{j-1,j-2} + \cdots + T_{2,1} + T_{10}.$$

By developing a proof along the lines of Theorem 5.2.1 it is easy to show that $T_{j,j-1}$ and $T_{j-1,j-2}$ are independent r.v.'s (see Exercise 9.3.1) and in general that $\{T_{i,i-1}, i \geq 1\}$ is a sequence of independent, and as earlier deduced, identically distributed r.v.'s. Thus, by Corollary 2.5.2B,

$$F_{j0}(s) = [F(s)]^j. \qquad j = 1, 2, \ldots. \tag{9.3.6}$$

From Theorem 5.1.6(b) we see that

$$f_{10}^{(1)} = k_0 \qquad \text{and} \qquad f_{10}^{(n)} = \sum_{l=1}^{\infty} k_l f_{10}^{(n-1)} \qquad (n \geq 2), \tag{9.3.7}$$

and thus that

$$
\begin{aligned}
F_{10}(s) &= k_0 s + \sum_{l=1}^{\infty} k_l s \sum_{n=2}^{\infty} f_{10}^{(n-1)} s^{n-1} \\
&= k_0 s + \sum_{l=1}^{\infty} k_l s F_{10}(s) \\
&= s\left[k_0 + \sum_{l=1}^{\infty} k_l \{F(s)\}^l \right] \qquad \text{[using Eq. (9.3.6)]} \\
&= sA(F(s)),
\end{aligned}
$$

and hence, from Eq. (9.3.5),

$$F(s) = sA(F(s)). \tag{9.3.8}$$

Under the assumption that $\rho \leq 1$, $F(1) = F_{00}(1) = f_{00} = 1$, so that $F(s)$ is a proper p.g.f. Thus, differentiating both sides of Eq. (9.3.8) we obtain, for $0 \leq s < 1$,

$$F^{(1)}(s) = A(F(s)) + sF^{(1)}(s)A^{(1)}(F(s)),$$

and hence that

$$F^{(1)}(s) = \frac{A(F(s))}{1 - sA^{(1)}(F(s))}, \qquad 0 \leq s < 1.$$

Now $F(s) \uparrow 1$ and $A(s) \uparrow 1$ as $s \uparrow 1$ and thus

$$\mu_{00} = \lim_{s \uparrow 1} F^{(1)}(s) = \begin{cases} \dfrac{1}{1 - \rho} & \text{if } \rho < 1, \\[2mm] \infty & \text{if } \rho = 1; \end{cases}$$

and the result follows from our earlier observation.

[We can establish the distinction between the transient and persistent cases from Eq. (9.3.8). Since $F(1) = F_{00}(1) = f_{00}$, Eq. (9.3.8) implies that f_{00} is a root of the equation $s = A(s)$. In fact, we shall establish that f_{00} is the smallest root α of this equation in the interval $[0, 1]$. To this end, observe that for any two independent nonnegative r.v.'s X and Y,

$$P\{X + Y \leq n\} \leq P\{(X \leq n) \cap (Y \leq n)\} = P(X \leq n)P(Y \leq n).$$

Thus

$$P\{T_{l_0} \leq n\} = P\{T_{l,l-1} + T_{l-1,0} \leq n\} \leq P\{T_{l,l-1} \leq n\}P\{T_{l-1,0} \leq n\},$$

and, hence, since $T_{l,l-1}$ is distributed as T_{10},

$$\begin{aligned} P\{T_{l0} \leq n\} &\leq P\{T_{10} \leq n\}P\{T_{l-1,0} \leq n\} \\ &\leq P\{T_{10} \leq n\}P\{T_{10} \leq n\}P\{T_{l-2,0} \leq n\} \\ &\vdots \\ &\leq (P\{T_{10} \leq n\})^l \qquad (l = 1, 2, \ldots). \end{aligned}$$

Now from Eq. (9.3.7) for $n \geq 0$,

$$\begin{aligned} P\{T_{10} \leq n + 1\} &= \sum_{k=1}^{n+1} f_{10}^{(k)} = k_0 + \sum_{l=1}^{\infty} k_l \sum_{k=2}^{n+1} f_{l0}^{(k-1)} \\ &= k_0 + \sum_{l=1}^{\infty} k_l P\{T_{l0} \leq n\} \\ &\leq k_0 + \sum_{l=1}^{\infty} k_l (P\{T_{10} \leq n\})^l = A(P\{T_{10} \leq n\}). \end{aligned}$$

Using an inductive proof, almost identical to that used in the proof of Theorem 8.1.4, we deduce that

$$P\{T_{10} \leq n\} \leq \alpha \qquad \text{for} \quad n \geq 0.$$

Thus, since $P\{T_{10} \leq n\} \uparrow f_{10} = f_{00}$ as $n \uparrow \infty$, $f_{00} \leq \alpha$ and hence that $f_{00} = \alpha$, the smallest root in $[0, 1]$ of $s = A(s)$.

Now if $\rho = A^{(1)}(1) \leq 1$ and $k_1 < 1$, which holds since the MC is irreducible, Lemma 8.1.3(b) shows that $s = A(s)$ has no root in $[0, 1]$ other than at 1, so that $f_{00} = 1$ and the MC is persistent.

If $\rho > 1$, then Lemmas 8.1.3(c) and 8.1.3(d) show that $\alpha < 1$ so that $f_{00} < 1$ and the MC is transient.] $\square$

Before we specialize to the case of the Geometric/G/1 model we shall derive some additional results for MC's $\{Q_n, n \geq 0\}$ having a transition matrix of the form P_Q. In particular we shall look at techniques for deriving the n-step transition probabilities and the stationary probabilities (when they exist).

*THEOREM 9.3.3: Let $\{Q_n, n \geq 0\}$ be a MC with transition matrix given by Eq. (9.3.1) where $\{k_j\}$ is a probability distribution with p.g.f. $A(s)$. If $p_{ij}^{(n)} = P\{Q_n = j | Q_0 = i\}$, then for $0 < s < 1$, $|z| \leq 1$,

$$\sum_{n=0}^{\infty} \sum_{j=0}^{\infty} p_{ij}^{(n)} s^n z^j = \frac{z^{i+1}[1 - F(s)] - (1 - z)sA(z)[F(s)]^i}{[1 - F(s)][z - sA(z)]}, \qquad (9.3.9)$$

where $z = F(s)$ is the unique positive root of $z = sA(z)$ for $s \in (0, 1]$.

*Proof: Let

$$P_i^{(n)}(z) \equiv \sum_{j=0}^{\infty} p_{ij}^{(n)} z^j = E(z^{Q_n} | Q_0 = i).$$

Since the mechanism generating the MC [cf. Eq. (9.3.3)] can be represented as $Q_{n+1} = A_{n+1} + \max(Q_n - 1, 0)$ where A_{n+1} has a probability distribution $\{k_j\}$,

$$P_i^{(n+1)}(z) = E(z^{A_{n+1} + \max(Q_n - 1, 0)} | Q_0 = i)$$
$$= E(z^{A_{n+1}}) E(z^{\max(Q_n - 1, 0)} | Q_0 = i),$$

because A_{n+1} is independent of Q_n and Q_0. Thus

$$P_i^{(n+1)}(z) = A(z) \left[p_{i0}^{(n)} + \sum_{l=1}^{\infty} p_{il}^{(n)} z^{l-1} \right]$$
$$= A(z) \left[p_{i0}^{(n)} + \frac{P_i^{(n)}(z) - p_{i0}^{(n)}}{z} \right].$$

Since $P_i^{(0)}(z) = z^i$, if we write

$$P_i(z, s) = \sum_{n=0}^{\infty} \sum_{j=0}^{\infty} p_{ij}^{(n)} z^j s^n$$

we have that

$$P_i(z, s) = \sum_{n=0}^{\infty} P_i^{(n)}(z) s^n = P_i^{(0)}(z) + \sum_{n=0}^{\infty} P_i^{(n+1)}(z) s^{n+1}$$
$$= z^i + sA(z) \left[\sum_{n=0}^{\infty} p_{i0}^{(n)} s^n + \frac{1}{z} \left\{ P_i(z, s) - \sum_{n=0}^{\infty} p_{i0}^{(n)} s^n \right\} \right].$$

Observe that $P_i(0, s) = \sum_{n=0}^{\infty} p_{i0}^{(n)} s^n$ so that solving for $P_i(z, s)$ yields

$$P_i(z, s) = \frac{z^{i+1} - s(1 - z)A(z)P_i(0, s)}{z - sA(z)}. \tag{9.3.10}$$

To completely specify $P_i(z, s)$ we need to find an expression for $P_i(0, s) \equiv P_{i0}(s)$. One way to find such an expression is to use the relationship between $F_{i0}(s)$ and $P_{i0}(s)$ (Theorem 6.2.5) and the result, established in the course of proving Theorem 9.3.2(c), that $F_{i0}(s) = [F(s)]^i$, where $F(s)$ satisfies the equation $F(s) = sA(F(s))$. (By working through the argument given in the proof of Theorem 8.1.8 it is seen that $F(s)$ is the unique positive root of $z = sA(z)$ for $s \in [0, 1]$.) Hence

$$P_{i0}(s) = \frac{F_{i0}(s)}{1 - F_{00}(s)} \quad (i \neq 0) \qquad \text{with} \quad P_{00}(s) = \frac{1}{1 - F_{00}(s)},$$

leading to

$$P_{i0}(s) = \frac{[F(s)]^i}{1 - F(s)} \qquad (i \geq 0). \tag{9.3.11}$$

Substitution of the expression given by Eq. (9.3.11) in Eq. (9.3.10) gives the required expression for $P_i(z, s)$, Eq. (9.3.9).

[An alternative argument, which utilizes complex variable theory, is to see that for each value of s such that $|s| < 1$ the function $P_i(z, s)$ is regular (differentiable) in z throughout $|z| < 1$. Now for each s, exactly one root of the equation $z = sA(z)$, i.e., $z = F(s)$, lies in the region $|z| < 1$. The regularity of $P_i(z, s)$ requires that the numerator of the fraction given by Eq. (9.3.10) must vanish whenever the denominator does. Thus, putting $z = F(s)$,

$$[F(s)]^i - s(1 - F(s))A(F(s))P_i(0, s) = 0$$

which leads to Eq. (9.3.9) using Eq. (9.3.8).] $\square$

THEOREM 9.3.4:   Let $\{Q_n, n \geq 0\}$ be an irreducible, persistent nonnull MC with transition matrix given by Eq. (9.3.1) where $\{k_j\}$ is a probability distribution, with p.g.f. $A(s)$ (and with $0 < k_0 \leq k_0 + k_1 < 1$ and $\rho = \sum_{j=1}^{\infty} jk_j < 1$).

For such a MC the stationary distribution $\{\pi_j\}$ with p.g.f. $\Pi(s)$ exists, where

$$\Pi(s) = \frac{(1 - \rho)(1 - s)A(s)}{A(s) - s}. \tag{9.3.12}$$

Furthermore, if $\sum_{l=i+1}^{\infty} k_l = r_i \ (i \geq 0)$, then

$$\pi_0 = 1 - \rho,$$

$$\pi_1 = (1 - \rho)r_0/k_0,$$ (9.3.13)

$$\pi_{j+1} = (1 - \rho) \sum_{l=1}^{j} \alpha_{jl}/k_0^{l+1} \qquad (j \geq 1),$$

where, for $j \geq l \geq 1$, $\alpha_{jl} = \sum_{\boldsymbol{\alpha} \in S_{jl}} r_{\alpha_1} r_{\alpha_2} \cdots r_{\alpha_l}$, $\boldsymbol{\alpha} = (\alpha_1, \ldots, \alpha_l)$, and S_{jl} is the set of all $\boldsymbol{\alpha}$ such that $\alpha_i \geq 1$ with $\alpha_1 + \cdots + \alpha_l = j$.

Proof: Under the stated conditions the existence of the stationary distribution is ensured and is given by the solution of the stationary equations

$$\pi_j = \sum_{i=0}^{\infty} \pi_i p_{ij} \quad (j \geq 0), \qquad \text{with} \quad \sum_{j=0}^{\infty} \pi_j = 1.$$

Substitution for the transition probabilities yields

$$\pi_j = \pi_0 k_j + \sum_{i=1}^{j+1} \pi_i k_{j-i+1} \qquad (j \geq 0). \qquad (9.3.14)$$

To solve these equations we use generating functions:

$$\Pi(s) = \sum_{j=0}^{\infty} \pi_j s^j = \pi_0 \sum_{j=0}^{\infty} k_j s^j + \sum_{j=0}^{\infty} s^j \sum_{i=1}^{j+1} \pi_i k_{j-i+1}$$

$$= \pi_0 A(s) + \frac{1}{s} \sum_{i=1}^{\infty} \pi_i s^i \sum_{j=i-1}^{\infty} k_{j-i+1} s^{j-i+1}$$

$$= \pi_0 A(s) + \frac{1}{s} [\Pi(s) - \pi_0] A(s),$$

implying that

$$\Pi(s) = \frac{\pi_0 A(s)(1 - s)}{A(s) - s}.$$

To find π_0 we use the fact that $\Pi(s)$ is a p.g.f. and thence take the limit as $s \uparrow 1$ to yield

$$1 = \pi_0 \lim_{s \uparrow 1} \frac{1 - s}{A(s) - s} = \pi_0 \frac{1}{1 - A^{(1)}(1)} = \frac{\pi_0}{1 - \rho},$$

using L'Hospital's rule. Thus $\pi_0 = 1 - \rho$ and hence Eq. (9.3.12) follows.

[An alternative proof can be based upon Theorem 9.3.3. Since the MC is irreducible, persistent nonnull, and aperiodic,

$$\lim_{n \to \infty} p_{ij}^{(n)} = \pi_j \qquad \text{for all} \quad i, j = 0, 1, 2, \ldots,$$

an application of Theorem 2.3.3 yields

$$\sum_{j=0}^{\infty} \pi_j z^j = \lim_{s\uparrow 1}(1-s)\sum_{n=0}^{\infty}\left(\sum_{j=0}^{\infty}p_{ij}^{(n)}z^j\right)s^n$$

$$= \lim_{s\uparrow 1}\left[\frac{(1-s)z^{i+1}}{z-sA(z)} - \frac{(1-s)(1-z)sA(z)[F(s)]^i}{[1-F(s)][z-sA(z)]}\right]$$

$$= \frac{(1-z)A(z)}{A(z)-z}\lim_{s\uparrow 1}\frac{1-s}{1-F(s)}$$

$$= \frac{(1-z)A(z)}{A(z)-z}(1-\rho)$$

as required, since

$$F(1) = 1 \quad \text{and} \quad \lim_{s\uparrow 1}\frac{1-s}{1-F(s)} = \frac{1}{F^{(1)}(1)} \quad \text{with} \quad F^{(1)}(1) = \frac{1}{1-\rho}$$

from the proof of Theorem 9.3.2(c).]

Our derivation of the explicit expressions for the π_j is based on a procedure developed by Çinlar (1975).

With $r_i = 1 - k_0 - \cdots - k_i$ and $\sum_{i=0}^{\infty} r_i = \rho$ (cf. the Theorem 2.4.3 using "tail probabilities"), adding the stationary equations (9.3.14) for $j = 0, 1, \ldots, l$, and solving for $\pi_{l+1}k_0$ yields, successively,

$$\pi_1 k_0 = \pi_0 r_0 \tag{9.3.15}$$

and, in general, for $l \geq 1$,

$$\pi_{l+1}k_0 = \pi_0 r_l + \sum_{i=1}^{l}\pi_i r_{l-i+1}. \tag{9.3.16}$$

Now adding Eqs. (9.3.15) and (9.3.16) for $l = 1, 2, \ldots$ gives

$$\left(\sum_{i=1}^{\infty}\pi_i\right)k_0 = \pi_0\rho + \sum_{l=1}^{\infty}\sum_{i=1}^{l}\pi_i r_{l-i+1}$$

$$= \pi_0\rho + \left(\sum_{i=1}^{\infty}\pi_i\right)\left(\sum_{l=i}^{\infty}r_{l-i+1}\right),$$

and hence that

$$(1-\pi_0)(1-r_0) = \pi_0\rho + (1-\pi_0)(\rho - r_0),$$

which implies that

$$\pi_0 = 1 - \rho.$$

From Eq. (9.3.15), $\pi_1 = (1-\rho)r_0/k_0$.

In addition, from Eq. (9.3.16) for $l = 1$,

$$\pi_2 = (1 - \rho)r_1/k_0^2 = (1 - \rho)\alpha_{11}/k_0^2.$$

Further, from Eq. (9.3.16) for $l = 2$,

$$\pi_3 = (1 - \rho)\left\{\frac{r_2}{k_0^2} + \frac{r_1^2}{k_0^3}\right\} = (1 - \rho)\left\{\frac{\alpha_{21}}{k_0^2} + \frac{\alpha_{23}}{k_0^3}\right\}.$$

The general expression for π_{j+1} given by Eq. (9.3.13) is established by induction. Assume the result true for $\pi_1, \pi_2, \ldots, \pi_j$ and show, using Eq. (9.3.16), that

$$\pi_{j+1} = (1 - \rho)\left[r_j/k_0^2 + \sum_{l=2}^{j}\left(\sum_{i=l}^{j}\alpha_{i-1,\,l-1}r_{j-i+1}\right)\Big/k_0^{l+1}\right].$$

The final conclusion follows by showing (Exercise 9.3.2) that $\alpha_{j1} = r_j$ and

$$\sum_{i=l}^{j}\alpha_{i-1,\,l-1}r_{j-i+1} = \alpha_{jl}. \qquad \square$$

Let us now return to the $\{Q_n, n \geq 0\}$ process embedded within the Geometric/G/1 queueing model. An application of the results derived in Theorems 9.3.2 and 9.3.4 to this process yields the following corollary of Theorem 9.3.1

COROLLARY 9.3.1A: Let $\{Q_n, n \geq 0\}$ be the MC with transition matrix given by Eq. (9.3.1) embedded in the Geometric/G/1 queueing model at instants following the departures from the system. Let $\{g_n, n \geq 1\}$ be the service time distribution with p.g.f. $G(s)$. The interarrival time distribution is geometric (parameter p).

(a)   The MC $\{Q_n, n \geq 0\}$ is irreducible if and only if $g_1 < 1$.
(b)   If $g_1 < 1$ and $\rho = p\sum_{n=1}^{\infty} ng_n$, then the MC is

   (i)   transient if $\rho > 1$,
   (ii) persistent null if $\rho = 1$, and
   (iii) persistent nonnull if $\rho < 1$.

(c)   When $g_1 < 1$ and $\rho < 1$, the p.g.f. of the stationary distribution $\Pi_Q(s)$ is given by

$$\Pi_Q(s) = \frac{(1 - \rho)(1 - s)G(\bar{p} + ps)}{G(\bar{p} + ps) - s}. \qquad (9.3.17)$$

Proof: (a)   If $g_1 < 1$, then for some $m \geq 2$, $g_m > 0$. Thus

$$k_0 = \sum_{k=1}^{\infty} g_k\bar{p}^k \geq g_m\bar{p}^m > 0$$

and

$$k_m = \sum_{k=m}^{\infty} g_k \binom{k}{m} p^m \bar{p}^{k-m} \geq g_m p^m > 0,$$

implying that $0 < k_0$ and $k_0 + k_1 < 1$ and thus the MC is irreducible by Theorem 9.3.2(a). Conversely, if $g_1 = 1$, $k_0 = \bar{p}$, $k_1 = p$, and $k_0 + k_1 = 1$, violating the condition for irreducibility.

(b) Since

$$\rho \equiv \sum_{j=1}^{\infty} jk_j = \mathsf{E}(A_n)$$

$$= A^{(1)}(1) = pG^{(1)}(\bar{p} + ps)|_{s \uparrow 1}$$

$$= pG^{(1)}(1) = p \sum_{n=1}^{\infty} ng_n,$$

the result follows from Theorem 9.3.2(c).

(c) The result follows from Eqs. (9.3.12) and (9.3.2). ☐

EXAMPLE 9.3.1: *The Geometric/D/1 Model.* A by-product of the analysis presented in Corollary 9.3.1A is that the Geometric/D/1 model, with late arrival system with delayed access, embedded at instants following departures, is not irreducible since $g_1 = 1$. In fact, a simple investigation of this model will show that once the system size reaches 0 or 1 it stays in those states (except possibly to state 2 just after an arrival and immediately prior to a departure). Here $k_0 = \bar{p}$ and $k_1 = p$ with $k_0 + k_1 = 1$ and $k_i = 0$ for $i \geq 2$. (Note that $\rho = p < 1$ does not imply irreducibility of the state space $\{0, 1, 2, \ldots\}$.)

However, if we restrict attention to the state space $\{0, 1\}$ and start the chain in one of these states, then the transition matrix becomes

$$P_Q = \begin{bmatrix} \bar{p} & p \\ \bar{p} & p \end{bmatrix},$$

which is clearly irreducible. Obviously this is a two-state MC corresponding to independent Bernoulli trials and thus, for all n,

$$\pi_0 = P[Q_n = 0] = \bar{p} \quad \text{and} \quad \pi_1 = P[Q_n = 1] = p. \quad \square$$

EXAMPLE 9.3.2: *The Geometric/Geometric/1 Model.* For the Geometric/Geometric/1 model, with late arrival system with delayed access, embedded at instants following departures, observe that the service time distribution is given by $g_n = \bar{r}^{n-1} r (n = 1, 2, \ldots)$ so that $g_1 = r < 1$ and thus, by Corollary 9.3.1A(a), the MC $\{Q_n, n \geq 0\}$ is irreducible.

Since

$$G(s) = \sum_{n=1}^{\infty} \bar{r}^{n-1} r s^n = \frac{rs}{1 - \bar{r}s},$$

with

$$G^{(1)}(s) = \frac{r}{(1 - \bar{r}s)^2},$$

we have that

$$\rho = pG^{(1)}(1) = p/r,$$

so that the MC is persistent nonnull if $p < r$. We shall show that in this case the stationary distribution is given by

$$\pi_0 = 1 - \rho,$$
$$\pi_j = \rho(1 - \alpha)\alpha^{j-1} \qquad (j = 1, 2, \ldots),$$

where $\alpha = p\bar{r}/\bar{p}r$.

Since

$$G(\bar{p} + ps) = \frac{r(\bar{p} + ps)}{1 - \bar{p}\bar{r} - p\bar{r}s},$$

substitution in Eq. (9.3.17) yields, after simplification,

$$\Pi_Q(s) = \frac{(1 - \rho)r(\bar{p} + ps)}{(\bar{p}r - p\bar{r}s)} = \frac{(1 - \rho)\left(1 + \dfrac{p}{\bar{p}}s\right)}{(1 - \alpha s)}$$

$$= (1 - \rho)\left(1 + \frac{p}{\bar{p}}s\right)\sum_{j=0}^{\infty} \alpha^j s^j.$$

Extracting the coefficient of s^j yields the expressions for π_j above by noting that $(1 - \rho)(\alpha + p/\bar{p}) = \rho(1 - \alpha)$.

Our presentation has thus far concentrated on the $\{Q_n, n \geq 0\}$ process. What changes are required for the $\{R_n, n \geq 0\}$ process when we examine the Geometric/G/1 system with an alternative arrival system? The required generalization of Theorem 9.3.1 is as follows:

THEOREM 9.3.5:   In the Geometric/G/1 queueing model with interarrival times generated by a geometric (parameter p) distribution according to the early arrival system (or the late arrival system with immediate access) and service times independently distributed with distribution $\{g_n, n \geq 1\}$ and p.g.f. $G(s)$, let

> R_n = Number of customers in the system just after the service
> completion of the nth customer.

Then $\{R_n, n \geq 0\}$ is a homogeneous MC with transition matrix given by

$$P_R = \begin{bmatrix} c_0 & c_1 & c_2 & c_3 & \cdots \\ k_0 & k_1 & k_2 & k_3 & \cdots \\ 0 & k_0 & k_1 & k_2 & \cdots \\ 0 & 0 & k_0 & k_1 & \cdots \\ \vdots & \vdots & \vdots & \vdots & \ddots \end{bmatrix}, \qquad (9.3.18)$$

where $k_j = P\{A_n = j\}$, $c_j = P\{B_n = j\}$ $(j \geq 0)$, with

> A_n = Number of customers entering the system during the
> servicing of the nth customer given that $R_{n-1} > 0$

and

> B_n = Number of customers entering the system during the
> servicing of the nth customer given that $R_{n-1} = 0$.

If $A(s) \equiv \sum_{j=0}^{\infty} k_j s^j$ and $B(s) \equiv \sum_{j=0}^{\infty} c_j s^j$, then

$$A(s) = G(\bar{p} + ps), \qquad (9.3.19)$$

$$B(s) = \frac{G(\bar{p} + ps)}{\bar{p} + ps}. \qquad (9.3.20)$$

Proof: Without loss of generality let us consider a typical sample path of the model under an early arrival system for the two cases $R_n > 0$ and $R_n = 0$, respectively. We shall assume that the service time of the $(n + 1)$th customer S_{n+1} takes three time slots and that $A_{n+1} = 2$ and $B_{n+1} = 2$. Figure 9.3.2 is the required generalization of Fig. 9.3.1.

When $R_n = 0$ observe that even though the service time takes three time slots there are only two time points available for arrivals rather than three, as in the $R_n > 0$ case. (For the late arrival system with immediate access, if $R_n = 0$ the initial service is regarded as taking 1 unit of time.)

Analogous to Eq. (9.3.3) we see that

$$R_{n+1} = \begin{cases} R_n - 1 + A_{n+1} & \text{if } R_n \geq 1, \\ B_{n+1}, & \text{if } R_n = 0, \end{cases}$$

where A_{n+1} is as in the proof of Theorem 9.3.1.

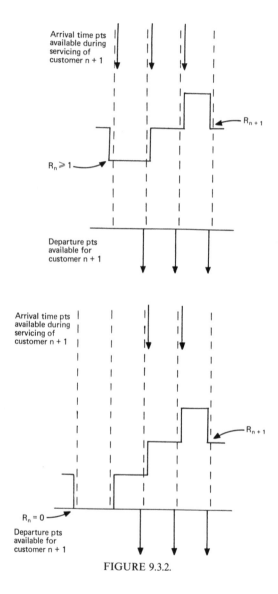

Arrival time pts available during servicing of customer n + 1

R_{n+1}

$R_n \geqslant 1$

Departure pts available for customer n + 1

Arrival time pts available during servicing of customer n + 1

R_{n+1}

$R_n = 0$

Departure pts available for customer n + 1

FIGURE 9.3.2.

This relationship gives the Markov property and implies that

$$p_{ij} = P[R_{n+1} = j | R_n = i]$$

$$= \begin{cases} P[A_{n+1} = j - i + 1] = k_{j-i+1} & (i \geq 1), \\ P[B_{n+1} = j] = c_j & (i = 0), \end{cases}$$

and the structure of the transition matrix given by Eq. (9.3.18) now follows. To derive the distribution of B_{n+1} observe that

$$c_j = P[B_{n+1} = j | R_n = 0]$$

$$= \sum_{k=1}^{\infty} P[B_{n+1} = j | R_n = 0, S_{n+1} = k] P[S_{n+1} = k | R_n = 0].$$

Given $R_n = 0$ and $S_{n+1} = k$, there are $k - 1$ Bernoulli trials available for arrivals and we require j successes (arrivals) during these trials with probability of success p $(0 \leq j \leq k - 1)$. Thus, since S_{n+1} is independent of R_n, for $j \geq 0$,

$$c_j = \sum_{k=j+1}^{\infty} \binom{k-1}{j} p^j \bar{p}^{k-1-j} g_k.$$

The p.g.f. of the distribution of B_{n+1} is thus given by

$$B(s) = \sum_{j=0}^{\infty} \sum_{k=j+1}^{\infty} \binom{k-1}{j} p^j \bar{p}^{k-1-j} g_k s^j$$

$$= \sum_{k=1}^{\infty} g_k \sum_{j=0}^{k-1} \binom{k-1}{j} (ps)^j \bar{p}^{k-1-j}$$

$$= \sum_{k=1}^{\infty} g_k (ps + \bar{p})^{k-1}$$

$$= \frac{G(ps + \bar{p})}{ps + \bar{p}}. \qquad \square$$

From Eqs. (9.3.19) and (9.3.20) observe that

$$G(\bar{p} + ps) = A(s) = (\bar{p} + ps)B(s) \qquad (9.3.21)$$

and thus, equating coefficients of s^j, we have that

$$k_0 = \bar{p} c_0$$
$$k_j = \bar{p} c_j + p c_{j-1} \qquad (j = 1, 2, \ldots). \qquad (9.3.22)$$

THEOREM 9.3.6: Let $\{R_n, n \geq 0\}$ be a homogeneous MC with transition matrix having the structure given by Eq. (9.3.18) where $\{k_j\}$ and $\{c_j\}$ are probability distributions satisfying the relationships given by Eqs. (9.3.22).

(a)   The MC $\{R_n, n \geq 0\}$ is irreducible iff $0 < k_0 \leq k_0 + k_1 < 1$.
(b)   If $\{R_n, n \geq 0\}$ is irreducible, it is aperiodic.
(c)   Under the conditions of irreducibility, the MC $\{R_n, n \geq 0\}$ is

(i)   transient if $\rho > 1$,
(ii)   persistent null if $\rho = 1$, and
(iii) persistent nonnull if $\rho < 1$,

where $\rho = \sum_{j=1}^{\infty} jk_j$.

Proof: We do not go into all the details, but the proof of this theorem parallels that of Theorem 9.3.2.

(a)   If $k_0 = 0$, then $p_{ij} = 0$ for all $j < i$ and the MC is not irreducible.
If $k_0 + k_1 = 1$, then $k_j = 0$ for all $j \geq 2$ and, hence, by Eq. (9.3.22), $\bar{p}c_j + pc_{j-1} = 0$ for all $j \geq 2$. By the nonnegativity of the c_i this then implies that $c_j = 0$ for all $j \geq 1$ and hence that $c_0 = 1$. Consequently, $p_{00} = 1$ and the MC is not irreducible.
Conversely, if $k_0 > 0$, then, from Eq. (9.3.22), $c_0 > 0$ while if $k_0 + k_1 < 1$ there exists an $r \geq 2$ such that $k_r > 0$. Consequently, $\bar{p}c_r + pc_{r-1} > 0$ so that there is an $s \geq 1$ ($s = r$ or $r - 1$) such that $c_s > 0$. From these observations, since $p_{k, k-1} = k_0$, $p_{0s} = c_s > 0$, $p_{l, l+r-1} = k_r > 0$ for $l \geq 1$ so that for all i, j it is possible to construct a chain $i \to i - 1 \to \cdots \to 0 \to s \to \cdots \to j$ and hence establish the irreducibility.

(b)   The aperiodicity follows using the proof given in Theorem 9.3.2.

(c)   The proof for the transience and persistence also follows as in Theorem 9.3.2(c) since the techniques used there for $\rho > 1$ and $\rho \leq 1$ do not depend on the form of the transition probabilities p_{ij} for $i = 0$. To distinguish between the $\rho = 1$ and $\rho < 1$ cases some modifications are required. In particular, it is no longer true that T_{00} and T_{10} have the same distribution. However, $F_{j0}(s) = [F_{10}(s)]^j$ for $j \geq 1$, as derived in Theorem 9.3.2.
Since

$$f_{00}^{(1)} = c_0 \qquad \text{and} \qquad f_{00}^{(n)} = \sum_{j=1}^{\infty} c_j f_{j0}^{(n-1)} \qquad (n \geq 2),$$

$$F_{00}(s) = c_0 s + \sum_{n=2}^{\infty} \sum_{j=1}^{\infty} c_j f_{j0}^{(n-1)} s^j$$

$$= c_0 s + \sum_{j=1}^{\infty} c_j s F_{j0}(s)$$

$$= s \left[c_0 + \sum_{j=1}^{\infty} c_j \{F_{10}(s)\}^j \right]$$

$$= s B(F_{10}(s)), \qquad (9.3.23)$$

where, as in the proof of Theorem 9.3.2(c), Eq. (9.3.8) gives

$$F_{10}(s) = sA(F_{10}(s)).$$ (9.3.24)

Under the assumption that $\rho \leq 1$, $F_{00}(1) = 1$ and thus from Eq. (9.3.23), $B(F_{10}(1)) = 1$ implying that $F_{10}(1) = 1$. Differentiation of Eq. (9.3.23) gives

$$F_{00}^{(1)}(s) = B(F_{10}(s)) + sF_{10}^{(1)}(s)B^{(1)}(F_{10}(s)).$$

Upon taking the limit as $s \uparrow 1$, we obtain

$$\mu_{00} = 1 + \mu_{10}B^{(1)}(1).$$ (9.3.25)

Now, using Eq. (9.3.24), it is easily seen that (cf. proof of Theorem 9.3.2)

$$\mu_{10} = \begin{cases} \dfrac{1}{1-\rho}, & \rho < 1, \\ \infty, & \rho = 1, \end{cases}$$ (9.3.26)

and, from Eq. (9.3.21),

$$A^{(1)}(s) = (\bar{p} + ps)B^{(1)}(s) + pB(s),$$

from which we derive that

$$B^{(1)}(1) = \rho - p.$$ (9.3.27)

Substituting the results given by Eqs. (9.3.26) and (9.3.27) into Eq. (9.3.25) yields

$$\mu_{00} = \begin{cases} \dfrac{\bar{p}}{1-\rho} < \infty, & \rho < 1, \\ \infty, & \rho = 1, \end{cases}$$

leading to the required distinction between the null and nonnull cases.   □

COROLLARY 9.3.5A: Let $\{R_n, n \geq 0\}$ be the MC with transition matrix given by Eq. (9.3.18) embedded in the Geometric/G/1 queueing model at instants following the departures from the system. Let $\{g_n, n \geq 1\}$ be the service time distribution with p.g.f. $G(s)$. The interarrival time distribution is geometric (parameter p).

(a)   The MC $\{R_n, n \geq 0\}$ is irreducible if and only if $g_1 < 1$.
(b)   If $g_1 < 1$ and $\rho = p \sum_{n=1}^{\infty} ng_n$, then the MC is

   (i)   transient if $\rho > 1$,
   (ii)   persistent null if $\rho = 1$, and
   (iii)   persistent nonnull if $\rho < 1$.

(c) When $g_1 < 1$ and $\rho < 1$, the p.g.f. of the stationary distribution $\Pi_R(s)$ is given by

$$\Pi_R(s) = \frac{(1 - \rho)(1 - s)G(\bar{p} + ps)}{(\bar{p} + ps)[G(\bar{p} + ps) - s]}. \tag{9.3.28}$$

Proof: The conclusions of (a) and (b) follow as in Corollary 9.3.1A by virtue of the identical conditions established in Theorems 9.3.2 and 9.3.6. To obtain the expression for the p.g.f. of the stationary distribution (which exists) we use the generating function method to solve the stationary equations:

$$\pi_j = \pi_0 c_j + \sum_{i=1}^{j+1} \pi_i k_{j-i+1}, \qquad (j \geq 0).$$

Paralleling the proof given in Theorem 9.3.4, we see that

$$\Pi_R(s) = \pi_0 B(s) + \frac{1}{s}[\Pi_R(s) - \pi_0]A(s).$$

Using the result given by Eq. (9.3.21) and solving gives

$$\Pi_R(s) = \pi_0 \left[\frac{A(s)(1 - s)}{A(s) - s} \right]\left[\frac{\bar{p}}{ps + \bar{p}} \right].$$

Taking the limit as $s \uparrow 1$, as in the proof of Theorem 9.3.4, gives

$$1 = \pi_0 \frac{\bar{p}}{1 - A^{(1)}(1)} = \frac{\pi_0 \bar{p}}{1 - \rho}.$$

Thus $\pi_0 = (1 - \rho)/\bar{p}$ and Eq. (9.3.28) follows after substitution for $A(s)$ by Eq. (9.3.19). $\square$

EXAMPLE 9.3.3: *The Geometric/Geometric/1 Model.* The p.g.f. for the stationary distribution for the MC $\{R_n, n \geq 0\}$ embedded in the Geometric/Geometric/1 model can be derived from Eq. (9.3.28). The service time distribution $\{g_n\}$ is given, as in Example 9.3.2, by $g_n = \bar{r}^{n-1}r$ $(n \geq 1)$, with p.g.f. $G(s) = rs/(1 - \bar{r}s)$. Substitution in Eq. (9.3.28) yields, after simplification and using the result that $1 - \rho = \bar{p}(1 - \alpha)$,

$$\Pi_R(s) = \frac{1 - \alpha}{1 - \alpha s}$$

and hence that $\pi_j = (1 - \alpha)\alpha^j, j = 0, 1, 2, \ldots.$ $\square$

Examples 9.3.2 and 9.3.3 give further evidence for the occurrence of geometric and modified geometric distributions in Geometric/Geometric/1 queueing models at different embedding points to those considered in Section 9.2. In the next section we shall reexamine this model, this time at different embedding points, and find that the same structure for the associated stationary distributions appears once again.

Expressions for the expected number of customers in the system immediately following a service completion can be deduced from our earlier results.

THEOREM 9.3.7: Under stationary conditions, for the Geometric/G/1 system,

$$EQ_n = \rho + \frac{p^2}{2(1-\rho)} E[S(S-1)], \tag{9.3.29}$$

$$ER_n = \rho - p + \frac{p^2}{2(1-\rho)} E[S(S-1)]. \tag{9.3.30}$$

where S is a service time r.v. with probability distribution $\{g_n\}$, $\rho = pES$, and p is the probability of an arrival during an interval.

Proof: We derive the expression given by Eq. (9.3.29) using the method given by Meisling (1958). Observe that if $A(s) = G(\bar{p} + ps)$, Eq. (9.3.17) becomes

$$\Pi_Q(s) = \frac{(1-\rho)(1-s)A(s)}{A(s) - s}.$$

Now $EQ_n = \Pi_Q^{(1)}(1)$, where $\Pi_Q^{(1)}(s) = (1-\rho)H(s)/K(s)$ with

$$H(s) = A(s)\{1 - A(s)\} - s(1-s)A^{(1)}(s)$$

and

$$K(s) = \{A(s) - s\}^2.$$

Since $H(1) = K(1) = 0$ we use L'Hospitals technique. But

$$H^{(1)}(s) = 2A^{(1)}(s)\{s - A(s)\} + s(1-s)A^{(2)}(s)$$

and

$$K^{(1)}(s) = 2\{A(s) - s\}\{A^{(1)}(s) - 1\},$$

implying that $H^{(1)}(1) = K^{(1)}(1) = 0$ so we need to consider second derivatives at $s = 1$.

$$H^{(2)}(s) = 2A^{(1)}(s)\{1 - A^{(1)}(s)\} - A^{(2)}(s)\{1 - 4s + 2A(s)\} - s(1-s)A^{(3)}(s),$$

$$K^{(2)}(s) = 2(A^{(1)}(s) - 1)^2 + 2\{A(s) - s\}A^{(2)}(s),$$

leading to

$$\Pi_Q^{(1)}(1) = \frac{(1-\rho)H^{(2)}(1)}{K^{(2)}(1)} = \frac{(1-\rho)[2A^{(1)}(1)\{1 - A^{(1)}(1)\} + A^{(2)}(1)]}{2\{1 - A^{(1)}(1)\}^2}.$$

Now, from Eq. (9.3.2), $A^{(1)}(1) = p\mathrm{E}S = \rho$ and $A^{(2)}(1) = p^2\mathrm{E}[S(S-1)]$. Substitution above leads to Eq. (9.3.29).

Furthermore, Eqs. (9.3.28) and (9.3.17) are related, since

$$\Pi_Q(s) = (\bar{p} + ps)\Pi_R(s),$$

so that

$$\Pi_Q^{(1)}(s) = p\Pi_R(s) + (\bar{p} + ps)\Pi_R^{(1)}(s),$$

implying that, by taking the limit as $s \uparrow 1$,

$$\mathrm{E}Q_n = p + \mathrm{E}R_n.$$

Equation (9.3.30) now follows from Eq. (9.3.29). □

Note that, in the notation of Theorem 9.3.5, from Eq. (9.3.27)

$$\mathrm{E}A_n = A^{(1)}(1) = \rho \qquad \text{and} \qquad \mathrm{E}B_n = B^{(1)}(1) = \rho - p$$

so that the initial terms in Eqs. (9.3.29) and (9.3.30) arise from the expected number of customers entering the system after a departure that has left the system empty.

In many queueing situations we are interested in the *queueing time* (the time spent in the system prior to servicing) or the *waiting time* (the total spent in the system, i.e., queueing time plus service time) of a typical customer.

Time in a discrete system has different interpretations. Concerning the waiting time we adopt the following conventions. For a system having late arrivals with immediate access we count the number of service time positions spent in the system, whereas for a system having late arrivals with delayed access we count the completed number of time slots spent in the system. For an early arrival system both of the above methods give the same waiting time. This means that in the delayed access case the service position immediately following an arrival (whether it arrives to an empty system or not) is not counted.

Let W_n denote the waiting time of the nth customer. Under "steady state" conditions let $w_k = \lim_{n \to \infty} P\{W_n = k\}$ so that $\{w_k\}$ $(k \geq 1)$ denotes the limiting probability distribution of the waiting time r.v.

THEOREM 9.3.8:   In the Geometric/G/1 model with interarrival times geometric (parameter p) and service times independently distributed with distribution g_n $(n \geq 1)$ and p.g.f. $G(s)$, the p.g.f. $W(s)$ of the limiting probability distribution of the waiting time is given by

$$W(s) = \frac{(1 - \rho)(1 - s)G(s)}{[pG(s) - s + \bar{p}]}, \tag{9.3.31}$$

provided $\rho < 1$ and $g_1 < 1$.

Proof: First observe that the theorem is stated without any restrictions on the arrival system. Initially, let us assume a late arrival system with delayed access.

We use an indirect argument. Observe that when a customer departs from the system he leaves behind him those customers that have arrived during his waiting time (because of the FIFO queue discipline). Conditioning upon W_n, we have that

$$P\{Q_n = k\} = \sum_{l=1}^{\infty} P\{Q_n = k \mid W_n = l\}P\{W_n = l\}, \qquad (9.3.32)$$

where

$$P\{Q_n = k \mid W_n = l\} = P\{k \text{ customers arrive during } l \text{ time slots}\}$$

$$= \binom{l}{k} p^k \bar{p}^{l-k} \qquad (k \le l),$$

since we have, effectively, l Bernoulli trials and we require the probability of k successes.

Under the conditions $\rho < 1$ and $g_1 < 1$, the MC $\{Q_n, n \ge 0\}$ is irreducible, aperiodic, and persistent nonnull (i.e., ergodic) so that $\lim_{n \to \infty} P\{Q_n = k\} = \pi_k$. Thus, taking the limit in Eq. (9.3.32),

$$\pi_k = \sum_{l=k}^{\infty} w_l \binom{l}{k} p^k \bar{p}^{l-k}.$$

Forming generating functions we obtain

$$\Pi_Q(z) = \sum_{k=0}^{\infty} \pi_k z^k = \sum_{k=0}^{\infty} \left(\sum_{l=k}^{\infty} w_l \binom{l}{k} p^k \bar{p}^{l-k} \right) z^k$$

$$= \sum_{l=0}^{\infty} w_l \left(\sum_{k=0}^{l} \binom{l}{k} (pz)^k \bar{p}^{l-k} \right)$$

$$= \sum_{l=0}^{\infty} w_l (\bar{p} + pz)^l = W(\bar{p} + pz).$$

Using Eq. (9.3.17) we deduce that

$$W(\bar{p} + pz) = \frac{(1 - \rho)(1 - z)G(\bar{p} + pz)}{G(\bar{p} + pz) - z},$$

and thus replacing $\bar{p} + pz$ by s, Eq. (9.3.31) follows after simplification.

To derive the results for the other arrival systems we have that

$$P\{R_n = k\} = \sum_{l=1}^{\infty} P\{R_n = k \mid W_n = l\}P\{W_n = l\},$$

where

$$P\{R_n = k \mid W_n = l\} = P\{k \text{ customers arrive during } l - 1 \text{ time slots}\}$$

$$= \binom{l-1}{k} p^k \bar{p}^{l-1-k} \qquad (k \le l - 1).$$

(A close examination of a sample path for the models under consideration will show that if customer n spends l service time positions in the system there are $l - 1$ possible arrival positions available.) Proceeding as earlier

$$\pi_k = \sum_{l=k+1}^{\infty} w_l \binom{l-1}{k} p^k \bar{p}^{l-1-k},$$

and forming generating functions (cf. the proof of Theorem 9.3.5) and using Eq. (9.3.28),

$$\Pi_R(z) = \frac{W(\bar{p} + pz)}{(\bar{p} + pz)} = \frac{(1-\rho)(1-z)G(\bar{p} + pz)}{(\bar{p} + pz)[G(\bar{p} + pz) - z]}.$$

The term $\bar{p} + pz$ cancels and the expression for $W(s)$ follows as before. $\square$

Since we have formally defined the waiting time as queueing time plus service time, it is easy to interpret the queueing time in the early arrival system and late arrival system with immediate access as the number of service positions spent in the system prior to servicing commencing. In the late arrival with delayed access case, however, a zero queueing time is assigned to a customer arriving to an empty system or to a customer arriving to a system containing only one unit which is on the verge of completing its service.

With such a convention, let $W_n^{(q)}$ be the queueing time of the nth customer and $w_k^{(q)} = \lim_{n \to \infty} P\{W_n^{(q)} = k\}$ the limiting distribution of the queueing time, under steady state conditions.

COROLLARY 9.3.8A: Under stationary conditions ($\rho < 1$, $g_1 < 1$) $w_k^{(q)} = \lim_{n \to \infty} P\{W_n^{(q)} = k\}$ exists and

$$W^{(q)}(s) \equiv \sum_{k=0}^{\infty} w_k^{(q)} s^k = \frac{(1-\rho)(1-s)}{pG(s) - s + \bar{p}}. \qquad (9.3.33)$$

Proof: $W_n = W_n^{(q)} + S_n$, where S_n is the service time of the nth customer, and since $W_n^{(q)}$ and S_n are independent r.v.'s

$$W(s) = \lim_{n \to \infty} E[s^{W_n}]$$

$$= \lim_{n \to \infty} E[s^{W_n^{(q)}}] E[s^{S_n}]$$

$$= W^{(q)}(s) G(s),$$

and Eq. (9.3.33) follows from Eq. (9.3.31). $\square$

COROLLARY 9.3.8B: Under stationary conditions

$$EW_n = ES + \frac{p}{2(1 - \rho)} E[S(S - 1)], \tag{9.3.34}$$

$$EW_n^{(q)} = \frac{p}{2(1 - \rho)} E[S(S - 1)], \tag{9.3.35}$$

where S is a service time r.v.

Proof: We need to separate out the two cases depending on which embedded process $\{Q_n\}$ or $\{R_n\}$ is appropriate. We leave the $\{R_n\}$ case as an exercise (Exercise 9.3.5).

Using the theory of conditional expectations

$$EQ_n = E[EQ_n|W_n = l],$$

where

$$E[Q_n|W_n = l] = \sum_{k=0}^{l} kP[Q_n = k|W_n = l]$$

$$= \sum_{k=0}^{l} k\binom{l}{k} p^k \bar{p}^{l-k} = lp,$$

so that

$$EQ_n = E(W_n p) = pEW_n.$$

Equation (9.3.34) now follows from Eq. (9.3.29) by noting $\rho = pES$.

Equation (9.3.35) follows from Eq. (9.3.34) by observing that $EW_n = EW_n^{(q)} + ES_n = EW_n^{(q)} + ES$.

(Alternatively, the expressions for the expectations can be found from their respective generating functions and a procedure similar to that used in deriving Theorem 9.3.7.) $\square$

EXAMPLE 9.3.4:   *The Geometric/Geometric/1 Model.* We make use of the results of Example 9.3.2 to derive expressions for the limiting distributions of the waiting time and queueing time r.v.'s and their expectations.

Since $G(s) = rs/(1 - \bar{r}s)$, it is easily seen that $E[S] = 1/r$, $E[S(S - 1)] = 2\bar{r}/r^2$, $\rho = p/r$. Let us use $\lambda \equiv \bar{r}/\bar{p}$ and $\alpha \equiv p\bar{r}/\bar{p}r = \lambda\rho$.

From Eqs. (9.3.31) and (9.3.33) we obtain

$$W(s) = \frac{(1 - \lambda)s}{1 - \lambda s} \quad \text{and} \quad W^{(q)}(s) = \frac{(1 - \lambda)(1 - \bar{r}s)}{(1 - \lambda s)r},$$

implying that

$$w_k = (1 - \lambda)\lambda^{k-1} \quad (k = 1, 2, \ldots),$$

showing that the waiting time is distributed, in the limit, as a geometric $(1 - \lambda)$ r.v., and that

$$w_k^{(q)} = \begin{cases} 1 - \lambda\rho & (k = 0), \\ \rho(1 - \lambda)\lambda^k & (k = 1, 2, \ldots). \end{cases}$$

Under stationary conditions, we also see that

$$\mathsf{E}W_n = \frac{\rho\bar{p}}{(1 - \rho)p} = \frac{1}{1 - \lambda} = \frac{1}{r(1 - \alpha)},$$

$$\mathsf{E}W_n^{(q)} = \frac{\rho\bar{r}}{(1 - \rho)r} = \frac{\alpha}{1 - \lambda} = \frac{\alpha}{r(1 - \alpha)}.$$

Note also that from Theorem 9.3.7 (or Examples 9.3.2 and 9.3.3)

$$\mathsf{E}Q_n = \frac{\rho\bar{p}}{1 - \rho} = \frac{p}{1 - \lambda} = \frac{\rho}{1 - \alpha},$$

$$\mathsf{E}R_n = \frac{\rho\bar{r}}{1 - \rho} = \frac{\lambda p}{1 - \lambda} = \frac{\alpha}{1 - \alpha}.$$

[Note: $1 - \rho = p(1 - \alpha)$ and $1 - \lambda = r(1 - \alpha)$.]   □

[Meisling (1958) derives similar expressions to $\mathsf{E}Q_n$ [Eq. (9.3.29)] and $\mathsf{E}W_n^{(q)}$ [Eq. (9.3.35)] and looks at the special case of geometric service times. However, his model permits a customer to be given a "zero" service on arrival—implying that such a customer, effectively, does not enter the service facility. Our models do not permit this phenomenon.]

We conclude this section with two interesting applications of branching chains to the determination of the essential characteristics of a *busy period*. A busy period starts when a customer arrives and finds the server free. It continues until the server eventually becomes free again, having served the initial customer and all others that have arrived during such an interval.

Our analysis assumes a Geometric/G/1 system with late arrivals having delayed access. It can be shown (see Exercise 9.3.6) that by making suitable modifications and redefinitions the results obtained can be adapted to models having alternative arrival systems. Figure 9.3.3 gives a typical sample path for a busy period for our model under consideration.

To examine the number of customers served in a busy period we construct a branching chain as follows. Suppose a customer arrives at time 0 to find the server free. Such an initial customer comprises the zeroth generation. The first generation consists of the customers arriving prior to or at the time of the termination of the initial customer's service time. If there are no such direct descendants the process stops. Otherwise the direct

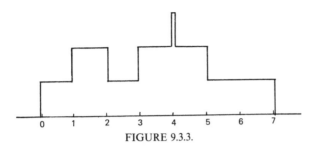

FIGURE 9.3.3.

descendants are served successively, and during their service times their direct descendants join the queue.

Figure 9.3.4 gives a realization for this branching chain constructed from the sample path given by Fig. 9.3.3.

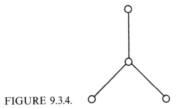

FIGURE 9.3.4.

Using the terminology of Section 8.1 observe that the offspring distribution has p.g.f. $P(s) = A(s) = G(ps + \bar{p})$, the p.g.f. for the number of customers that arrive during the servicing of a typical customer. Now, from Corollary 9.3.1A, $\mu = pE(S) = \rho$ and thus, from Corollary 8.1.4A, the branching chain is certain to die out, or equivalently the busy period is certain to terminate only if $\rho \le 1$. The total progeny consists of all the customers (including the initial customer) arriving during the busy period and has finite expectation only if $\rho < 1$. In other words, congestion is guaranteed when $\rho = 1$ and long queues will occur unless ρ is much less than 1, being consistent with the conditions for transience and persistence as given by Corollary 9.3.1A.

Another branching chain can be constructed by considering time points as elements of the process. Such an elegant device, which is due to Good (1951), enables us to look at the actual duration of the busy period. We say that time point n has no descendants if no customer arrives at that time point. If such a customer arrives and his service lasts for r time units, then time points $n + 1, \ldots, n + r$ are counted as direct descendants of the time point. Suppose that at time 0 the server is free.

A moment's reflection will show that Fig. 9.3.5 gives a realization for this chain for the busy period presented in Fig. 9.3.3.

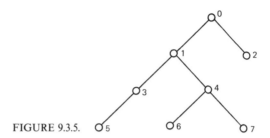

FIGURE 9.3.5.

At any time point we have no arrival, with probability $\bar{p}$, and thus no descendant, or an arrival occurs, with probability p, and such an arrival generates r descendents with probability g_r. Consequently the offspring distribution has p.g.f. $P(s) = \bar{p} + pG(s)$. The root ρ of $s = P(s)$ gives the probability of a termination of the busy period (which gives the same results as the aforementioned branching chain). The duration of the busy period consists of the total progeny (not including the first) and thus from Eq. (8.1.37) has p.g.f. $B(s)$ satisfying $B(s) = \bar{p} + pG(sB(s))$.

By embedding the Geometric/G/1 queue, with late arrivals having delayed access, using the $\{Z_n, n \geq 0\}$ process (i.e., the system size at time points $n+$), Bhat (1968) finds expressions for the joint probability distribution of $N_i = \min\{n: Z_n = 0 | Z_0 = i\}$ $(i \geq 1)$, and D_n, the number of customers served during an interval of length n. In particular he shows that

$$P\{N_i = n, D_n = k\} = \frac{i}{k} p^{k-i}\bar{p}^{n-k+i} \binom{n}{k-i} g_n^{(k)},$$

where $\{g_n^{(k)}\} = \{g_n\}^{k*}$ with $\{g_n^{(0)}\} \equiv \{\delta_{n0}\}$.

From this result the joint distribution and hence the marginal distributions of the length of and the number of customers served in a busy period can be obtained by setting $i = 1$.

Bhat also shows that, for $i \geq 0$,

$$P\{Z_n = 0 | Z_0 = i\} = \sum_{k=i}^{\infty} p^{k-i}\bar{p}^{n-k+i} \frac{1}{k-i} \binom{n-1}{k-i-1}$$

$$\times \sum_{n=1}^{\infty} \left[n - m\left(1 - \frac{i}{k}\right) \right] g_m^{(k)}.$$

Exercises 9.3

1. In the notation of Theorem 9.3.2 give a formal proof of the result that $T_{j,j-1}$ and $T_{j-1,j-2}$ are independent r.v.'s.

2. In the notation of Theorem 9.3.4, show that

$$\sum_{i=l}^{j} \alpha_{i-1,l-1} r_{j-i+1} = \alpha_{jl}$$

and hence fill in the details of the required inductive proof of that theorem.

3. Equation (9.3.3) can be expressed as $Q_{n+1} = Q_n + A_{n+1} - C_n$ where C_n is 0 or 1, according as Q_n is 0 or positive.

(a) By taking expectations of this equation show, under stationary conditions (in which case $EQ_{n+1} = EQ_n$) that

$$\pi_0 = P[Q_n = 0] = 1 - \rho.$$

(b) By squaring the equation and taking expectations derive (under stationary conditions) the expression for EQ_n given by Eq. (9.3.29).

4. Consider a Geometric/G/1/N + 1 queue where we impose the restriction of a finite waiting room of capacity N. Then all customers who arrive to find the waiting room full (namely $N + 1$ customers in the system) leave the system never to return. We shall assume that we have a late arrival system with delayed access. Let Q_n be the number of customers in the system just after the nth departure.

(a) Using the relationship [cf. Eq. (9.3.3) when $N = \infty$]

$$Q_{n+1} = \begin{cases} \min(Q_n - 1 + A_{n+1}, N), & Q_n = 1, 2, \ldots, N, \\ \min(A_{n+1}, N), & Q_n = 0, \end{cases}$$

show that $\{Q_n, n \geq 0\}$ is a MC with state space $S = \{0, 1, \ldots, N\}$ and transition matrix given by

$$P = \begin{bmatrix} k_0 & k_1 & k_2 & \cdots & k_{N-1} & r_{N-1} \\ k_0 & k_1 & k_2 & \cdots & k_{N-1} & r_{N-1} \\ 0 & k_0 & k_1 & \cdots & k_{N-2} & r_{N-2} \\ \vdots & \vdots & \vdots & & \vdots & \vdots \\ 0 & 0 & 0 & \cdots & k_0 & k_1 & r_1 \\ 0 & 0 & 0 & \cdots & 0 & k_0 & r_0 \end{bmatrix}$$

where k_n is the probability that exactly n arrivals occur during a service time and where $r_n = k_{n+1} + k_{n+2} + \cdots$ [cf. Eq. (9.3.1) when $N = \infty$].

(b) For a MC with transition matrix of the form given by P show that it is irreducible and aperiodic if $0 < k_0 \leq k_0 + k_1 < 1$ and that a stationary distribution exists irrespective of the value of $\rho = \sum_{j=1}^{\infty} j k_j$ (cf. Theorem 9.3.2).

(c)   Let $\pi_i^{(N)}$ ($i = 0, 1, \ldots, N$) be the stationary probabilities and let $\{\pi_i\}$ be the stationary distribution for the unlimited waiting room case. Show that

$$\pi_i^{(N)} = \frac{\pi_i}{\sum_{k=0}^{N} \pi_k}, \qquad i = 0, 1, \ldots, N.$$

Furthermore, $b_i \equiv \pi_i^{(N)}/\pi_0^{(N)} = \pi_i/\pi_0$ is independent of N and if $B(s) = \sum_{i=0}^{\infty} b_i s^i$ show that

$$B(s) = \frac{(1 - s)A(s)}{A(s) - s},$$

where

$$A(s) = \sum_{i=0}^{\infty} k_i s^i \qquad (= G(sp + \bar{p})).$$

(d)   Discuss the changes required to the above analysis when we have different arrival systems (cf. Theorem 9.3.5).

5. Prove Corollary 9.3.8B for the case of a system with early arrivals or late arrivals with immediate access.

6. By redefining a "busy period" and/or the service time distribution show how the analysis of Section 9.3 pertaining to busy periods can be adapted to the Geometric/G/1 model with (a) early arrival system and (b) late arrival system with immediate access.

7. Let $R(s)$ be the g.f. of the total number of customers served during a busy period (initiated by a single customer) in a Geometric/G/1 queue with late arrivals having delayed access. If the service time distribution has p.g.f. $G(s)$ show using
(a)   branching chain results and
(b)   first passage time arguments that

$$R(s) = sG(\bar{p} + pR(s)).$$

8. The following is a simple model for the storage of a disposable commodity. The system commences with an initial supply of items. At each time instant the item at the head of the line in the store either remains at that position with probability p or is disposed of with probability $\bar{p} = 1 - p$. At the time instant following the complete depletion of the stock, j items ($j = 0, 1, 2, \ldots$) are added to the store with probability c_j.
(a)   Show that this system can be represented as an irreducible MC with transition probabilities given by $p_{0j} = c_j$ ($j = 0, 1, 2, \ldots$), $p_{ii} = p(i = 1, 2, \ldots)$ and $p_{i, i-1} = \bar{p}$ ($i = 1, 2, \ldots$).
(b)   Find the g.f. $F_{i0}(s)$ for the first passage time T_{i0} from state i to state 0.

(c)   Show that the MC is persistent.

(d)   Show that the MC is persistent nonnull or persistent null according to whether the mean of the distribution $\{c_j\}$ is finite or infinite.

9.4   The GI/Geometric/1 Model

In this section we consider a queueing model that is, in some way, the "dual" to the Geometric/G/1 model. By interchanging the role played by the geometric distribution as a service time distribution (as in the Geometric/G/1 model) to an interarrival time distribution and by observing the system at instants just prior to an arrival joining the system instead of just after the departure of a customer we are able to, once again, detect the presence of an embedded homogeneous Markov chain. It is, of course, the property that geometric service times can be regarded as being generated by a sequence of independent Bernoulli trials, with its associated lack of memory property, that enables us to make such a deduction. For any other service pattern the number of service completions during an interarrival interval will depend on the amount of service time already expended by the item (if any) in the service facility at the time of arrival of the customer initiating such an interval.

Our development will parallel that presented in Section 9.3 with appropriate changes and modifications.

Let

$$U_n \equiv \text{Number of customers in the GI/Geometric/1 system}$$
$$\text{immediately } prior \text{ to the arrival of the } n\text{th customer.}$$

The sample path for such an embedded process will depend on the type of arrival system we are assuming. If Fig. 9.2.2 gives typical sample paths for a GI/Geometric/1 system, realizations of the $\{U_n, n \geq 0\}$ process for the model under the various arrival type assumptions (appropriately identified) are as follows:

Early arrival system $\{U_n^{(e)}\}$:	0 1 0 0 1 1 0
Late arrival system with immediate access $\{U_n^{(i)}\}$:	0 1 0 0 1 1 0
Late arrival system with delayed access $\{U_n^{(d)}\}$:	0 1 1 0 1 2 0

The $\{U_n^{(e)}\}$ and $\{U_n^{(i)}\}$ processes have the same sample paths and hence can be regarded as equivalent processes, different than the $\{U_n^{(d)}\}$ process. To eliminate the superscripts, henceforth we shall use

$$U_n \equiv U_n^{(e)} = U_n^{(i)} \quad \text{and} \quad V_n \equiv U_n^{(d)}.$$

We shall first examine the $\{U_n, n \geq 0\}$ process.

THEOREM 9.4.1:   In the GI/Geometric/1 queueing model let the inter-arrival times have probability distribution $\{f_n\}$ $(n \geq 1)$ and let the service times be distributed independently according to a geometric (parameter r) distribution. Let the arrivals be generated according to an early arrival system or a late arrival system with immediate access and let

$$U_n = \text{Number of customers in the system just prior to}$$
$$\text{the arrival of the } n\text{th customer.}$$

Then $\{U_n, n \geq 0\}$ is a homogeneous MC with transition matrix given by

$$P_U = \begin{bmatrix} r_0 & k_0 & 0 & 0 & \cdots \\ r_1 & k_1 & k_0 & 0 & \cdots \\ r_2 & k_2 & k_1 & k_0 & \cdots \\ r_3 & k_3 & k_2 & k_1 & \cdots \\ \vdots & \vdots & \vdots & \vdots & \ddots \end{bmatrix}, \tag{9.4.1}$$

where, if $\bar{r} = 1 - r$, for $j = 0, 1, 2, \ldots$

$$k_j = \sum_{k=j}^{\infty} f_k \binom{k}{j} r^j \bar{r}^{k-j} \tag{9.4.2}$$

and

$$r_j = 1 - \sum_{i=0}^{j} k_i.$$

Furthermore, $\{k_j\}$ forms a probability distribution with probability generating function given by

$$C(s) = \sum_{j=0}^{\infty} k_j s^j = F(rs + \bar{r}),$$

where $F(s)$ is the p.g.f. of the interarrival times $\{T_i\}$.

Proof: First observe that, irrespective of the arrival system used,

$$U_{n+1} = U_n + 1 - C_n \qquad (U_n \geq 0, 0 \leq C_n \leq U_n + 1)$$

where C_n is the number of customers served during the interarrival time between the nth and the $(n + 1)$th arrivals. Thus the Markov property holds and if

$$p_{ij} = P\{U_{n+1} = j | U_n = i\},$$

then

$$p_{ij} = \begin{cases} P\{C_n = i - j + 1 | U_n = i\}, & i \geq 0, 0 \leq j \leq i + 1, \\ 0, & \text{otherwise.} \end{cases}$$

To evaluate these probabilities we have to take into consideration the arrival system being used, and in what follows we shall assume an early arrival system. A moments reflection will show that the same analysis will hold for the late arrival system with immediate access.

Note also that we need to treat the cases $j = 0$ and $j > 0$ separately. When $U_{n+1} = 0$ the system is empty just prior to the arrival of the $(n + 1)$th customer so that the server may have been idle for a portion of the interarrival interval between the nth and $(n + 1)$th customers (of duration T_n). Consequently it is not sufficient to say that $i - j + 1 = i + 1$ are served during an interval of length T_n as they could have been served in less time than T_n.

In Fig. 9.4.1 we give typical sample paths for this queueing model during an interarrival interval with $T_n = 3$ and under the assumption that $U_{n+1} > 0$. We consider separately the cases $U_n \geq 1$ and $U_n = 0$.

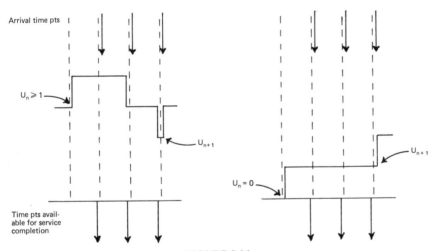

FIGURE 9.4.1.

Observe that in both cases ($U_n = 0$ and $U_n \geq 1$) if the interarrival interval consists of k units of time and the server is busy during this entire interval (equivalent to $U_{n+1} > 0$), then there are k servicing positions available.

Let us use C_n^+ to denote the r.v. C_n under the assumption that the server is operating continuously during the entire interarrival interval.

Thus, by the property of geometric service times, services are generated by a Bernoulli sequence, and if we are given that $T_n = k$ the probability that $C_n^+ = l$ is equivalent to the probability of l successes in k Bernoulli trials with probability of success r. Hence for $l = 0, 1, 2, \ldots$

$$k_l = P\{C_n^+ = l\}$$

$$= \sum_{k=l}^{\infty} P\{C_n^+ = l \mid T_n = k\} P\{T_n = k\}$$

$$= \sum_{k=l}^{\infty} f_k \binom{k}{l} r^l \bar{r}^{k-l},$$

leading to Eq. (9.4.2). The expression $C(s)$ for the g.f. of the sequence $\{k_j\}$ follows by an analogous proof to that given in Theorem 9.3.1 for $A(s)$ but with p replaced by r and g_k by f_k. Thus

$$C(s) = F(1 - r + rs),$$

and since $C(1) = F(1) = 1$, $\{k_j\}$ forms a probability distribution.

Now, for $0 < j \le i + 1$, $i \ge 0$ (which implies $U_{n+1} > 0$),

$$p_{ij} = P\{C_n = i - j + 1 \mid U_n = i\} = P\{C_n^+ = i - j + 1\} = k_{i-j+1}.$$

We have thus determined all the transition probabilities except those when $U_{n+1} = 0$, namely p_{i0} for $i \ge 0$.

Quite simply, however,

$$p_{i0} = 1 - \sum_{j=1}^{i+1} p_{ij} = 1 - \sum_{j=1}^{i+1} k_{i-j+1} = r_i,$$

and the transition matrix given by Eq. (9.4.1) follows. (For a direct evaluation of p_{i0} see Exercise 9.4.1.)   □

Let us now examine the structure of MC's having a transition matrix of the form given by Eq. (9.4.1). The results given by Theorem 9.4.2, which follows, are similar to those given in Theorem 9.3.2 for the "dual" model. Note, however, the different definition for ρ.

THEOREM 9.4.2:   Let $\{U_n, n \ge 0\}$ be a homogeneous MC with transition matrix given by Eq. (9.4.1) where $\{k_j\}$ is a probability distribution and $r_j = 1 - \sum_{i=0}^{j} k_i$.

(a) $\{U_n, n \ge 0\}$ is an irreducible MC if and only if $0 < k_0 \le k_0 + k_1 < 1$.
(b)   If $\{U_n, n \ge 0\}$ is irreducible, it is aperiodic.
(c)   Under the conditions of irreducibility (and aperiodicity) the MC $\{U_n, n \ge 0\}$ is

   (i)   transient if $\rho > 1$,
   (ii)   persistent null if $\rho = 1$, and
   (iii) persistent nonnull if $\rho < 1$,

where $\rho^{-1} = \sum_{j=1}^{\infty} j k_j$.

Proof: (a)   If $k_0 = 0$, then $r_0 = 1$ and state 0 is absorbing.

If $k_0 + k_1 = 1$, then $k_i = 0$ for $i \geq 2$ and $r_i = 0$ for $i \geq 1$, implying that state j is never reached from state i for all $0 \leq i \leq j$.

Conversely, if $k_0 > 0$ and $k_0 + k_1 < 1$, then, for some $r \geq 2$, $k_r > 0$. These conditions are sufficient to establish the irreducibility. In particular $k_0 > 0$ implies that for $j > i$ we can construct a chain $i \to i + 1 \to \cdots \to j$ with positive probability $(\geq k_0^{j-i})$, and state j can be reached from state i. Similarly we can construct suitable paths to show that this same conclusion holds for $i > j$ and $i = j$. (See Exercise 9.4.2.)

(b)   Since the irreducibility conditions imply that $r_0 > 0$, state 0 is obviously aperiodic and hence the chain is aperiodic.

(c)   We first use Theorem 7.1.10 to show that if $\rho < 1$ the chain is persistent nonnull (and hence ergodic). Consider the equations $\sum_{i=0}^{\infty} x_i p_{ij} = x_j$ $(j \geq 0)$ and let $x_i = s^i$. Then, for $j \geq 1$,

$$\sum_{i=j-1}^{\infty} s^i k_{i-j+1} = s^j$$

or

$$\sum_{i=j-1}^{\infty} s^{i-j+1} k_{i-j+1} = s$$

or

$$C(s) \equiv \sum_{n=0}^{\infty} k_n s^n = s. \tag{9.4.3}$$

For the case $j = 0$, if $0 < s < 1$,

$$\begin{aligned}
\sum_{i=0}^{\infty} s^i p_{i0} &= \sum_{i=0}^{\infty} \left(\sum_{j=i+1}^{\infty} k_j \right) s^i = \sum_{j=1}^{\infty} \sum_{i=0}^{j-1} k_j s^i \\
&= \sum_{j=1}^{\infty} k_j \left(\frac{1 - s^j}{1 - s} \right) = \frac{1}{1 - s} \left(1 - k_0 - \sum_{j=1}^{\infty} k_j s^j \right) \\
&= \frac{1}{1 - s} (1 - k_0 - C(s) + k_0) = \frac{1 - C(s)}{1 - s}. \tag{9.4.4}
\end{aligned}$$

Now consider Eq. (9.4.3). Since $C(0) = k_0 > 0$ and $C(1) = 1$, if $C^{(1)}(1) = \sum_{n=1}^{\infty} n k_n = \rho^{-1} > 1$, by Lemma 8.1.3(d) there exists an α $(0 < \alpha < 1)$ such that $C(\alpha) = \alpha$. Consequently Eq. (9.4.3) is satisfied by taking $s = \alpha$ as is Eq. (9.4.4) since $0 < \alpha < 1$. Furthermore $x_i \neq 0$ and $\sum_{i=0}^{\infty} |x_i| = \sum_{i=0}^{\infty} \alpha^i = (1 - \alpha)^{-1} < \infty$ and the result follows by Theorem 7.1.10 (Foster's theorem).

We now use the same theorem to show that if $\rho \geq 1$, then the system is either transient or persistent null. Under this condition, $x_i \equiv 1$ is a solution of the inequalities $\sum_{i=0}^{\infty} x_i p_{ij} \leq x_j$ $(j \geq 0)$, since [cf. the derivation of Eqs.

(9.4.3) and (9.4.4)],

$$\sum_{i=0}^{\infty} p_{ij} = \sum_{i=j-1}^{\infty} k_{i-j+1} = \sum_{n=0}^{\infty} k_n = 1 \leq 1 \qquad (j \geq 1),$$

$$\sum_{i=0}^{\infty} p_{i0} = \sum_{j=1}^{\infty} \sum_{i=0}^{j-1} k_j = \sum_{j=1}^{\infty} j k_j = \rho \leq 1 \qquad (j = 0).$$

Since $\sum x_i = \infty$, Theorem 7.1.10 implies that the MC cannot be ergodic and hence it must be either transient or persistent null. Consequently the MC is persistent nonnull if and only if $\rho < 1$.

The required conclusion of (c) will follow if we establish that the system is transient if and only if $\rho > 1$. To make such a deduction we shall use Theorem 7.1.11. Thus we now prove that the system of equations $\sum_{j=0}^{\infty} p_{ij} y_j = y_i$ $(i \geq 1)$ has a nonconstant bounded solution if and only if $\rho > 1$. Since this system obviously admits a constant solution (e.g., $y_j = 1$) we may, without loss of generality, suppose that $y_0 = 0$. Under this assumption the system reduces to

$$\sum_{j=1}^{i+1} k_{i-j+1} y_j = \sum_{j=0}^{i+1} k_{i+1-j} y_j = y_i \qquad (i \geq 1)$$

or, in array form,

$$k_2 y_0 + k_1 y_1 + k_0 y_2 = y_1$$

$$k_3 y_0 + k_2 y_1 + k_1 y_2 + k_0 y_3 = y_2$$

$$\vdots$$

$$k_{i+1} y_0 + k_i y_1 + \cdots + k_1 y_i + k_0 y_{i+1} = y_i$$

$$\vdots$$

Define $Y(s) = \sum_{i=0}^{\infty} y_i s^i$ and $C(s) = \sum_{i=0}^{\infty} k_i s^i$; then, by observing the convolution form of the system from the array, if we multiply the ith equation by s^{i+1} and sum we obtain

$$Y(s)C(s) - sk_0 y_1 = sY(s)$$

or that

$$Y(s) = \frac{sk_0 y_1}{C(s) - s}, \qquad (9.4.5)$$

provided $C(s) \neq s$.

Note that the irreducibility conditions require $k_1 < 1$ and $k_0 > 0$, and since $C^{(1)}(1) = \sum_{n=0}^{\infty} n k_n = \rho^{-1}$ Lemmas 8.1.3(b) and 8.1.3(d) imply that if $\rho < 1$, $C(s) = s$ for some $s \in (0, 1)$, while if $\rho \geq 1$, $C(s) \neq s$ for $s \in [0, 1)$.

(Thus if $\rho < 1$, $Y(s)$ cannot have bounded coefficients since in this case $Y(s)$ would then converge for every $s \in [0, 1]$. Consequently the system is persistent, by Theorem 7.1.11.)

Under the assumption that $\rho \geq 1$,

$$C(s) - s = (1 - s)\left[1 - \frac{1 - C(s)}{1 - s}\right] \equiv (1 - s)[1 - R(s)],$$

where, from Theorem 2.4.1, $R(s) = \sum_{n=0}^{\infty} r_n s^n$ with $r_n = \sum_{i=n+1}^{\infty} k_i = 1 - \sum_{i=0}^{n} k_i$ ("tail" probabilities for the $\{k_n\}$ sequence). Furthermore, from Theorem 2.4.3,

$$\sum_{n=0}^{\infty} r_n = \sum_{n=0}^{\infty} n k_n = \rho^{-1} \leq 1.$$

Thus, for $|s| < 1$,

$$\left|\frac{1 - C(s)}{1 - s}\right| = |R(s)| \leq \sum_{n=0}^{\infty} r_n \leq 1,$$

implying that $|C(s) - s| \neq 0$ and that the power series expansion

$$\frac{1 - s}{C(s) - s} = \frac{1}{1 - R(s)} = \sum_{n=0}^{\infty} \{R(s)\}^n \equiv U(s) = \sum_{n=0}^{\infty} u_n s^n$$

is valid for $|s| < 1$ with coefficients $u_n \geq 0$. Application of Theorem 2.3.2 (Abel's theorem) now yields

$$\sum_{n=0}^{\infty} u_n = \lim_{s \uparrow 1} \frac{1 - s}{C(s) - s} = \lim_{s \uparrow 1} \frac{1}{1 - R(s)} = \frac{1}{1 - \rho^{-1}} = \begin{cases} \dfrac{\rho}{\rho - 1}, & \rho > 1, \\ \infty, & \rho = 1. \end{cases}$$

Now, from Eq. (9.4.5), for $\rho > 1$,

$$Y(s) = \frac{s k_0 y_1}{(1 - s)(1 - R(s))} = \frac{s k_0 y_1 U(s)}{(1 - s)} \equiv s k_0 y_1 V(s),$$

where, if

$$V(s) = \sum_{n=0}^{\infty} v_n s^n,$$

then

$$v_n = \sum_{k=0}^{n} u_k \leq \sum_{k=0}^{\infty} u_k = \frac{\rho}{\rho - 1}.$$

Thus $y_{n+1} = k_0 y_1 v_n$ and the sequence $\{y_n\}$ is bounded and nonconstant (since $y_0 = 0$, y_1 arbitrary) if and only if $\rho > 1$. Theorem 7.1.11 now gives the required conclusion.

The approach we have used to classify the states has been based upon indirect arguments. The traditional approach would be to concentrate on one particular state, say 0, and deduce the results from the sequence $\{f_{00}^{(n)}\}$. However, the derivation of this first passage time distribution is a little complicated. We outline a derivation below (with the reader to fill in extra details in Exercise 9.4.3) using the concept of "taboo probabilities" as briefly mentioned in Exercise 6.2.4.

Let $l_{ij}^{(n)} = P\{U_n = j, U_k \neq i$ for $k = 1, \ldots, n-1 | U_0 = i\}$ $(n \geq 1)$, and in particular consider $l_{0j}^{(n)}$, the "zero avoiding" transition probabilities. We first deduce the following results:

(i) $\{l_{i,i+1}^{(n)}\} = \{l_{01}^{(n)}\} = \{f_{10}^{(n)}\}$ $(i \geq 0)$,

(ii) $\{l_{i,i+2}^{(n)}\} = \{l_{i,i+1}^{(n)}\} * \{l_{i+1,i+2}^{(n)}\}$ $(i \geq 0)$,

(iii) $\{l_{0j}^{(n)}\} = \{l_{01}^{(n)}\}^{j*}$ $(j \geq 1)$,

(iv) $\{l_{0j}^{(n)}\} = \{f_{j0}^{(n)}\}$ $(j \geq 1)$,

where $\{f_{j0}^{(n)}\}$ $(j \geq 1)$ is the n-step first passage time distribution for the queueing chain $\{Q_n, n \geq 0\}$, as given in Theorem 9.3.2.

To establish (i) observe that state $i + 1$ can never be approached from state i (except initially) because state i is to be avoided. Thus

$$l_{i,i+1}^{(1)} = p_{i,i+1} = k_0,$$

$$l_{i,i+1}^{(2)} = p_{i,i+1}p_{i+1,i+1} = k_0 k_1,$$

and for $n \geq 3$,

$$l_{i,i+1}^{(n)} = P\{U_1 = i+1, U_2 \geq i+1, \ldots, U_{n-1} \geq i+1, U_n = i+1 | U_0 = i\},$$

$$= \sum_{r_1 \geq i+1} \cdots \sum_{r_{n-2} \geq i+1} p_{i,i+1}p_{i+1,r_1} \cdots p_{r_{n-3},r_{n-2}}p_{r_{n-2},i+1},$$

$$= \sum_{r_1 \geq i+1} \cdots \sum_{r_{n-2} \geq i+1} k_0 k_{i+1-r_1+1} \cdots k_{r_{n-3}-r_{n-2}+1}k_{r_{n-2}-i},$$

$$= \sum_{t_1 \geq 0} \cdots \sum_{t_{n-2} \geq 0} k_0 k_{1-t_1} \cdots k_{t_{n-3}-t_{n-2}+1}k_{t_{n-2}+1},$$

$$= \sum_{s_1 \geq 0} \cdots \sum_{s_{n-2} \geq 0} k_0 k_{1-s_{n-2}} \cdots k_{s_2-s_1+1}k_{s_1+1}, \tag{9.4.6}$$

where, successively $t_k = r_k - i - 1$ and $s_k = t_{n-k-1}$.

Observe that Eq. (9.4.6) does not depend on i and is identical in form to Eq. (9.3.4) of Theorem 9.3.2.

Result (ii) is left as an exercise with the hint that since $i + 2$ must be reached by passing through $i + 1$ we can condition upon the last time state $i + 1$ is visited. The convolution form follows by the Markovian structure.

Results (iii) and (iv) are now direct consequences of the earlier results by considering the passage from 0 to j via states $1, 2, \ldots, j-1$.

To find $\{f_{00}^{(n)}\}$ for the $\{U_n, n \geq 0\}$ MC observe now that

$$f_{00}^{(1)} = p_{00} = r_0,$$

$$f_{00}^{(n+1)} = \sum_{j=1}^{\infty} l_{0j}^{(n)} p_{j0} = \sum_{j=1}^{\infty} r_j l_{0j}^{(n)} \qquad (n \geq 1).$$

Using result (iv) and Eq. (9.3.6) we have that $\sum_{n=1}^{\infty} l_{0j}^{(n)} s^n = [F(s)]^j$, where $F(s)$ is the generating function of the sequence $\{l_{01}^{(n)}\}$. Thus

$$
\begin{aligned}
F_{00}(s) &= r_0 s + \sum_{n=1}^{\infty} \left(\sum_{j=1}^{\infty} r_j l_{0j}^{(n)} \right) s^{n+1} \\
&= r_0 s + s \sum_{j=1}^{\infty} r_j [F(s)]^j \\
&= s R(F(s)) \\
&= \frac{s[1 - C(F(s))]}{1 - F(s)} = \frac{s - F(s)}{1 - F(s)},
\end{aligned}
\tag{9.4.7}
$$

where we have used the result that $F(s) = sC(F(s))$, as given by Eq. (9.3.8) with $A(s)$ taken as $C(s)$.

Concerning the properties of $F(s)$ observe that replacing ρ by $\rho^{-1} = \sum n k_n$ we deduce from Theorem 9.3.2 that

$$
F(1) = \begin{cases} 1, & \rho \geq 1, \\ \alpha, & \rho < 1, \end{cases}
\tag{9.4.8}
$$

where α is the smallest root (< 1) of the equation $s = C(s)$ in the interval $[0, 1]$. Furthermore, for $\rho \geq 1$, we have that

$$
F^{(1)}(1) = \begin{cases} \dfrac{\rho}{\rho - 1}, & \rho > 1, \\[2mm] \infty, & \rho = 1. \end{cases}
\tag{9.4.9}
$$

With these results we have, from Eq. (9.4.7),

$$
f_{00} = \lim_{s \uparrow 1} F_{00}(s) = \begin{cases} \dfrac{1 - F(1)}{1 - F(1)} = 1 & \text{if } F(1) < 1, \\[3mm] \lim_{s \uparrow 1} \dfrac{1 - F^{(1)}(s)}{-F^{(1)}(s)} & \text{if } F(1) = 1. \end{cases}
$$

Using Eqs. (9.4.8) and (9.4.9) we have, after simplification, that

$$f_{00} = \begin{cases} 1, & \rho \leq 1, \\ \dfrac{1}{\rho}, & \rho > 1, \end{cases}$$

showing that state 0, and hence the MC, is transient if $\rho > 1$ and persistent if $\rho \leq 1$. In this latter case the nature of the mean recurrence time of state 0, μ_{00}, will give the required distinction between the null and nonnull cases.

From Theorems 2.4.1 and 2.4.3, and Eq. (9.4.7),

$$\mu_{00} = \lim_{s\uparrow 1} \frac{1 - F_{00}(s)}{1 - s} = \lim_{s\uparrow 1} \frac{1}{1 - F(s)} = \begin{cases} \dfrac{1}{1 - \alpha}, & \rho < 1, \\ \infty, & \rho = 1, \end{cases}$$

and the classification of the MC is complete. □

The following two theorems hold for MC's $\{U_n, n \geq 0\}$ having a transition matrix P_U as given by Eq. (9.4.1). They are analogs to Theorems 9.3.3 and 9.3.4.

*THEOREM 9.4.3: Let $\{U_n, n \geq 0\}$ be a MC with transition matrix given by Eq. (9.4.1) where $\{k_j\}$ is a probability distribution with p.g.f. $C(s)$ and $r_j = \sum_{i=j+1}^{\infty} k_i$ $(j \geq 0)$.

If $p_{ij}^{(n)} = P\{U_n = j \mid U_0 = i\}$, then for $0 < s < 1$, $|z| \leq 1$,

$$\sum_{n=0}^{\infty} \sum_{i=0}^{\infty} p_{ij}^{(n)} s^n z^i = \frac{(1-s)(1-z)z^{j+1} + s(z - C(z))(1 - F(s))[F(s)]^j}{(1-s)(1-z)(z - sC(z))}, \quad (9.4.10)$$

where $z = F(s)$ is the unique positive root of $z = sC(z)$ for $s \in (0, 1]$.

*Proof: By Theorem 5.1.2(b),

$$p_{ij}^{(n+1)} = \sum_{l=0}^{\infty} p_{il} p_{lj}^{(n)} = \sum_{l=1}^{i+1} k_{i-l+1} p_{lj}^{(n)} + r_i p_{0j}^{(n)}$$

$$= \sum_{m=0}^{i} k_m p_{i+1-m,j}^{(n)} + r_i p_{0j}^{(n)}. \quad (9.4.11)$$

For $|z| \leq 1$, let $P_j^{(n)}(z) = \sum_{i=0}^{\infty} p_{ij}^{(n)} z^i$. Then from Eq. (9.4.11) we have, for $n \geq 0$,

$$z P_j^{(n+1)}(z) - C(z) P_j^{(n)}(z) = \frac{\{z - C(z)\}}{1 - z} p_{0j}^{(n)}, \quad (9.4.12)$$

where, clearly, $P_j^{(0)}(z) = z^j$. Further, if we introduce the g.f.

$$P_j(z, s) = \sum_{n=0}^{\infty} P_j^{(n)}(z) s^n = \sum_{n=0}^{\infty} \sum_{i=0}^{\infty} p_{ij}^{(n)} z^n s^i,$$

which is convergent for $|z| \leq 1$ and $|s| < 1$, then by Eq. (9.4.12) we obtain that

$$P_j(z, s) = \left(z^{j+1} + s\frac{z - C(z)}{1 - z}\sum_{n=0}^{\infty} p_{0j}^{(n)}s^n\right)\Big/(z - sC(z)). \qquad (9.4.13)$$

Observe that $\sum_{n=0}^{\infty} p_{0j}^{(n)}s^n = P_{0j}(s)$ $(j \geq 0)$. From Theorem 6.2.5 and Eq. (9.4.7) we have that

$$P_{00}(s) = \frac{1}{1 - F_{00}(s)} = \frac{1 - F(s)}{1 - s}.$$

By using the result of Exercise 6.2.4(d), or by showing $\{p_{0j}^{(n)}\} = \{p_{00}^{(n)}\} * \{l_{0j}^{(n)}\}$, we have that

$$P_{0j}(s) = P_{00}(s)L_{0j}(s) = \frac{1 - F(s)}{1 - s}[F(s)]^j. \qquad (9.4.14)$$

Substitution of Eq. (9.4.14) in Eq. (9.4.13) leads to the stated result, Eq. (9.4.10).

[Alternatively, $P_j(z, s)$ is a differentiable function of z if $|z| < 1$ and $|s| < 1$. The denominator of Eq. (9.4.13) has one and only one root $z = F(s)$ in the region $|z| < 1$. Consequently $z = F(s)$ must be a root of the numerator of Eq. (9.4.13) too. This observation gives a further derivation of Eq. (9.4.14).]   □

THEOREM 9.4.4:   Let $\{U_n, n \geq 0\}$ be an irreducible persistent nonnull MC with transition matrix given by Eq. (9.4.1), where $\{k_j\}$ is a probability distribution with p.g.f. $C(s)$ (and with $0 < k_0 \leq k_0 + k_1 < 1$, $r_j = \sum_{i=j+1}^{\infty} k_i$ and $\rho^{-1} = \sum_{j=1}^{\infty} jk_j > 1$.)

For such a MC the stationary distribution $\{\pi_j\}$ exists and is given by

$$\pi_j = (1 - \alpha)\alpha^j, \qquad j \geq 0, \qquad (9.4.15)$$

where α $(0 < \alpha < 1)$ is the unique root of the equation $s = C(s)$, which lies in the region $(0, 1)$.

Proof: By Theorem 9.4.2, the conditions are necessary and sufficient for the existence of a stationary distribution $\{\pi_j\}$, which is the unique solution of the equations $\sum_{i=0}^{\infty} \pi_i p_{ij} = \pi_j$ $(j \geq 0)$, subject to $\sum_{j=0}^{\infty} \pi_j = 1$.

In the proof of Theorem 9.4.2(c) we showed that, for $\rho < 1$, the equations $\sum_{i=0}^{\infty} x_i p_{ij} = x_j$ $(j \geq 0)$ have a solution $x_i = \alpha^i$ where α is as stated above. Thus $\pi_j = A\alpha^j$ where A is determined so that $\{\pi_j\}$ is a probability distribution, i.e., $A = 1 - \alpha$. Since there cannot be a second solution, Eq. (9.4.15) gives the required stationary distribution.   □

An interesting observation is that for all MC's having a transition matrix with the structure as given by Eq. (9.4.1) the stationary distribution (when it exists) is always a geometric distribution. The particular form of the p.g.f. of $\{k_j\}$, $C(s)$, determines the parameter of the distribution.

Let us now return to the $\{U_n, n \geq 0\}$ process embedded in the GI/Geometric/1 model and summarize the results so far derived as they pertain to this system.

COROLLARY 9.4.1A: Let $\{U_n, n \geq 0\}$ be the MC with transition matrix given by Eq. (9.4.1) embedded in the GI/Geometric/1 queueing model at instants just prior to the arrival of successive customers. Let $\{f_n\}$ $(n \geq 1)$ be the interarrival time distribution with p.g.f. $F(s)$. The service time distribution is geometric (parameter r).

(a)   The MC $\{U_n, n \geq 0\}$ is irreducible if and only if $f_1 < 1$.
(b)   If $f_1 < 1$ and $\rho^{-1} = r \sum_{n=1}^{\infty} nf_n$, then the MC is

   (i)   transient if $\rho > 1$,
   (ii)   persistent null if $\rho = 1$, and
   (iii)   persistent nonnull if $\rho < 1$.

(c)   When $f_1 < 1$ and $\rho < 1$, the stationary distribution $\{\pi_j\}$ is given by

$$\pi_j = (1 - \alpha)\alpha^j, \qquad j \geq 0, \tag{9.4.16}$$

where α $(0 < \alpha < 1)$ is the unique root of the equation $s = F(rs + \bar{r})$ that lies in the region $(0, 1)$.

Proof: (a)   Observe that Eq. (9.4.2) giving expressions for k_j are identical to the expressions for k_j as given in Theorem 9.3.1 with g_k replaced by f_k and p replaced by r. Since the irreducibility conditions for the $\{Q_n, n \geq 0\}$ MC of the type embedded in the Geometric/G/1 (as in Theorem 9.3.2) are identical to the conditions given in Theorem 9.4.2, the proof of Corollary 9.3.1A(a) can be adapted to give the required conditions.
   (b) $\rho^{-1} = \sum_{n=1}^{\infty} nk_n$ and the result follows as in the proof of Corollary 9.3.1A(b).
   (c)   This follows Theorems 9.4.4 and 9.4.1 since $C(s) = F(rs + \bar{r})$.   □

EXAMPLE 9.4.1:   *The Geometric/Geometric/1 Model.*   We examine the $\{U_n, n \geq 0\}$ process for the Geometric/Geometric/1 model with early arrival system or late arrival system with immediate access.
   In this case $f_n = \bar{p}^{n-1}p$ $(n \geq 1)$ so that if $f_1 = p < 1$ the MC is irreducible.
   Since $\sum_{n=1}^{\infty} nf_n = 1/p$, $\rho^{-1} = r/p$, or $\rho = p/r$. Thus if $p < r$, a stationary distribution exists and is given by $\pi_j = (1 - \alpha)\alpha^j$ $(j \geq 0)$, where α is the unique root (< 1) of $s = F(sr + \bar{r})$.
   Since

$$F(s) = \frac{ps}{1 - \bar{p}s},$$

$$F(sr + \bar{r}) = \frac{p(sr + \bar{r})}{1 - \bar{p}r - \bar{p}rs},$$

and thus $s = F(sr + \bar{r})$ if and only if $(\bar{p}rs - p\bar{r})(s - 1) = 0$, implying that $\alpha = p\bar{r}/\bar{p}r$.

Thus the $\{U_n, n \geq 0\}$ MC has the same stationary distribution as the $\{Y_n, n \geq 0\}$ process considered for the Geometric/Geometric/1 model in Corollary 9.2.2A. Observe, however, that whereas the $\{Y_n, n \geq 0\}$ process is observed systematically unit time slots apart, the $\{U_n, n \geq 0\}$ process is observed only just prior to an arrival joining the system. □

To round out this section we need to examine the $\{V_n, n \geq 0\}$ process that appears in the GI/Geometric/1 model with late arrival system having a delayed access.

THEOREM 9.4.5: In the GI/Geometric/1 queueing model with interarrival times having a probability distribution $\{f_n\}$ ($n \geq 1$) and being generated by a late arrival system with delayed access and service times distributed independently according to a geometric (parameter r) distribution let

$$V_n = \text{Number of customers in the system just prior to}$$
$$\text{the arrival of the } n\text{th customer.}$$

Then $\{V_n, n \geq 0\}$ is a homogeneous MC with transition matrix given by

$$P_V = \begin{bmatrix} s_0 & c_0 & 0 & 0 & \cdots \\ r_1 & k_1 & k_0 & 0 & \cdots \\ r_2 & k_2 & k_1 & k_0 & \cdots \\ r_3 & k_3 & k_2 & k_1 & \cdots \\ \vdots & \vdots & \vdots & \vdots & \ddots \end{bmatrix}, \tag{9.4.17}$$

where, if $\bar{r} = 1 - r$, for $j \geq 0$,

$$k_j = \sum_{k=j}^{\infty} f_k \binom{k}{j} r^j \bar{r}^{k-j},$$

$$r_j = 1 - \sum_{i=0}^{j} k_i,$$

and

$$c_0 = k_0/\bar{r}, \quad s_0 = 1 - c_0.$$

Proof: As observed in the proof of Theorem 9.4.1

$$V_{n+1} = V_n + 1 - D_n \qquad (V_n \geq 0, 0 \leq D_n \leq V_{n+1}),$$

where D_n is the number of customers served during the interarrival time between the nth and the $(n + 1)$th arrivals.

To illustrate the changes required to Theorem 9.4.1 let us give in Fig. 9.4.2 typical sample paths for this model. As before let us assume that

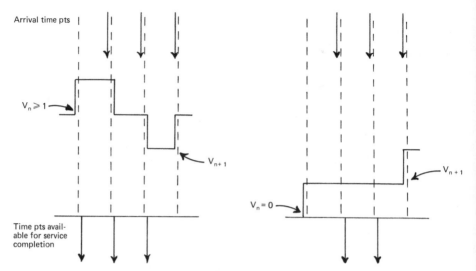

FIGURE 9.4.2.

T_n, the interarrival time between the nth and $(n + 1)$th arrivals, takes on the value 3 and that $V_{n+1} > 0$. We consider the cases $V_n \geq 1$ and $V_n = 0$ separately.

If the interarrival interval consists of k units of time and the server is busy during this entire interval, then, if $V_n \geq 1$ there are k positions available for service completion whereas if $V_n = 0$ there are only $k - 1$ positions available because a possible initial service position is blocked due to the arrival mechanism that is imposed upon the model.

This means that if $V_n \geq 1$, then D_n is distributed the same as C_n, as given in Theorem 9.4.1. Consequently the transition probabilities p_{ij} for $i \geq 1$ are identical to those obtained for the $\{U_n, n \geq 0\}$ chain.

The theorem will follow upon the derivation of the transition probability p_{01}, which is given by

$$p_{01} = P\{V_{n+1} = 1 \,|\, V_n = 0\} = P\{D_n = 0 \,|\, V_n = 0\}$$

$$= \sum_{k=1}^{\infty} P\{D_n = 0 \,|\, V_n = 0, \, T_n = k\} P\{T_n = k\},$$

since T_n does not depend on V_n. Observe that when $V_n = 0$ and $T_n = 1$ no service is possible and thus $D_n = 0$ with probability 1. Also, when $V_n = 0$ and $T_n = k \geq 1$ then $D_n = 0$ occurs iff at each of $k - 1$ independent trials there are no service completions (each with probability $\bar{r}$). Thus

$$c_0 \equiv p_{01} = 1 \cdot f_1 + \sum_{k=2}^{\infty} \bar{r}^{k-1} f_k = \sum_{k=1}^{\infty} \bar{r}^{k-1} f_k.$$

Since $k_0 = \sum_{k=1}^{\infty} f_k \bar{r}^k$ we have that $k_0 = c_0 \bar{r}$ and the transition matrix P_V follows. $\square$

Paralleling Theorem 9.3.6 we have the analogous results concerning chains with transition matrix structure P_V. The proof of the theorem is left as an exercise (Exercise 9.4.4).

THEOREM 9.4.6: Let $\{V_n, n \geq 0\}$ be a homogeneous MC with transition matrix given by Eq. (9.4.17) where $\{k_j\}$ is a probability distribution with $c_0 = k_0/\bar{r}$ for some $\bar{r}$.
(a) $\{V_n, n \geq 0\}$ is an irreducible MC if and only if $0 < k_0 \leq k_0 + k_1 < 1$.
(b) If the MC is irreducible and $c_0 < 1$, then the MC is aperiodic.
(c) Under the conditions of irreducibility the MC $\{V_n, n \geq 0\}$ is

  (i) transient if $\rho > 1$,
  (ii) persistent null if $\rho = 1$, and
  (iii) persistent nonnull if $\rho < 1$,

where $\rho^{-1} = \sum_{j=1}^{\infty} j k_j$. $\square$

COROLLARY 9.4.5A: Let $\{V_n, n \geq 0\}$ be the MC with transition matrix given by Eq. (9.4.17) embedded in the GI/Geometric/1 queueing model at instants just prior to the arrival of successive customers. Let $\{f_n\}$ $(n \geq 1)$ be the interarrival time distribution with p.g.f. $F(s)$. The service time distribution is geometric (parameter r).
(a) The MC $\{V_n, n \geq 0\}$ is irreducible (and aperiodic) if and only if $f_1 < 1$.
(b) If $f_1 < 1$ and $\rho^{-1} = r \sum_{n=1}^{\infty} n f_n$, then the MC is

  (i) transient if $\rho > 1$,
  (ii) persistent null if $\rho = 1$, and
  (iii) persistent nonnull if $\rho < 1$.

(c) When $f_1 < 1$ and $\rho < 1$, the stationary distribution $\{\pi_j\}$ is given by

$$\pi_j = \begin{cases} 1 - \lambda, & j = 0, \\ \lambda(1 - \alpha)\alpha^{j-1}, & j \geq 1, \end{cases} \tag{9.4.18}$$

where α $(0 < \alpha < 1)$ is the unique root of the equation $s = F(rs + \bar{r})$, which lies in the region $(0, 1)$ and

$$\lambda = \alpha/(\alpha r + \bar{r}). \tag{9.4.19}$$

Proof: (a) The irreducibility conditions follow as in Corollary 9.4.1A. Note, however, that to establish the aperiodicity one needs to show that $c_0 = k_0/\bar{r} < 1$. Now $k_0 = \sum_{n=1}^{\infty} f_n \bar{r}^n = F(\bar{r})$ so that $c_0 = 1$ iff $F(\bar{r}) = \bar{r}$. Let $\mu = \sum n f_n$ and since $f_0 = 0$ and $f_0 < 1$, Lemma 8.1.3(b) shows for the $\mu \leq 1$

case while Lemma 8.1.3(c) shows for the $\mu > 1$ case that $F(s) = s$ has no roots in $(0, 1)$ and since $0 < \bar{r} < 1$, $F(\bar{r}) \neq \bar{r}$. Consequently $c_0 < 1$.

(b) Follows, as for Corollary 9.4.1A.

(c) The existence of a stationary distribution follows from Theorem 9.4.6. The stationary equations, $\pi_j = \sum_{i=0}^{\infty} \pi_i p_{ij}$, are given by

$$\pi_0 = s_0 \pi_0 + \sum_{i=1}^{\infty} r_i \pi_i, \qquad (9.4.20)$$

$$\pi_1 = c_0 \pi_0 + \sum_{i=1}^{\infty} k_i \pi_i, \qquad (9.4.21)$$

$$\pi_j = \sum_{i=j-1}^{\infty} \pi_i k_{i-j+1} \qquad (j \geq 2). \qquad (9.4.22)$$

Observe that $\pi_j = A\alpha^j$ ($j \geq 1$) satisfy Eq. (9.4.22) as found in the course of establishing Theorem 9.4.4. With such a substitution and taking $\pi_0 = B$, Eqs. (9.4.20) and (9.4.21) become, respectively,

$$Bc_0 = A \sum_{i=1}^{\infty} r_i \alpha^i, \qquad (9.4.23)$$

$$A\alpha = Bc_0 + A \sum_{i=1}^{\infty} k_i \alpha^i. \qquad (9.4.24)$$

Since $C(\alpha) = \alpha$ where $C(s)$ is the p.g.f. of the $\{k_i\}$ distribution, Eq. (9.4.24) simplifies to give

$$B = (k_0/c_0)A = \bar{r}A. \qquad (9.4.25)$$

Furthermore, if $r_0 = 1 - k_0$ and $R(s) = \sum_{i=0}^{\infty} r_i s^i$, then $R(s) = [1 - C(s)]/(1 - s)$, implying that $R(\alpha) = 1$ and Eq. (9.4.23) also simplifies to Eq. (9.4.25).

To determine A (and hence B) we use the fact that $\sum_{j=0}^{\infty} \pi_j = 1$, i.e.,

$$B + A \sum_{j=1}^{\infty} \alpha^j = 1,$$

implying that

$$B + A\alpha/(1 - \alpha) = 1. \qquad (9.4.26)$$

Solving Eqs. (9.4.25) and (9.4.26) yields

$$A = \frac{1 - \alpha}{\alpha r + \bar{r}}, \qquad B = \frac{\bar{r}(1 - \alpha)}{\alpha r + \bar{r}}.$$

By defining $\lambda = 1 - B = \alpha/(\alpha r + \bar{r})$, Eq. (9.4.18) follows by noting that $\lambda(1 - \alpha) = A\alpha$. $\square$

EXAMPLE 9.4.2:   *The Geometric/Geometric/1 Model.*   For a Geometric/Geometric/1 model having a late arrival system with delayed access the

$\{V_n, n \geq 0\}$ MC gives the system size embedded at time points just prior to the arrival of successive customers.

As in Example 9.4.1, $f_n = \bar{p}^{n-1}p$ and the MC is irreducible and aperiodic if $f_1 < 1$. Furthermore, in this example, $\rho = p/r$, $\alpha = p\bar{r}/\bar{p}r$, and $\lambda = p/r$.

Consequently if $p < r$ a stationary distribution exists and is given by

$$\pi_j = \begin{cases} 1 - \rho, & j = 0, \\ \rho(1 - \alpha)\alpha^{j-1}, & j \geq 1, \end{cases}$$

a modified geometric distribution. $\square$

We have thus seen that for the Geometric/Geometric/1 model stationary distributions of either geometric or modified geometric form exist when the system is embedded at a variety of time points. The actual stationary distribution that arises depends also on the arrival system that is imposed upon the model. To summarize our findings derived in Sections 9.2, 9.3, and 9.4 we give a table of stationary distributions covering all the situations considered for a Geometric/Geometric/1 model. The arrival parameter is r and the service parameter p and we assume that $\rho = p/r \; (< 1)$, $\alpha = p\bar{r}/\bar{p}r \; (< 1)$.

In Table 9.4.1 an entry "Geometric" means that the stationary distribution is of the form $\pi_i = (1 - \alpha)\alpha^i \; (i \geq 0)$, and "Modified Geometric" means $\pi_0 = 1 - \rho$, $\pi_i = \rho(1 - \alpha)\alpha^{i-1} \; (i \geq 1)$.

TABLE 9.4.1

Embedding points	Early arrival system	Late arrival system with immediate access	Late arrival system with delayed access
$n-$	$\{X_n\}$ Modified Geometric (Corollary 9.2.1A)	$\{X_n^{(i)}\} = \{Y_n\}$ Geometric (Corollary 9.2.2A)	$\{X_n^{(d)}\} = \{X_n\}$ Modified Geometric (Corollary 9.2.1A)
n	$\{Y_n\}$ Geometric (Corollary 9.2.2A)	$\{Y_n^{(i)}\} = \{X_{n+1}\}$ Modified Geometric (Corollary 9.2.1A)	$\{Y_n^{(d)}\}$ —
$n+$	$\{Z_n\} = \{X_{n+1}\}$ Modified Geometric (Corollary 9.2.1A)	$\{Z_n^{(i)}\} = \{Y_{n+1}\}$ Geometric (Corollary 9.2.2A)	$\{Z_n^{(d)}\} = \{X_{n+1}\}$ Modified Geometric (Corollary 9.2.1A)
Just after a departure from the system	$\{Q_n^{(e)}\} = \{R_n\}$ Geometric (Example 9.3.3)	$\{Q_n^{(i)}\} = \{R_n\}$ Geometric (Example 9.3.3)	$\{Q_n^{(d)}\} = \{Q_n\}$ Modified Geometric (Example 9.3.2)
Just before an arrival joins the system	$\{U_n^{(e)}\} = \{U_n\}$ Geometric (Example 9.4.1)	$\{U_n^{(i)}\} = \{U_n\}$ Geometric (Example 9.4.1)	$\{U_n^{(d)}\} = \{V_n\}$ Modified Geometric (Example 9.4.2)

Table 9.4.1 reinforces what we have said earlier. Much care must be taken when modeling discrete time stochastic systems to be absolutely certain what process is being examined. Often a sketch of a typical sample path is required to clarify what is going on.

We conclude this section with a derivation of the limiting waiting time and queueing time distributions for a typical customer in a GI/Geometric/1 model. We use the same definitions for W_n and $W_n^{(q)}$, the waiting time and queueing time of the nth customer, as established in Section 9.3.

THEOREM 9.4.7:   Let us assume that we are considering a GI/Geometric/1 queueing system having either an early arrival system or a late arrival system with immediate access. The interarrival time distribution is $\{f_n\}$ $(n \geq 1)$ with p.g.f. $F(s)$ and the service time distribution is geometric (parameter r). If $f_1 < 1$ and $\rho < 1$ where $\rho^{-1} = r \sum_{n=1}^{\infty} nf_n$, then W_n and $W_n^{(q)}$ have limiting probability distributions $\{w_n\}$ and $\{w_n^{(q)}\}$, respectively, given by

$$w_n = (1 - \lambda)\lambda^{n-1} \qquad (n \geq 1) \qquad\qquad (9.4.27)$$

and

$$w_n^{(q)} = \begin{cases} 1 - \alpha & (n = 0), \\ \alpha(1 - \lambda)\lambda^{n-1} & (n \geq 1), \end{cases} \qquad (9.4.28)$$

where α $(0 < \alpha < 1)$ is the unique root of the equation $s = F(rs + \bar{r})$, which lies in the region $(0, 1)$ and $\lambda = \alpha r + \bar{r}$.

Proof: With the conventions established earlier for the particular arrival systems being used, W_n is the number of time slots spent in the system by the nth customer while $W_n^{(q)}$ is the number of time slots spent before servicing commences by the nth customer.

First observe that under the stated conditions if U_n is the number of customers in the system just prior to the arrival of the nth customer, then $\{U_n, n \geq 0\}$ has a limiting distribution $\{\pi_j\}$ given by $\pi_j = (1 - \alpha)\alpha^j$ $(j \geq 0)$.

Now $W_n^{(q)} = 0$ if and only if $U_n = 0$ so that

$$w_0^{(q)} = \lim_{n \to \infty} P\{W_n^{(q)} = 0\} = \lim_{n \to \infty} P\{U_n = 0\} = \pi_0 = 1 - \alpha.$$

If $U_n = k$ (≥ 1) there are k customers in the system whose services must be completed before the nth customer enters the service facility. Thus $W_n^{(q)} = S_1' + S_2 + \cdots + S_k$ where S_1' is the remaining service time of customer in the service facility and S_i $(2 \leq i \leq k)$, the service times of the customers to be serviced. By the lack of memory property of the geometric distribution, $S_1' \equiv S_1$, distributed as geometric (parameter r), as is $S_2, \ldots, S_k$. Hence

$$P\{S_i = l\} = \bar{r}^{l-1}r \qquad (l \geq 1). \qquad\qquad (9.4.29)$$

Therefore, for $l \geq 1$,

$$P\{W_n^{(q)} = l\} = \sum_{k=1}^{l} P\{S_1 + \cdots + S_k = l | U_n = k\} P\{U_n = k\}$$

$$= \sum_{k=1}^{l} \binom{l-1}{k-1} r^{k-1} \bar{r}^{l-k} r P\{U_n = k\}, \qquad (9.4.30)$$

since we require the probability of k services during l slots with a service on the last slot, i.e., $k - 1$ successes in $l - 1$ Bernoulli trials together with a success on the last trial.

Taking the limit as n tends to ∞ in Eq. (9.4.30) we obtain, for $l \geq 1$,

$$w_l^{(q)} = \sum_{k=1}^{l} \binom{l-1}{k-1} r^k \bar{r}^{l-k} \pi_k$$

$$= (1 - \alpha) \bar{r}^l \sum_{k=1}^{l} \binom{l-1}{k-1} \left(\frac{r\alpha}{\bar{r}}\right)^k$$

$$= (1 - \alpha) \bar{r}^l \left(\frac{r\alpha}{\bar{r}}\right) \sum_{m=0}^{l-1} \binom{l-1}{m} \left(\frac{r\alpha}{\bar{r}}\right)^m$$

$$= (1 - \alpha) r\alpha \bar{r}^{l-1} \left(1 + \frac{r\alpha}{\bar{r}}\right)^{l-1}$$

$$= \alpha(1 - \lambda)\lambda^{l-1},$$

concluding the derivation of Eq. (9.4.28).

To obtain the limiting distribution of W_n observe that $W_n = W_n^{(q)} + S_n$ where S_n is the service time of the nth customer. Since $W_n^{(q)}$ and S_n are independent, if $W(s)$ and $W^{(q)}(s)$ are the p.g.f.'s of the limiting probability distribution of W_n and $W_n^{(q)}$, respectively, and $G(s)$ is the p.g.f. of the service time r.v., then

$$W(s) = W^{(q)}(s)G(s). \qquad (9.4.31)$$

Using the results that $1 - \lambda = (1 - \alpha)r$ and $\lambda - \alpha = (1 - \alpha)\bar{r}$ we have from Eq. (9.4.28) that

$$W^{(q)}(s) = \frac{(1 - \alpha)(1 - \bar{r}s)}{1 - \lambda s}$$

and from Eq. (9.4.29) that $G(s) = rs/(1 - \bar{r}s)$. Substitution into Eq. (9.4.31) yields

$$W(s) = \frac{(1 - \alpha)rs}{1 - \lambda s} = \sum_{n=1}^{\infty} (1 - \lambda)\lambda^{n-1} s^n,$$

from which Eq. (9.4.27) follows. $\square$

We have left the case when the model has a late arrival system with delayed access as an exercise. Care needs to be taken in this instance because of the convention adopted for the queueing time r.v. $W_n^{(q)}$ (see Section 9.3).

EXAMPLE 9.4.3: *The Geometric/Geometric/1 Model.* From Example 9.4.1 we have that $f_n = \bar{p}^{n-1} p (n \geq 1)$ and that $\alpha = p\bar{r}/\bar{p}r$. Thus $\lambda = \bar{r}/\bar{p}$ and the limiting probability distributions $\{w_n\}$ and $\{w_n^{(q)}\}$ as given by Eqs. (9.4.27) and (9.4.28) are the same as those derived by the methods of Section 9.3 and given in Example 9.3.4. □

We do not discuss the derivation of busy period distributions in a GI/Geometric/1 queue other than to say that if $g_k(n)$ is the probability that k customers are served in and that a busy period initiated by one customer is of length n, then for a system having a late arrival system with delayed access Bhat (1968) shows, using arguments involving zero avoiding transitions, that

$$g_1(n) = \bar{r}^{n-1} r \sum_{l=n+1}^{\infty} f_l,$$

$$g_k(n) = r^k \bar{r}^{n-k} \frac{1}{k-1} \binom{n-2}{k-2} \sum_{m=0}^{n-1} \sum_{l=n-m}^{\infty} (n-m-1) f_m^{(k-1)} f_l \qquad (k \geq 2),$$

where $\{f_n^{(k)}\} = \{f_n\}^{k*}$.

Exercises 9.4

1. In Theorem 9.4.1 we deduced expressions for the transition probabilities p_{i0} by using the result that the row sums of a transition matrix are unity. We show that the same expressions can be deduced from first principles

   (a) $p_{i0} = P\{C_n^0 = i + 1\}$ $(i \geq 0)$,

   where C_n^0 is C_n (as defined in the proof of Theorem 9.4.1) under the assumption that the server may be unoccupied for a portion of the interarrival interval.

   (b) Show

   $$p_{i0} = \sum_{k=i+1}^{\infty} f_k P\{S_1 + \cdots + S_{i+1} \leq k\},$$

   where S_j is the service time of the jth customer served in the interarrival interval.

(c) Show, using generating functions or otherwise, that

$$P\{S_1 + \cdots + S_{i+1} = l\} = \binom{l-1}{l-1-i} r^{i+1} \bar{r}^{l-1-i}$$

and hence deduce that

$$p_{i0} = \sum_{k=i+1}^{\infty} f_k \sum_{l=i+1}^{k} \binom{l-1}{l-1-i} r^{i+1} \bar{r}^{l-1-i}.$$

(d) Using the result of (c), show that

$$p_{i0} = \sum_{k=i+1}^{\infty} f_k \sum_{l=i+1}^{k} \binom{k}{l} r^l \bar{r}^{k-l}$$

$$= \sum_{j=i+1}^{\infty} k_j$$

$$= r_i,$$

since $\{k_j\}$ is a probability distribution. [*Note:* The proof (d) rests upon establishing the result that if

$$A_k = \sum_{l=i+1}^{k} \binom{l-1}{i} r^{i+1} \bar{r}^{l-1-i} \quad \text{and} \quad B_k = \sum_{l=i+1}^{k} \binom{k}{l} r^l \bar{r}^{k-l},$$

then $A_k = B_k$ for $k = i + 2, \ldots$. There are various ways of establishing this result:

(i) Use induction and the result that

$$B_k = \frac{k!}{i!(k-i-1)!} \int_0^r x^i (1-x)^{k-i-1} \, dx$$

or (ii) $B_k = P\{Y_k \geq i + 1\}$ where Y_k is the number of successes in k Bernoulli trials with probability of success r. A_k can be deduced by conditioning this probability with the event "the $(i + 1)$th success occurs at trial number l."]

*2. Fill in the details of the proof of Theorem 9.4.2(a) for the cases $i = j$ and $i > j$.

*3. Fill in the details of the direct proof of Theorem 9.4.2(c). In particular, establish result (ii) that

$$\{l^{(n)}_{i,\,i+2}\} = \{l^{(n)}_{i,\,i+1}\} * \{l^{(n)}_{i+1,\,i+2}\} \qquad (i \geq 0).$$

*4. Prove Theorem 9.4.6. [In proving (c) observe that the transient conditions follow by using, as in Theorem 9.4.2, Theorem 7.1.11. Using Theorem 7.1.10 for the $\rho < 1$ case consider taking $x_0 = \bar{r}$ and $x_i = \alpha^i$ $(i \geq 1)$

while for the $\rho \geq 1$ case consider taking $x_0 = \bar{r}$ and $x_i = 1$ $(i \geq 1)$. Modifications to the direct arguments give

(i) $$\sum_{n=1}^{\infty} l_{0j}^{(n)} s^n = \frac{1}{\bar{r}} [F(s)]^j \qquad (j \geq 1),$$

(ii) $$F_{00}(s) = \frac{1}{\bar{r}} \left[\frac{s - F(s)}{1 - F(s)} - sr \right],$$

(iii) $$f_{00} = \begin{cases} 1, & \rho \leq 1, \\ \dfrac{1 - \rho r}{\rho - \rho r}, & \rho > 1, \end{cases}$$

(iv) $$\mu_{00} = \begin{cases} \dfrac{\bar{r} + r\alpha}{\bar{r}(1 - \alpha)}, & \rho < 1, \\ \infty, & \rho = 1, \end{cases}$$

where $F(s)$ and α are as in the proof of Theorem 9.4.2.]

5. Consider a GI/Geometric/1/N queue with a waiting room of capacity $N - 1$. A customer arriving and finding N customers in the system cannot join and leaves, never to return. Let us assume that we have an early arrival system or late arrival system with immediate access. Let U_n be the number in the system just prior to the nth arrival.

(a) Show that $\{U_n, n \geq 0\}$ is a MC with state space $S = \{0, 1, \ldots, N\}$ and transition matrix

$$P = \begin{bmatrix} r_0 & k_0 & 0 & 0 & \cdots & 0 & 0 \\ r_1 & k_1 & k_0 & 0 & \cdots & 0 & 0 \\ r_2 & k_2 & k_1 & k_0 & \cdots & 0 & 0 \\ \vdots & \vdots & \vdots & \vdots & & \vdots & \vdots \\ r_{N-1} & k_{N-1} & k_{N-2} & & \cdots & k_1 & k_0 \\ r_{N-1} & k_{N-1} & k_{N-2} & & \cdots & k_1 & k_0 \end{bmatrix},$$

where k_n is the probability that exactly n customers are served during an interarrival time (provided the server is operating continuously) and $r_n = k_{n+1} + k_{n+2} + \cdots$.

(b) Note that if the states are relabeled so that $\bar{0} = N$, $\bar{1} = N - 1, \ldots,$ $\bar{N} = 0$, then the transition matrix $\bar{P}$ corresponding to this labeling is the transition matrix P given in Exercise 9.3.4(a).

(c) Conclude from this that the limiting distribution $\{\pi_{i,U}^{(N)}\}$ of the MC $\{U_n\}$ is related to the limiting distribution $\{\pi_{i,Q}^{(N)}\}$ of Exercise 9.3.4(c) through

$$\pi_{i,Q}^{(N)} = \pi_{N-i,U}^{(N)} \qquad (i = 0, 1, \ldots, N)$$

for specified probability distribution $\{k_n\}$.

6. Develop analogous results to those given in Theorem 9.4.7 for the limiting distribution of the waiting time and queueing time r.v.'s for a GI/Geometric/1 model with late arrivals having delayed access.

*9.5   Further Models and Generalizations

We conclude our study of discrete time queueing models by taking a look at some generalizations of our earlier models together with special cases and additional models that have been presented in the literature.

Recall that in Section 9.1 we saw that compound or bulk arrivals (or, for that matter, batch servicing) could be regarded as Geometric$^{(A)}$ type arrivals (or services). Such patterns can in fact replace the Geometric arrivals (or services) used in the models of Sections 9.3 and 9.4.

EXAMPLE 9.5.1: *The Geometric$^{(A)}$/G/1 Model.* The Geometric$^{(A)}$ arrival pattern is simply portrayed as a sequence of i.i.d. nonnegative integer-valued r.v.'s $\{A_n\}$ with p.g.f. $A(s)$, where A_n is the number of arrivals that occur in the nth time slot.

[Note that if we assume that at each arrival time point no arrival occurs, with probability $\bar{p}$, or a batch occurs with probability p and size j (≥ 1), with conditional probability b_j, then

$$A(s) = \sum_{j=0}^{\infty} P(A_n = j)s^j = \bar{p} + p \sum_{j=1}^{\infty} b_j s^j \equiv \bar{p} + pB(s).\Big]$$

One way to analyze the behavior of this model is to parallel the analysis given in Section 9.3 for the Geometric/G/1 model and attempt to find an embedded MC. To this end, let Q_n be the number of customers in the system just after the service completion of the nth customer and let C_n be the number of customers that arrive during the service time of the nth customer. Then, as in the proof of Theorem 9.3.1, under the assumption of a late arrival system with delayed access,

$$Q_{n+1} = \begin{cases} Q_n - 1 + C_{n+1} & \text{if } Q_n \geq 1, \\ C_{n+1} & \text{if } Q_n = 0, \end{cases}$$

i.e.,

$$Q_{n+1} = \max(Q_n - 1 + C_{n+1}, C_{n+1}). \tag{9.5.1}$$

Equation (9.5.1) holds for any single server system, but it only leads to a homogeneous MC $\{Q_n, n \geq 0\}$ under the assumption that $\{C_n\}$ is a sequence of independent and identically distributed r.v.'s. Such a property holds when we have Geometric$^{(A)}$ arrivals and a general service time distribution $\{g_n\}$ ($n \geq 1$).

Let $C(s)$ be the p.g.f. of a typical C_n r.v. By conditioning upon S_n, the service time of the nth customer, as in the proof of Theorem 9.3.1, or by using Theorem 2.6.1, it can be easily shown that $C(s) = G(A(s))$ where $G(s)$ is the p.g.f. of S_n.

Since $C(s) = \sum_{j=0}^{\infty} k_j s^j$ where $k_j = P[C_n = j]$, the analysis of the queueing system now follows via the MC $\{Q_n, n \geq 0\}$ with transition matrix P_Q given by Eq. (9.3.1). The classification of the chain is given by Theorem 9.3.2 and, by Theorem 9.3.4, we conclude that if $\rho = \sum_{j=1}^{\infty} jk_j = C^{(1)}(1) < 1$ and $0 < k_0 \leq k_0 + k_1 < 1$, a stationary distribution $\{\pi_j\}$, which is also the limiting distribution of Q_n, exists with p.g.f. $\Pi_Q(s)$ given by

$$\Pi_Q(s) = \frac{(1 - \rho)(1 - s)G(A(s))}{G(A(s)) - s}. \tag{9.5.2}$$

The p.g.f. of the limiting distribution $\Pi_Q(s)$ for a geometric (p) input follows as a special case of Eq. (9.5.2) since, from our observations above concerning batch arrivals, in this case $b_1 = 1$ with $B(s) = s$ implying $A(s) = \bar{p} + ps$. Equation (9.5.2) now reduces to Eq. (9.3.17), as given in Corollary 9.3.1A.

We saw in Section 9.3 that the system size r.v. in the Geometric/G/1 model when examined unit time points apart does not lead to a homogeneous MC. However, by augmenting this r.v. with a supplementary r.v., we can introduce Markovian structure. Such an idea was exploited by Dafermos and Neuts (1971) when they examined the Geometric$^{(A)}$/G/1 model in terms of stochastic processes embedded at each successive time point. To be specific, in terms of our notation, they considered a late arrival system with delayed access with service time distribution $\{g_n\}$ $(n \geq 1)$ with A_n arrivals occurring in $(n-, n)$ having a probability distribution $\{a_n\}$ $(n > 0)$ with $a_0 > 0$, $a_1 > 0$. The Markovian structure is introduced by considering a sequence $\{(Z_n, W_n), n \geq 0\}$ of pairs of r.v.'s where Z_n is the number of customers in the system at time $n+$ and W_n is the additional number of units of service time required by the customer in service at time $n+$. If $Z_n = 0$, then $W_n = 0$.

The pairs (Z_n, W_n) are recursively related by

$$Z_{n+1} = Z_n + A_{n+1}, \qquad W_{n+1} = W_n - 1 \quad \text{for} \quad Z_n \geq 1, W_n > 1;$$

$$Z_{n+1} = Z_n + A_{n+1} - 1, \qquad W_{n+1} = S \qquad \text{for} \quad Z_n \geq 1, W_n = 1,$$
$$Z_n + A_{n+1} > 1;$$

$$Z_{n+1} = W_{n+1} = 0, \qquad\qquad\qquad\quad \text{for} \quad Z_n = W_n = 0, A_{n+1} = 0,$$
$$\text{or} \quad Z_n = W_n = 1, A_{n+1} = 0;$$

$$Z_{n+1} = A_{n+1}, \qquad W_{n+1} = S \qquad \text{for} \quad Z_n = W_n = 0, A_{n+1} \geq 1;$$

where S denotes the amount of service requested by the customer who begins service at time $n + 1$. This shows that the pairs (Z_n, W_n), $n \geq 0$, form a MC with state space consisting of $(0, 0)$ and all integer pairs (i,j) with $i \geq 1$, $1 \leq j \leq J$, where $J = \max\{j : g_j > 0\}$. Moreover, it is easy to see that this MC is irreducible and aperiodic.

Let $p_{i_0 j_0, ij}^{(n)} \equiv P\{Z_n = i, W_n = j \mid Z_0 = i_0, W_0 = j_0\}$ be the n-step transition probabilities for the bivariate MC. For simplicity, if we suppress the dependence on i_0, j_0, we shall write $p_{ij}^{(n)}$ for $p_{i_0 j_0, ij}^{(n)}$. With such a convention the basic recurrence relations can be derived from above to give

$$p_{00}^{(n+1)} = a_0 p_{00}^{(n)} + p_0 p_{11}^{(n)},$$

$$p_{ij}^{(n+1)} = \sum_{l=1}^{i} a_{i-l} p_{l,j+1}^{(n)} + \sum_{l=0}^{i} a_{i-l} g_j p_{l+1,1}^{(n)} + a_i g_j p_{00}^{(n)} \tag{9.5.3}$$

for $i \geq 1, j \geq 1, n \geq 0$.

By using Eqs. (9.5.3) as a starting point for the theoretical discussion, generating functions are introduced and an expression for

$$R(z, w, \xi) = \sum_{n=0}^{\infty} \sum_{i=0}^{\infty} \sum_{j=0}^{\infty} \xi^n z^i w^j p_{ij}^{(n)}$$

is developed in terms of $A(s)$, $G(s)$ and the initial conditions.

The irreducibility and aperiodicity imply that the limits

$$\pi_{ij} = \lim_{n \to \infty} P\{Z_n = i, W_n = j \mid Z_0 = i_0, W_0 = j_0\}$$

exist and if

$$\Pi(z, w) = \sum_{i=0}^{\infty} \sum_{j=0}^{\infty} z^i w^j \pi_{ij},$$

then it can be shown that

$$\Pi(z, w)[w - A(z)][z - G(A(z))] = \pi_{00}[\{G(A(z)) - z\}\{A(z) - w\} + wz\{A(z) - 1\}\{G(w) - G(A(z))\}],$$

where, if $\rho = A^{(1)}(1)G^{(1)}(1)$, $\pi_{00} = 0$ when $\rho \geq 1$ and $\pi_{00} = 1 - \rho$ when $\rho < 1$.

If we are only concerned with the queue length, set $w = 1$ and we obtain as the generating function of the limiting marginal distribution of Z_n, when $\rho < 1$,

$$\Pi(z, 1) = \frac{(1 - \rho)(1 - z)G(A(z))}{G(A(z)) - z}. \tag{9.5.4}$$

Comparison of Eqs. (9.5.2) and (9.5.4) leads to the interesting result that the $\{Q_n, n \geq 0\}$ and the $\{Z_n, n \geq 0\}$ processes both have the same limiting distribution. For the late arrival system with delayed access, the $\{Z_n\}$ process includes all the Q_n points together with those time points following a possible departure point where no departure actually occurs but they both have the same limiting distribution. A further discussion of this phenomenon is given in Example 9.5.3 where we specialize to the case of constant service times.

The reader is asked to consider the derivation of the limiting distribution of the $\{Q_n, n \geq 0\}$ process when we impose either an early arrival system or a late arrival system with immediate access (see Exercise 9.5.1).

In two further papers (Klimko and Neuts, 1973; Heimann and Neuts, 1973) a discussion of the computation of the distribution and moments of the busy period of the above queueing model is presented. In particular, if $\beta_n = P\{$Busy period initiated by a single customer lasts for exactly n units$\}$ $(n \geq 1)$, then $\{\beta_n\}$ satisfies the nonlinear recurrence relation

$$\beta_n = \sum_{l=1}^{n} g_l \sum_{k=0}^{\infty} a_k^{(l)} \beta_{n-l}^{(k)} \qquad (n \geq 1), \tag{9.5.5}$$

where $\{a_n^{(k)}\} = \{a_n\}^{k*}$ and $\{\beta_n^{(k)}\} = \{\beta_n\}^{k*}$.

Equation (9.5.5) is obtained by application of the law of total probability. The probability that the first service lasts for l units of time annd that k customers arrive during it is $g_l a_k^{(l)}$. If $k \neq 0$, these k customers may be considered as the initial customers of k independent busy periods. The probability that these k busy periods together take up exactly the $n - l$ remaining units of time is given by $\beta_{n-l}^{(k)}$.

If we define $\beta(s) = \sum_{n=1}^{\infty} \beta_n s^n$, then from Eq. (9.5.5)

$$\beta(s) = \sum_{l=1}^{\infty} \sum_{n=l}^{\infty} g_l s^n \sum_{k=0}^{\infty} a_k^{(n)} \beta_{n-l}^{(n)}$$

$$= \sum_{l=1}^{\infty} g_l s^l \sum_{v=0}^{\infty} s^v \sum_{k=0}^{\infty} a_k^{(l)} \beta_v^{(k)}$$

$$= \sum_{l=1}^{\infty} g_l s^l \left(\sum_{v=0}^{\infty} a_k^{(l)} \beta^k(s) \right)$$

$$= \sum_{l=1}^{\infty} g_l s^l A^l(\beta(s)),$$

implying

$$\beta(s) = G(sA(\beta(s))), \qquad |s| \leq 1. \tag{9.5.6}$$

The functional Eq. (9.5.6) is examined in detail in the Dafermos and Neuts (1971) paper where it is shown that $\beta(1) = \sum_{n=1}^{\infty} \beta_n = 1$ iff $\rho \leq 1$ and is < 1 if $\rho < 1$. Thus the busy period distribution is a proper distribution when $\rho \leq 1$. Furthermore, when $\rho = 1$ the mean of the busy period is infinite and when $\rho < 1$ the mean of the busy period is given by $A^{(1)}(1)/(1 - \rho)$.

(*Note*:

$$\rho = A^{(1)}(1)G^{(1)}(1) = \frac{\text{mean number of arrivals for unit time}}{\text{mean service time}},$$

which must be <1 for stability of the system.)

The Klimko and Neuts (1973) paper deals primarily with the numerical problems arising in the computation of higher moments of the busy period. A moment g.f. version of the classical functional Eq. (9.5.6) is used and the higher order derivatives at zero of the m.g.f. are computed by repeated use of the classical differentiation formula of Fáa di Bruno (1876). Moments of order up to fifty may be computed in this manner.

In the Heimann and Neuts (1973) paper explicit expressions for the probabilities β_n are derived by applying Lagrange's inversion formula to Eq. (9.5.6) together with a further application of Fáa di Bruno's formula. The expressions derived are not suitable for numerical computation. However, a monotone iterative procedure is developed for solving the non-linear difference Eqs. (9.5.5) numerically. $\square$

EXAMPLE 9.5.2: *The Geometric$^{(A)}$/G/1/N + 1 Model.* The limiting distribution of the number of customers Q_n in the system immediately following a departure is discussed for the Geometric/G/1/N + 1 model in Exercise 9.3.4 under the assumption of late arrivals having delayed access. The analysis presented therein can be adapted along similar lines to that given in Example 9.5.1 for the case of Geometric$^{(A)}$ arrivals.

The key properties of this model are examined in some detail in a sequence of papers by Neuts (1973b), Neuts and Klimko (1973), and the Heimann and Neuts (1973) paper mentioned earlier. Stress is placed upon the computational aspects since this model serves as a useful description for many discrete time queues of practical interest. Neuts and his co-researchers show that a wealth of numerical information may be obtained by relatively unsophisticated methods.

The assumptions inherent in the models considered in these afore-mentioned papers is that of Geometric$^{(A)}$ arrivals with the $\{A_n\}$ having a probability distribution $\{a_n\}$ $(0 \le n \le K)$, a service time distribution $\{g_n\}$ $(1 \le n \le L)$, and a maximum waiting room capacity of N. The arrival pattern is effectively that of late arrivals having delayed access (although by introducing a quantity g_0 that a service time is equal to zero one can, with suitable modifications, consider the immediate access case).

(In setting up the computer programs it is assumed that $1 \le K < N$, a restriction that is not essential but is usually satisfied in practice. $L = N = 100$ appear to be practical upper limits to the computer implementation presented.)

In the initial paper (Neuts, 1973b), the bivariate MC $\{(Z_n, W_n), n \geq 0\}$ where Z_n is the system size at time $n+$ and W_n is the number of additional units of service time required by the customer in service at time n is examined. Recurrence relations analogous to those of Eq. (9.5.3) but involving the various boundary conditions are derived and from them an iterative scheme for finding $\{p_{ij}^{(n)}\}$ ($1 \leq i \leq N$, $1 \leq j \leq L$, and $i, j = 0, 0$), the conditional joint probability distribution of the Z_n and W_n, is developed. From this joint probability distribution various "derived" queue features are obtained. In particular, the queue length distribution of Z_n, $\{p_{00}^{(n)}, \sum_{j=1}^{L} p_{ij}^{(n)}\}$, and its moments follow routinely.

The waiting time at time n is defined to be zero if and only if $Z_n = W_n = 0$. For $Z_n = 1$, $W_n = j$ the waiting time equals j. For $Z_n > 1$, $W_n = j$ the waiting time is the sum of j and $Z_n - 1$ independent service times. The waiting time is thus an integer-valued r.v. with value between 0 and NL. Expressions for the probability distribution of the waiting time can thus be found from $\{p_{ij}^{(n)}\}$ as can expressions for its moments.

All the above results are derived for fixed values of n. In a direct sequel (Neuts and Klimko, 1973), a computational procedure for finding the stationary probabilities π_{ij} ($i = 1, \ldots, N; j = 1, \ldots, L$ and $i, j = 0, 0$) by solving the stationary equations is presented. The system of equations contains $LN + 1$ independent linear equations in $LN + 1$ unknowns. It is shown that by introducing the vectors $\pi_j' = (\pi_{1j}, \ldots, \pi_{Nj})$ ($1 \leq j \leq L$), the solution may be expressed in terms of the solution of a homogeneous system of N equations in N unknowns. Moreover, this latter system has a "nearly upper triangular" structure, which makes the solution particularly simple.

From the expressions for the $\{\pi_{ij}\}$ various other stationary distributions can be found, for example, the waiting time distribution and the number of lost customers. The stationary distribution of the system size r.v. following departures is also obtained (using auxiliary quantities that are computed in the process of evaluating the π_{ij}). A comparison of these two stationary distributions shows that a very substantial difference may exist between them. In fact, the authors point out that the asymptotic distribution of the system size may be of limited practical value since it is possible for realizations to exhibit very substantial fluctuations that are not reflected in the asymptotic distributions.

The Heimann and Neuts (1973) paper is concerned with procedures for finding the probability distribution of the busy period. In this bounded waiting room case it is found by a standard investigation of the absorption time distribution in a finite MC. (Actually, the iterative procedure for determining the probability distribution of the busy period in the unbounded case uses as a starting solution the probability distribution found for this bounded case.) □

EXAMPLE 9.5.3: *The Geometric$^{(A)}/D/1$ Model and Its Variants.* We consider the model discussed in Example 9.5.1 but with the service time distribution taken as deterministic with service times assumed, without loss of generality, to be of unit duration. Since $g_1 = 1$, the analysis of Example 9.5.1 follows but with $G(s) = s$ and thus with $C(s) = A(s)$.

Recall that the MC $\{Q_n, n \geq 0\}$ embedded in this model arises when we consider the system size at time points immediately following departure (for a late arrival system with delayed access). What happens, however, when we embed the system at different time points and with different arrival systems?

In accordance with the notation established in Section 9.2, let $X_n = X(n-)$, $Y_n = X(n)$, and $Z_n = X(n+)$ where $X(t)$ refers the *system size* r.v. at time t. Furthermore, let $Q(t)$ be the *queue size* r.v. at time defined as $Q(t) \equiv \max(X(t) - 1, 0)$ and take $X_n^{(q)} \equiv Q(n-)$, $Y_n^{(q)} \equiv Q(n)$, and $Z_n^{(q)} \equiv Q(n+)$.

We shall investigate the behavior of these r.v.'s for each of the arrival systems considered earlier. We shall adopt the convention that all the arrivals that may occur in an interval occur at one time at the arrival point specified by the model under consideration. For example, in the early arrival system we have A_n arrivals occurring in $(n, n+)$ while in the late arrival systems we have A_n arrivals occurring in $(n-, n)$. Note also that in an early arrival system a service occurs in $(n-, n)$ iff $X_n \geq 1$, whereas for the late arrival systems a service occurs in $(n, n+)$ for the immediate case if $X_n + A_n \geq 1$ and in the delayed case iff $X_n \geq 1$.

We set up the following recurrence relationships:

$$Q_{n+1} = \begin{cases} Q_n - 1 + A_{n+1}, & Q_n \geq 1, \\ A_{n+1}, & Q_n = 0, \end{cases} \tag{9.5.7}$$

and

$$P_{n+1} = \begin{cases} P_n - 1 + A_n, & P_n + A_n \geq 1, \\ 0, & P_n + A_n = 0. \end{cases} \tag{9.5.8}$$

Observe that these two relationships can also be expressed as

(a) $\qquad\qquad Q_{n+1} = \max(Q_n - 1, 0) + A_{n+1},$ $\qquad$ (9.5.7)

(b) $\qquad\qquad P_{n+1} = \max(P_n + A_n - 1, 0).$ $\qquad$ (9.5.8)

Consider further the following two relationships:

(c) $\qquad\qquad P_{n+1} = \max(Q_n - 1, 0),$ $\qquad$ (9.5.9)

(d) $\qquad\qquad Q_n = P_n + A_n.$ $\qquad$ (9.5.10)

It is an interesting exercise to examine the interrelationships that exist between (a), (b), (c), and (d) above. In fact, one can develop the following "algebra of queues": For sequences $\{A_n\}$, $\{P_n\}$, and $\{Q_n\}$ that satisfy any two of Eqs. (9.5.7) to (9.5.10) the remaining two relationships are automatically satisfied. (See Exercise 9.5.2.) In particular, with the labels (a)–(d) as given above observe that:

If $\{A_n, Q_n\}$ satisfies (a) and $\{P_n\}$ is defined by (c) [or (d)], then $\{A_n, P_n, Q_n\}$ satisfies (b) and (d) [or (b) and (c)].

If $\{A_n, P_n\}$ satisfies (b) and $\{Q_n\}$ is defined by (c) [or (d)], then $\{A_n, P_n, Q_n\}$ satisfies (a) and (d) [or (a) and (c)].

In terms of application if $\{Q_n\}$ is a relationship between *system size* r.v.'s at time points n, then $\{P_{n+1}\}$ gives, by virtue of Eq. (9.5.9), the relationship between the *queue size* r.v.'s at the same time points.

Further, if $\{P_n\}$ is a relationship between *system size* r.v.'s at time points n, the $\{Q_n\}$ gives the relationship between the *system size* r.v.'s at the instant following an additional input A_n (before a service completion).

If we now look at the Geometric$^{(A)}$/D/1 system at the various embedding points and look for relationships satisfied recursively by the system size and queue size r.v.'s we obtain the results of Table 9.5.1 for a given sequence $\{A_n\}$ of arrival r.v.'s.

TABLE 9.5.1

	Early arrival system	Late arrival system with immediate access	Late arrival system with delayed access
$\{X_n\} \equiv$	$\{Q_{n-1}\}$	$\{P_n\}$	$\{Q_{n-1}\}$
$\{Y_n\} \equiv$	$\{P_n\}$	$\{Q_n\}$	$\{Q_{n-1} + A_n\}$
$\{Z_n\} \equiv$	$\{Q_n\}$	$\{P_{n+1}\}$	$\{Q_n\}$
$\{X_n^{(q)}\} \equiv$	$\{P_n\}$	$\{\max(P_n - 1, 0)\}$	$\{P_n\}$
$\{Y_n^{(q)}\} \equiv$	$\{\max(P_n - 1, 0)\}$	$\{P_{n+1}\}$	$\{\max(Q_{n-1} + A_n - 1, 0)\}$
$\{Z_n^{(q)}\} \equiv$	$\{P_{n+1}\}$	$\{\max(P_{n+1} - 1, 0)\}$	$\{P_{n+1}\}$

The establishment of the relationships given in Table 9.5.1 is left as an exercise (Exercise 9.5.2).

Note that the above queueing models are also intimately connected with the following *storage model* (or *dam model*): Consider a storage facility (dam) with unlimited capacity. During each interval $(n-1, n)$ there is an input of A_n units. At the end of each time interval (at time n) a unit release occurs, provided there is at least one unit in the storage facility. Let $S(t)$ be the storage in the dam at time t (measured in discrete units $0, 1, 2, \ldots$). Then

$U_n \equiv S(n-)$ is distributed as Q_n, in accordance with the relationship given by Eq. (9.5.7) and $V_n \equiv S((n-1)+)$ is distributed as P_n, in accordance with the relationship given by Eq. (9.5.8).

It has been traditional to examine the stochastic behavior of the queueing model in terms of the $\{Q_n, n \geq 0\}$ chain (c.f. Prabhu, 1958; Boudreau et al., 1962) and the behavior of the storage model in terms of the $\{P_n, n \geq 0\}$ chain (cf. Moran, 1956; Gani, 1957; Prabhu, 1964).

Since the input $\{A_n\}$ consists of a sequence of i.i.d. nonnegative integer-valued r.v.'s, Eqs. (9.5.7) and (9.5.8) imply $\{Q_n, n \geq 0\}$ and $\{P_n, n \geq 0\}$ are both homogeneous MC's. We have examined MC's of the form $\{Q_n, n \geq 0\}$ in some considerable detail in Section 9.3 and the results given there still hold for the cases considered in this example. In particular, under the conditions of Example 9.5.1 with $g_1 = 1$,

$$\Pi_Q(s) = \frac{(1-\rho)(1-s)A(s)}{A(s) - s} \tag{9.5.11}$$

gives the p.g.f. of the limiting distribution of $\{Q_n\}$ when $\rho = A^{(1)}(1) < 1$.

An analysis of the MC $\{P_n, n \geq 0\}$ is discussed in Exercise 9.5.4. Needless to say, results concerning its limiting distribution can be obtained from Eq. (9.5.11) and the relationship given by Eq. (9.5.9) or (9.5.10).

An interesting observation concerns the MC $\{Q_n, n \geq 0\}$, which, as we saw earlier, arises in the queueing model with late arrivals having delayed access for system size r.v. embedded just after departures. This same MC also appears for the $\{Z_n\}$ embedding, i.e., after every departure time point n that in effect augments the $\{Q_n\}$ process by including those time points when no departure actually occurs. The reason that this should be so is that the number of customers who arrive during a unit length of time also has the same distribution as the number of customers who arrive during the servicing of an arbitrary individual.   □

EXAMPLE 9.5.4:   *The Geometric$^{(A)}/D/1/N + 1$ Model and Its Variants.* We now consider the imposition of a waiting room of finite capacity N (or system capacity of $N + 1$) on the model presented in the previous example. With the notation of Example 9.5.3 and under the restriction of a late arrival system with delayed access, it is easily seen that the $\{Q_n, n \geq 0\}$ process has a state space $\{0, 1, \ldots, N\}$ and the $\{P_n, n \geq 0\}$ process has a state space $\{0, 1, \ldots, N - 1\}$. (For the other arrival systems the results that follow all hold but with N replaced by $N + 1$.)

Analogous to Eqs. (9.5.7) to (9.5.10) we have that

$$Q_{n+1} = \begin{cases} \min(Q_n - 1 + A_{n+1}, N), & Q_n = 1, 2, \ldots, N, \\ \min(A_{n+1}, N), & Q_n = 0, \end{cases} \tag{9.5.12}$$

$$P_{n+1} = \begin{cases} N-1, & P_n + A_n \geq N, \\ P_n + A_n - 1, & P_n + A_n = 1, 2, \ldots, N-1, \\ 0, & P_n + A_n = 0, \end{cases} \quad (9.5.13)$$

$$P_{n+1} = \begin{cases} Q_n - 1, & Q_n = 1, 2, \ldots, N, \\ 0, & Q_n = 0, \end{cases} \quad (9.5.14)$$

$$Q_n = \min(P_n + A_n, N). \quad (9.5.15)$$

The queueing system $\{Q_n, n \geq 0\}$ satisfying Eq. (9.5.12) has already been considered in Exercise 9.3.4 and the results presented therein are still valid here under the assumption that $\{k_i\}$ is the probability distribution of A_{n+1}.

The $\{P_n, n \geq 0\}$ process can be examined directly from Eq. (9.5.13) or indirectly from the result of the $\{Q_n, n \geq 0\}$ process and Eq. (9.5.14) or (9.5.15). The transition matrix for the $\{P_n, n \geq 0\}$ process is given by

$$P = \begin{bmatrix} k_0 + k_1 & k_2 & k_3 & \cdots & & k_{N-1} & r_N \\ k_0 & k_1 & k_2 & \cdots & & k_{N-2} & r_{N-1} \\ 0 & k_0 & k_1 & \cdots & & k_{N-3} & r_{N-2} \\ \vdots & \vdots & \vdots & & & \vdots & \vdots \\ 0 & 0 & 0 & \cdots & k_0 & k_1 & r_2 \\ 0 & 0 & 0 & \cdots & 0 & k_0 & r_1 \end{bmatrix}$$

where $r_n = k_n + k_{n+1} + \cdots$.

If $0 < k_0 \leq k_0 + k_1 < 1$, this MC is irreducible and aperiodic. Prabhu (1958) attributes to Moran (1956) the result that if $\pi_{i,P}^{(N)}$ $(i = 0, 1, \ldots, N-1)$ is the stationary probability distribution of this MC, the ratios $b_i = \pi_{i,P}^{(N)}/\pi_{0,P}^{(N)}$ are independent of N, where the b_i can be found as the coefficients of s^i in $B(s)$ where

$$B(s) = \frac{k_0(1-s)}{A(s) - s}, \qquad \left(A(s) = \sum_{i=0}^{\infty} k_i s^i \right). \quad \square$$

In the examples considered in this section thus far we have looked at models that have been generalizations of the Geometric/G/1 model considered in Section 9.3. We now consider extensions to the GI/Geometric/1 model.

EXAMPLE 9.5.5:   *The GI/Geometric/N Model.*   Chan and Maa (1978) examine the GI/Geometric/N model, i.e., the model of Section 9.4 but with N servers. Their presentation assumes, effectively, either an early arrival system or late arrival system with immediate access. Arrivals are assumed to occur at time epochs $\tau_1, \tau_2, \ldots$ where $T_n = \tau_{n+1} - \tau_n$ $(n \geq 1)$ are the inter-arrival times each i.i.d. nonnegative integer-valued r.v.'s with probability distribution $\{f_n\}$ $(n \geq 1)$. The sequence of service times $\{S_i\}$ are i.i.d. geometric r.v.'s with $P\{S_i = n\} = \bar{r}^{n-1} r (n \geq 1)$. It is assumed that each unit time slot can contain at most one service time point with arrivals occurring at integer-valued time points. The model of Theorem 9.4.1 still holds and thus

$$U_{n+1} = U_n + 1 - C_n \qquad (U_n \geq 0, 0 \leq C_n \leq U_n + 1),$$

where U_n is the number of customers in the system at the instant just before an arrival epoch and C_n is the r.v. denoting the number of departures during the interarrival time T_n.

It is clear that $\{U_n, n \geq 0\}$ is a homogeneous MC. However, the derivation of the transition probabilities p_{ij} are slightly more complicated in this case. Chan and Maa (1978) show that

$$p_{ij} = \begin{cases} a_{ij}, & 0 \leq i < N, \quad j \leq i+1 < N, \\ b_{ij}, & i \geq N, \quad j \leq N, \\ c_{ij}, & i \geq N, \quad N+1 \leq j \leq i+1, \\ 0, & \text{otherwise}, \end{cases} \qquad (9.5.16)$$

where

$$a_{ij} = \sum_{n=1}^{\infty} \binom{i+1}{j} \bar{r}^{nj} (1 - \bar{r}^n)^{i+1-j} f_n,$$

$$b_{ij} = \sum_{n=1}^{\infty} \left[\sum_{k=1}^{n} \sum_{l=j}^{N} \sum_{m=N-l+1}^{N} \binom{Nk-N}{i+1-m-l} \binom{N}{m} \binom{l}{j} \right.$$
$$\left. \times r^{i+1-l} \bar{r}^{Nk-i-1+l+(n-k)j} (1 - \bar{r}^{n-k})^{l-j} \right] f_n,$$

$$c_{ij} = \sum_{n=1}^{\infty} \binom{Nn}{i+1-j} r^{i+1-j} \bar{r}^{Nn-(i+1-j)} f_n.$$

(The properties of the hypergeometric distribution enable one to show that $c_{iN} = b_{iN}$.)

It is easy to verify that when $N = 1$ the transition probabilities reduce to those specified by Eqs. (9.4.1) and (9.4.2).

The stationary distribution $\{\pi_i\}$ $(i \geq 0)$ is found by solving the stationary equations. The proof of Corollary 9.4.1A needs to be modified but it can be shown that if $Nr \sum_{n=1}^{\infty} nf_n > 1$ the stationary distribution exists and is given by

$$\pi_j = \begin{cases} \sum_{l=j}^{N-1} (-1)^{l-j} \binom{l}{j} u_l, & j = 0, 1, \ldots, N-1, \\ C\alpha^{j-N}, & j \geq N, \end{cases}$$

where α $(0 < \alpha < 1)$ is the unique root of the equation $s = F((rs + \bar{r})^N)$ with $F(s)$ the p.g.f. of the $\{f_n\}$. Furthermore,

$$u_l = Cb_l \sum_{j=l+1}^{N} \frac{h_j}{b_j(1 - a_j)},$$

where

$$C = \left[\frac{1}{1 - \alpha} + \sum_{j=1}^{N} \frac{h_j}{b_j(1 - a_j)} \right]^{-1},$$

$$b_j = \prod_{k=1}^{j} \frac{a_k}{1 - a_k}, \qquad (j \geq 1) \quad \text{with} \quad b_0 = 1,$$

$$a_j = \sum_{k=1}^{j} \bar{r}^{kj} f_k = F(\bar{r}^j),$$

and

$$h_j = \binom{N}{j} - \frac{\bar{r}^N}{\alpha^{N+1}} \frac{(\alpha - a_j)}{\{\bar{r} + r\alpha\}^N - \bar{r}^j} \sum_{m=1}^{N} \binom{N}{m} \left[\frac{r\alpha}{\bar{r}} \right]^m \sum_{k=N+1-m}^{N} \binom{k}{j} \alpha^k.$$

Chan and Maa also derive expressions for the limiting distribution of the queueing time r.v. $W^{(q)}$ $(= \lim_{n \to \infty} W_n^{(q)})$. In particular

$$P\{W^{(q)} > k\} = \sum_{j=0}^{\infty} \pi_j P\{W^{(q)} > k \,|\, \text{the arrival finds } j \text{ in the system}\}$$

$$= \sum_{j=N}^{\infty} \pi_j P\{W^{(q)} > k \,|\, \text{the arrival finds } j \text{ in the system}\}$$

$$= \sum_{j=N}^{\infty} \pi_j \sum_{l=0}^{j-N} \binom{Nk}{l} r^l \bar{r}^{Nk-l}$$

$$= \sum_{j=N}^{\infty} \sum_{l=0}^{j-N} \binom{Nk}{l} r^l \bar{r}^{Nk-l} C\alpha^{j-N}$$

$$= \frac{C}{1 - \alpha} (\bar{r} + r\alpha)^{Nk}. \tag{9.5.17}$$

(The key step above follows by observing that if the arriving customer finds $j \geq N$ in the system, the N servers are busy and there are $j - N$ waiting at that time. If $W^{(q)}$ is to exceed k, then in k time units, equivalent to Nk Bernoulli trials, the number of departures must not exceed $j - N$.)

The mean queueing time may be found by virtue of Eq. (9.5.17):

$$EW^{(q)} = \frac{C}{1 - \alpha} \{1 - (\bar{r} + r\alpha)^N\}.$$

Finally, the authors also examine the GI/Geometric/N/N system (the "loss system" where, if all the servers are busy, an arriving customer leaves without receiving service). For this model the transition probabilities are given by

$$p_{ij} = \begin{cases} a_{ij}, & 0 \le i < N, \quad 0 \le j \le i + 1 \le N, \\ a_{N-1,j}, & i = N, \qquad 0 \le j \le N, \end{cases}$$

where the $\{a_{ij}\}$ are given by Eq. (9.5.16).

The stationary distribution $\{\pi_j\}$ ($j = 0, 1, \ldots, N$) is given by

$$\pi_j = \sum_{l=j}^{N} (-1)^{l-j} \binom{l}{j} v_l,$$

where

$$v_l = b_j \sum_{k=j}^{N} \binom{N}{k} \frac{1}{b_k} \bigg/ \sum_{k=0}^{N} \binom{N}{k} \frac{1}{b_k},$$

with a_j and b_j as defined above. $\square$

EXAMPLE 9.5.6: *The GI/Geometric$^{(B)}$/1 Model.* The model of Section 9.4 can be further generalized by allowing the server to still have geometric service times but with batch departures (of fixed or random size) with D_n departures during the nth slot.

We do not intend to elaborate on all the details since the analysis parallels the generalization of Example 9.5.1.

For the model with early arrival or late arrival with immediate access system, Theorem 9.4.1 generalizes as follows:

$$U_{n+1} = U_n + 1 - C_n \qquad (U_n \ge 0, 0 \le C_n \le U_n + 1),$$

where, as before, U_n is the number of customers in the system just prior to the arrival of the nth customer and C_n, the number of customers served during the interarrival time between the nth and $(n + 1)$th customer. The transition matrix given by Eq. (9.4.1) still holds but with the probability distribution $\{k_j\}$ having p.g.f. $C(s) = F(D(s))$ where $D(s)$ is the p.g.f. of the number of customers served in an interval of unit length (cf. Example 9.5.1), i.e., $D(s) = \sum_{j=0}^{\infty} P(D_n = j)s^j$.

Theorems 9.4.2, 9.4.3, and 9.4.4 still hold but with the changed form of the distribution $\{k_j\}$. In particular, when $0 < k_0 \le k_0 + k_1 < 1$ and $\rho^{-1} = \sum_{j=1}^{\infty} jk_j = D^{(1)}(1)F^{(1)}(1) > 1$, the stationary distribution exists and is given by

$$\pi_j = (1 - \alpha)\alpha^j, \qquad j \ge 0,$$

where α ($0 < \alpha < 1$) is the unique root of the equation $s = F(D(s))$. $\square$

EXAMPLE 9.5.7:   *The D/Geometric$^{(B)}$/1 Model.*   By assuming that we have a deterministic input of, say, 1 unit each time slot the analysis given in Example 9.5.6 holds with $f_1 = 1$, $F(s) = s$ and thus $C(s) = D(s)$.

Let us parallel the treatment given in Example 9.5.3 where we considered the Geometric$^{(A)}$/D/1 model. With the notation established there, i.e., X_n, Y_n, Z_n and $X_n^{(q)}$, $Y_n^{(q)}$, $Z_n^{(q)}$ for the appropriate system size and queue size r.v.'s at time points $n-$, n, and $n+$, we can develop recurrence relations for each of these r.v.'s for the model under prescribed arrival patterns.

We shall assume that D_n departures take place at time n or more particularly in $(n-, n)$ for the early arrival system and in $(n, n+)$ for the late arrival systems.

We set up the following recurrence relationships:

(a) $U_{n+1} = U_n + 1 - D_n$ $(U_n \geq 0, 0 \leq D_n \leq U_n + 1)$,   (9.5.18)

(b) $V_{n+1} = V_n + 1 - D_n$ $(V_n \geq 1, 0 \leq D_n \leq V_n)$,   (9.5.19)

(c) $V_n = U_n + 1$, $(U_n \geq 0)$,   (9.5.20)

(d) $U_{n+1} = V_n - D_n$ $(V_n \geq 1, 0 \leq D_n \leq V_n)$.   (9.5.21)

Once again we can develop an "algebra of queues" (cf. Example 9.5.3).

In particular, with the labels (a)–(d) as assigned:

If $\{D_n, U_n\}$ satisfies (a) and $\{V_n\}$ is defined by (c) [or (d)], then $\{D_n, U_n, V_n\}$ satisfies (b) and (d) [or (b) and (c)].

If $\{D_n, V_n\}$ satisfies (b) and $\{U_n\}$ is defined by (c) [or (d)], then $\{D_n, U_n, V_n\}$ satisfies (a) and (d) [or (a) and (c)].

TABLE 9.5.2

	Early arrival system	Late arrival system with immediate access	Late arrival system with delayed access
$\{X_n\} \equiv$	$\{V_n\}$	$\{U_n\}$	$\{V_n\}$
$\{Y_n\} \equiv$	$\{U_{n+1}\}$	$\{V_n\}$	$\{V_n + 1\}$
$\{Z_n\} \equiv$	$\{V_{n+1}\}$	$\{U_{n+1}\}$	$\{V_{n+1}\}$
$\{X_n^{(q)}\} \equiv$	$\{U_n\}$	$\{\max(U_n - 1, 0)\}$	$\{U_n\}$
$\{Y_n^{(q)}\} \equiv$	$\{\max(U_{n+1} - 1, 0)\}$	$\{U_n\}$	$\{V_n\}$
$\{Z_n^{(q)}\} \equiv$	$\{U_{n+1}\}$	$\{\max(U_{n+1} - 1, 0)\}$	$\{U_{n+1}\}$

Analogous to Table 9.5.1 we have Table 9.5.2 giving the relationships satisfied by the system size and queue size r.v.'s at the specified embedding points under the assumption of a given sequence $\{D_n\}$ of departure r.v.'s.

Observe that if V_n is a system size r.v. at n, then $V_n \geq 1$ and thus U_n is the associated queue size r.v. at n. Also, if U_n is a system size r.v. at n and the next event to occur is an arrival, then V_n is the system size r.v. immediately following that event.

(Note that the state space for the $\{X_n, n \geq 0\}$ process for the case of early arrivals or late arrivals with delayed access is effectively the positive integers since if $X_0 = 0$, then $X_n \geq 1$ for all $n \geq 1$.)

The relationships given by Eqs. (9.5.18) and (9.5.19) imply that if $\{D_n\}$ is a sequence of i.i.d. r.v.'s, then $\{U_n, n \geq 0\}$ and $\{V_n, n \geq 0\}$ are both homogeneous MC's. The $\{U_n, n \geq 0\}$ MC has been examined in detail before in Section 9.4 and by Eq. (9.5.20) the $\{V_n, n \geq 0\}$ MC has identical characteristics but with a relabeling of the states. □

In all the examples considered so far we have seen systems where we have, at the same time, imposed tight conditions on one of the arrival or service components (e.g., Geometric or Deterministic) and relaxed conditions on the other component (e.g., Geometric $^{(A)}$ or General Independent). We now conclude this section by looking briefly at what can be said about discrete time systems when we attempt to put minimal conditions on both the arrival and service patterns.

EXAMPLE 9.5.8: *The GI/G/1 Model.* We assume that the interarrival times $\{T_n\}$ are i.i.d. r.v.'s with probability distribution $\{f_k\}$ and that the service time r.v.'s $\{S_n\}$ are i.i.d. with probability distribution $\{g_k\}$. For this model we cannot construct a MC for the system size r.v. at prechosen embedding points. Neither of the embeddings "just before an arrival" or "just after a departure" are suitable because of the absence of any lack of memory in the service time distribution or interarrival time distribution. A similar comment follows regards any embedding unit times apart.

One of the interesting problems is that of finding the stationary distribution of the queueing time of the nth customer. Konheim (1975) gives a simple method for calculating this distribution which only requires the ability to factor polynomials. He makes the assumption that the $\{f_k\}$ and $\{g_k\}$ distributions have compact support, i.e., $f_k = 0$ for $k > K_1$ and $g_k = 0$ for $k > K_2$. (Konheim also permits f_0 and g_0 to be positive. If $f_0 = g_0 = 0$, then the model assumptions imply an early arrival system.)

Let $W_n^{(q)}$ be the queueing time of the nth customer. Then $W_1^{(q)} = 0$ and

$$W_{n+1}^{(q)} = \begin{cases} W_n^{(q)} + S_n - T_{n+1}, & W_n^{(q)} + S_n \geq T_{n+1}, \\ 0, & \text{otherwise,} \end{cases}$$

which we may write as

$$W_{n+1}^{(q)} = \max(W_n^{(q)} + S_n - T_{n+1}, 0). \tag{9.5.22}$$

Let us define $R_n \equiv S_n - T_{n+1}$. Then since S_n and T_{n+1} are independent, R_n has a probability distribution $\{r_m\}$ $(m = -K_1, \ldots, -1, 0, 1, 2, \ldots, K_2)$

given by

$$r_m = \sum_l P\{S_n = l + m, T_{n+1} = l\} = \sum_l f_l g_{l+m},$$

where the summation extends over $\{l : 0 \le l \le K_1, 0 \le l + m \le K_2\}$.

Observe that if $\{f_k\}$ has p.g.f. $F(s)$ and $\{g_k\}$ has p.g.f. $G(s)$, then $\{r_m\}$ has g.f. $R(s) = F(1/s)G(s)$ and the distribution function of R_n, $\{\sum_{m \le k} r_m\} \equiv \{h_k\}$ has g.f. $H(s) = R(s)/(1 - s)$.

Since $W_n^{(q)}$ and R_n are independent, from Eq. (9.5.22) we deduce for $j \ge 0$,

$$P\{W_{n+1}^{(q)} \le j\} = P\{W_n^{(q)} + R_n \le j\} = \sum_l P\{W_n^{(q)} = l\} P\{R_n \le j - l\}. \quad (9.5.23)$$

It can be shown (Lindley, 1952) that if $ET_n > ES_n$, then $\lim_{n \to \infty} P\{W_n^{(q)} = k\}$ exists and $= w_k^{(q)}$ say. Under this assumption we have from Eq. (9.5.23) that for $j \ge 0$,

$$\sum_{k=0}^{j} w_k^{(q)} = \sum_{l=0}^{\infty} w_l^{(q)} h_{j-l}. \quad (9.5.24)$$

To solve Eq. (9.5.24) Konheim (1975) introduces an auxiliary sequence. We do not give the details. However the net result is the following procedure. Form the rational function

$$S(s) = [1 - F(1/s)G(s)]/(1 - s)$$

and factor $S(s)$ as $S(s) = S^+(s)S^-(s)$ with

$$S^+(s) = C_1 \Pi(s - \eta_i^+)^{\alpha_i}, \qquad 1 < |\eta_i^+|,$$

$$S^-(s) = C_2 s^{-C_3} \Pi(s - \eta_i^-)^{\beta_i}, \qquad 0 < |\eta_i^-| < 1,$$

choosing the constants C_1 and C_2 so that $S^+(1) = 1$. Finally

$$\sum_{k=0}^{\infty} w_k^{(q)} s^k = \frac{1}{S^+(s)}, \quad (9.5.25)$$

so that the determination of the queueing time distribution requires just (!) the ability to factor polynomials. The assumption that F and G are polynomials is of course unnecessary. The general case may be obtained by approximating an actual distribution by one of finite support.

If one is interested in the limiting waiting time distribution $\{w_k\}$, then since $W_n = W_n^{(q)} + S_n$, the sum of two independent r.v.'s, we have from Eq. (9.5.25) that

$$\sum_{k=0}^{\infty} w_k s^k = \frac{G(s)}{S^+(s)}. \quad \square \quad (9.5.26)$$

EXAMPLE 9.5.9: *The Geometric$^{(A)}$/Geometric$^{(B)}$/1 Model.* In some ways this model is a "macroscopic" version of the GI/G/1 model since in each unit time slot we permit multiple arrivals and multiple services whereas in the GI/G/1 model we can have at most one arrival and/or one service completion. There is, however, considerable similarity between the two models in the derivation of the system size distributions for the Geometric$^{(A)}$/Geometric$^{(B)}$/1 model and the queueing and waiting time distributions of the GI/G/1 model.

For the early arrival system assume A_n arrivals in $(n, n+)$ and D_n departures in $(n-, n)$. For the late arrival system we assume A_n arrivals in $(n-, n)$ with D_n departures in $(n, n+)$ for the delayed access case and D_{n+1} departures in $(n, n+)$ for the immediate access case.

We assume that $\{A_n\}$ is a sequence of i.i.d. r.v.'s with probability distribution $\{f_k\}$ and $\{D_n\}$ r.v.'s have probability distribution $\{g_k\}$.

Now consider the following recurrence relationships:

$$F_{n+1} = \max(F_n + A_n - D_{n+1}, 0), \tag{9.5.27}$$

$$G_{n+1} = \max(G_n - D_{n+1}, 0) + A_{n+1}, \tag{9.5.28}$$

$$G_n = F_n + A_n, \tag{9.5.29}$$

$$F_{n+1} = \max(G_n - D_{n+1}, 0). \tag{9.5.30}$$

As in earlier related models any one of Eqs. (9.5.27) and (9.5.28) together with any one of Eqs. (9.5.29) and (9.5.30) imply that all the relationships are satisfied.

With X_n, Y_n, and Z_n as the system size r.v.'s at $n-$, n, and $n+$, Table 9.5.3 gives an identification with the appropriate $\{F_n\}$ and $\{G_n\}$ process as defined above.

TABLE 9.5.3

	Early arrival system	Late arrival system with immediate access	Late arrival system with delayed access
$\{X_n\} \equiv$	$\{G_{n-1}\}$	$\{F_n\}$	$\{G_{n-1}\}$
$\{Y_n\} \equiv$	$\{F_n\}$	$\{G_n\}$	$\{G_{n-1} + A_n\}$
$\{Z_n\} \equiv$	$\{G_n\}$	$\{F_{n+1}\}$	$\{G_n\}$

If one examines Example 9.5.8, then Eq. (9.5.22) implies that $\{W_n^{(q)}\}$ is an $\{F_n\}$ sequence but with $\{A_n\}$ replaced by $\{S_n\}$ and $\{T_n\}$ replaced by $\{D_n\}$. Furthermore, since $W_n = W_n^{(q)} + S_n$, as application of Eq. (9.5.29) shows that $\{W_n\}$ is a $\{G_n\}$ sequence.

The outcome of the above analysis is that we can use the results of Example 9.5.8 to analyze the system size distributions of interest in the Geometric$^{(A)}$/Geometric$^{(B)}$/1 model.

Note also that the models considered in Examples 9.5.3 and 9.5.7 are special cases of the above model. □

EXAMPLE 9.5.10:   *The Geometric$^{(A)}$/Geometric$^{(B)}$/1/N Model.*   This single server queueing model with limited waiting room has, as one of its main applications, the study of buffer storage. At each time instant n ($n \geq 0$), A_n units of data arrive for ultimate transmission over a channel or some other processor. Immediately after this the channel will accept D_{n+1} data units for transmission. If we assume that with positive probability $D_{n+1} < A_n$, then at least for some of the time the channel will not be able to accept all the available data and a buffer is imposed between the source and the channel to store this excess. Let N be the capacity of the buffer (system size). Occasionally the buffer will be full and at these times the excess $A_n - D_{n+1}$ units must be discarded and considered as lost. The primary problem is what average fraction of data is lost due to the "buffer overflow" or what is the probability that the system is found at maximum size? The buffer's capacity and storage allocation strategy is of great importance for the cost-effective design of a computer-communication system.

Let X_n represent the system size at time $n-$ where for convenience we have taken a late arrival system with immediate access. Then

$$X_n = \begin{cases} 0 & \text{if } X_{n-1} + A_n - D_{n+1} < 0, \\ X_{n-1} + A_n - D_{n+1} & \text{if } 0 < X_{n-1} + A_n - D_{n+1} \leq N, \\ N & \text{if } X_{n-1} + A_n - D_{n+1} > N, \end{cases}$$

or

$$X_n = \min[\max(X_{n-1} + A_n - D_{n+1}, 0), N].$$

Let $p_N(k) = \lim_{n \to \infty} P\{X_n = k\}$. Wyner (1974) obtained the following results, based upon quite elaborate arguments, which we will not reproduce here:

(i)   When $EA_n = ED_{n+1}$, there exist finite constants c_0 and c_1 such that

$$p_N(N) = \frac{C(N)}{N} \quad \text{with} \quad c_0 \leq C(N) \leq c_1.$$

(ii)   When $EA_n < ED_{n+1}$, there exist finite constants d_0 and d_1 such that

$$p_N(N) = D(N)s_0^{-N} \quad \text{with} \quad d_0 \leq D(N) \leq d_1,$$

where $s_0(> 1)$ is the unique positive real root of the equation $H(s) \equiv F(s)G(s^{-1}) = 1$ with $F(s)$ and $G(s)$ being the p.g.f.'s of the r.v.'s A_n and D_{n+1}, respectively. Wyner obtained these results under the assumption that the r.v.'s A_n and D_{n+1} have compact support.   □

Our discussion of discrete time queueing models has centered mainly around single server systems. With the upsurge of interest in computer communication networks, recent research has focused on the extension of the results of this chapter to discrete time queueing networks. For a general review of the application of queueing models in computer communications systems analysis the reader is referred to the survey article by Kobayashi and Konheim (1977). For a brief consideration of some simpler models— feedback queues and simple networks of queues—refer to Bharath-Kumar (1980). For a discussion of tandem queues—where the output of one queue forms the input of another queue—see the papers of Pearce (1968), Hsu and Burke (1976), and Morrison (1979). An examination of two queues or storage facilities in parallel with correlated inputs is examined by Hunter (1981).

Exercises 9.5

1. Derive an expression for the p.g.f. of the limiting system size r.v., imme-diately following a service completion, in the Geometric$^{(A)}$/G/1 model when the arrival pattern is generated by an early arrival system or a late arrival system with immediate access (cf. the $\{R_n, n \geq 0\}$ process as discussed in Theorem 9.3.5).

2. Prove that for sequences $\{A_n\}$, $\{P_n\}$, and $\{Q_n\}$ that satisfy any two of Eqs. (9.5.7) to (9.5.10) the remaining two relationships are automatically satisfied.

3. Verify the entries given in Table 9.5.1.

4. Let $\{A_n\}$ be a sequence of independent and identically distributed non-negative integer-valued r.v.'s with probability distribution $\{k_j\}$. Let $\{P_n, n \geq 0\}$ be a sequence of r.v.'s satisfying the relationship

$$P_{n+1} = \max(P_n + A_n - 1, 0)$$

(a) Show that $\{P_n, n \geq 0\}$ is a homogeneous MC with transition matrix given by

$$p = \begin{bmatrix} k_0 + k_1 & k_2 & k_3 & \cdots \\ k_0 & k_1 & k_2 & \cdots \\ 0 & k_0 & k_1 & \cdots \\ \vdots & \vdots & \vdots & \ddots \end{bmatrix}.$$

(b) Show that if $0 < k_0 \leq k_0 + k_1 < 1$, then the MC $\{P_n, n \geq 0\}$ is irre-ducible and aperiodic. What happens if either $k_0 = 0$ or $k_0 + k_1 = 1$?

(c) Show that if the MC is irreducible (and aperiodic) and $\rho \equiv \sum_{n=1}^{\infty} nk_n < 1$, the MC is ergodic. In such a case the MC has a stationary distribution $\{\pi_{i,P}\}$, which is also the limiting distribution. If $\Pi_P(s)$ is the p.g.f. of this stationary distribution and $A(s)$ is the p.g.f. of the probability distribution $\{k_n\}$ show, using the stationary equations of the MC

$\{P_n, n \geq 0\}$, that

$$\Pi_P(s) = \frac{(1 - \rho)(1 - s)}{A(s) - s}.$$

(d)   Using the relationship given by Eq. (9.5.9), $P_{n+1} = \max(Q_n - 1, 0)$, derive the result of (c) from the expression for the p.g.f. of the stationary distribution of the MC $\{Q_n, n \geq 0\}$, i.e.,

$$\Pi_Q(s) = \frac{(1 - \rho)(1 - s)A(s)}{A(s) - s}.$$

(e)   Using the relationship given by Eq. (9.5.10), viz., $Q_n = P_n + A_n$, derive the expression for $\Pi_Q(s)$ [as given in (d)] by using the expression for $\Pi_P(s)$ [as given in (c)].

(f)   If $\{\pi_{i,Q}\}$ are the stationary probabilities of the $\{Q_n, n \geq 0\}$ MC, show that

(i) $\pi_{i,Q} = \displaystyle\sum_{j=0}^{i} \pi_{j,P} k_{i-j}.$

(ii) $\pi_{i,P} = \begin{cases} \pi_{0,Q} + \pi_{1,Q}, & i = 0, \\ \pi_{i+1,Q}, & i \geq 1. \end{cases}$

5.   Prove the results concerning the $\{P_n, n \geq 0\}$ process as stated in Example 9.5.4.

References

Aitken, A. C. (1956). *Determinants and Matrices*, 9th ed. Oliver & Boyd, Edinburgh.

Ayres, F., Jr. (1962). *Theory and Problems of Matrices.* Schaum, New York.

Bailey, N. T. J. (1964). *The Elements of Stochastic Processes with Applications to the Natural Sciences.* Wiley, New York.

Ben-Israel, A., and Greville, T. N. E. (1974). *Generalized Inverses: Theory and Applications.* Wiley (Interscience), New York.

Bharath-Kumar, K. (1980). Discrete time queueing systems and their networks, *IEEE Trans. Comm.* **COM-28**, 260–263.

Bhat, U. N. (1968). *A Study of the Queueing Systems M/G/1 and GI/M/1.* Springer-Verlag, Berlin and New York.

Bhat, U. N. (1972). *Elements of Applied Stochastic Processes.* Wiley, New York.

Bizley, M. T. L. (1957). *Probability: An Intermediate Textbook.* Cambridge Univ. Press, London and New York.

Bizley, M. T. L. (1962). Patterns in repeated trials, *J. Inst. Actuaries* **88,** 360–366.

Boudreau, P. E., Griffen, J. S., Jr., and Kac, M. (1962). An elementary queueing problem, *Amer. Math. Monthly* **69,** 713–724.

Boullion, T. L., and Odell, P. L. (1971). *Generalised Inverse Matrices.* Wiley (Interscience), New York.

Buck, R. C. (1956). *Advanced Calculus.* McGraw-Hill, New York.

Campbell, S. L., and Meyer, C. D., Jr. (1979). *Generalised Inverses of Linear Transformations.* Pitman, London.

Chan, W. C., and Maa, D. Y. (1978). The GI/Geom/N queue in discrete time, *INFOR—Canad. J. Oper. Res. Inform. Process.* **16,** 232–252.

Chung, K. L. (1960). *Markov Chains with Stationary Transition Probabilities*. Springer-Verlag, Berlin and New York.

Çinlar, E. (1975). *Introduction to Stochastic Processes*. Prentice-Hall, Englewood Cliffs, New Jersey.

Cooke, R. G. (1950). *Infinite Matrices and Sequence Spaces*. Macmillan, New York.

Cox, D. R., and Miller, H. D. (1965). *The Theory of Stochastic Processes*. Methuen, London.

Dafermos, S. C., and Neuts, M. F. (1971). A single server queue in discrete time, *Cahiers Centre Études Rech. Opér.* **13**, 23–40.

Debreu, G., and Herstein, I. N. (1953). Nonnegative square matrices, *Econometrica* **21**, 597–607.

Decell, H. P., Jr., and Odell, P. L. (1967). On the fixed point probability vector of regular or ergodic transition matrices, *J. Amer. Statist. Assoc.* **62**, 600–602.

Erdélyi, I. (1967). On the matrix equation $Ax = \lambda Bx$, *J. Math. Anal. Appl.* **17**, 119–132.

Fáa di Bruno, C. (1876). *Theorie des Formes Binaires*. Librairie Brero, Succr. de P. Marietti, Turin.

Faddeeva, V. (1959). *Computational Methods of Linear Algebra*. Dover, New York.

Feller, W. (1949). Fluctuation theory of recurrent events, *Trans. Amer. Math. Soc.* **67**, 98–119.

Feller, W. (1957). *An Introduction to Probability Theory and Its Applications*, 2nd ed., Vol. 1. Wiley, New York.

Feller, W. (1968). *An Introduction to Probability Theory and Its Applications*, 3rd ed., Vol. 1. Wiley, New York.

Fisher, R. A. (1922). On the dominance ratio, *Proc. Roy. Soc. Edinburgh* **42**, 321–341.

Fisher, R. A. (1930). *The Genetical Theory of Natural Selection*. Dover, New York.

Foster, F. G. (1953). On the stochastic matrices associated with certain queueing processes, *Ann. Math. Statist.* **22**, 355–360.

Frechet, M. (1950). *Recherches Theoriques Modernes sur le Calcul des Probabilitiés*, 2nd ed., Vol. 2. Gauthier-Villars, Paris.

Frobenius, G. (1908). Über Matrizen aus positiven Elementen, *Sitzungsber. K. Preuss. Akad. Wiss.*, pp. 471–476.

Frobenius, G. (1909). Über Matrizen aus positiven Elementen II, *Sitzungsber. K. Preuss. Akad. Wiss.*, pp. 514–518.

Frobenius, G. (1912). Über Matrizen aus nicht-negativen Elementen, *Sitzungsber K. Preuss. Akad Wiss.*, pp. 456–477.

Gabriel, K. R. (1959). The distribution of the number of successes in a sequence of dependent trials, *Biometrika* **46**, 454–460.

Galton, F. A., and Watson, H. W. (1874). On the probability of extinction of families, *J. Anthropol. Inst. Great Britain Ireland* **4**, 138–144.

Gani, J. (1957). Problems in the probability theory of storage systems, *J. Roy. Statist. Soc. Ser. B.* **19**, 181–206.

Gantmacher, F. R. (1959). *Applications of the Theory of Matrices*. Wiley (Interscience), New York.

Gihmann, I. I., and Skorohod, A. V. (1974). *The Theory of Stochastic Processes I*. Springer-Verlag, Berlin and New York.

Good, I. J. (1949). The number of individuals in a cascade process, *Proc. Cambridge Philos. Soc.* **45**, 360–363.

Good, I. J. (1951). Discussion to some problems in the theory of queues, *J. Roy. Statist. Soc. Ser. B* **13**, 173–185.

Gray, J. R. (1967). *Probability*. Oliver & Boyd, Edinburgh.

Haldane, J. B. S. (1927). A mathematical theory of natural and artificial selection, Part V: Selection and mutation, *Proc. Cambridge Philos. Soc.* **23**, 838–844.

Hardy, G. H. (1949). *Divergent Series*. Oxford Univ. Press (Clarendon), London and New York.

Heathcote, C. R. (1971). *Probability: Elements of the Mathematical Theory*. Allen & Unwin, London.

Heimann, D., and Neuts, M. F. (1973). The single server queue in discrete time—numerical analysis, IV, *Naval Res. Logist. Quart.* **20**, 753–766.

Helgert, H. J. (1970). On sums of random variables defined on a two-state Markov chain, *J. Appl. Probab.* **7**, 761–765.

Howard, R. A. (1971). *Dynamic Probabilistic Systems*, Vol. 1. Wiley, New York.

Hsu, J., and Burke, P. J. (1976). Behaviour of tandem buffers with geometric input and Markovian output, *IEEE Trans. Comm.* **COM-24**, 358–360.

Hunter, J. J. (1969). On the moments of Markov renewal processes, *Adv. in Appl. Probab.* **1**, 188–210.

Hunter, J. J. (1973). On the occurrence of the sequence SF in Markov dependent Bernoulli trials, *Math. Chronicle* **2**, 131–136.

Hunter, J. J. (1981). Queueing and storage systems in parallel with correlated inputs, *New Zealand Oper. Res.* **9**, 119–136.

Hunter, J. J. (1982). Generalised inverses and their application to applied probability problems, *Linear Algebra Appl.* **45**, 157–198.

Hyslop, J. M. (1959). *Infinite Series*, 5th ed. Oliver & Boyd, Edinburgh.

Jordan, C. (1947). *Calculus of Finite Differences*, 2nd ed. Chelsea, Bronx, New York.

Karlin, S. (1966). *A First Course in Stochastic Processes*. Academic Press, New York.

Kemeny, J. G. (1981). Generalisation of a fundamental matrix, *Linear Algebra Appl.* **38**, 193–206.

Kemeny, J. G., and Snell, J. L. (1960). *Finite Markov Chains*. Van Nostrand, Princeton, New Jersey.

Kemeny, J. G., Snell, J. L., and Knapp, A. W. (1966). *Denumerable Markov Chains*. Van Nostrand, Princeton, New Jersey.

Klimko, M. F., and Neuts, M. F. (1973). The single server queue in discrete time—numerical analysis, II, *Naval Res. Logist. Quart.* **20**, 304–319.

Kobayashi, H., and Konheim, A. G. (1977). Queueing models for computer communications system analysis, *IEEE Trans. Comm.* **COM-25**, 1–29.

Konheim, A. G. (1975). An elementary solution of the queueing system G/G/1, *SIAM J. Comput.* **4**, 540–545.

Lancaster, P. (1969). *Theory of Matrices*. Academic Press, New York.

Levinson, N. (1965). An elementary proof of the stationary distribution for an irreducible Markov chain, *Amer. Math. Monthly* **72**, 366–369.

Lindley, D. V. (1952). The theory of queues with a single server, *Proc. Cambridge Philos. Soc.* **48**, 277–289.

Lipschutz, S. (1968). *Theory and Problems of Linear Algebra*. Schaum, New York.

Liutikas, V. (1965). On the m.g.f. of the number of renewals of a discrete renewal process, *Litovsk Mat. Sb.* **5**, 421–425.

Lotka, A. J. (1931). The extinction of families, *J. Wash. Acad. Sci.* **21**, 377–380, 453–459.

Malécot, G. (1944). Sur un problème de probabilités en chaine que pose la génétique, *C. R. Acad. Sci. Paris Sér. A-B* **219**, 379–381.

Meisling, T. (1958). Discrete time queueing theory, *Oper. Res.* **6**, 96–105.

Meyer, C. D., Jr. (1975). The role of the group generalised inverse in the theory of finite Markov chains, *SIAM Rev.* **17**, 443–464.

Meyer, C. D., Jr. (1978). An alternative expression for the mean first passage time matrix, *Linear Algebra Appl.* **22**, 41–47.

Mirsky, L. (1955). *An Introduction to Linear Algebra*. Oxford Univ. Press, London and New York.

Moore, E. H. (1935). *General Analysis*. Amer. Philos. Soc., Philadelphia. Pennsylvania.

Moran, P. A. P. (1956). A probability theory of dams with a continuous release, *Quart. J. Math. Oxford* (2) **7**, 130–137.

Moran, P. A. P. (1968). *An Introduction to Probability Theory.* Oxford Univ. Press (Clarendon), London and New York.

Morrison, J. A. (1979). Two discrete time queues in tandem, *IEEE Trans. Comm.* **COM-27**, 563–573.

Neuts, M. F. (1973a). *Probability.* Allyn & Bacon, Boston, Massachusetts.

Neuts, M. F. (1973b). The single server queue in discrete time—numerical analysis, I, *Naval Res. Logist. Quart.* **20**, 297–304.

Neuts, M. F., and Klimko, E. M. (1973). The single server queue in discrete time—numerical analysis, III, *Naval Res. Logist. Quart.* **20**, 557–567.

Paige, C. C., Styan, G. P. H., and Wachter, P. G. (1975). Computation of the stationary distribution of a Markov chain, *J. Statist. Comput. Simulation* **4**, 173–186.

Parzen, E. (1962). *Stochastic Processes.* Holden-Day, San Francisco, California.

Pearce, C. (1968). On the joint equilibrium queue length distribution in a series queue, *J. Canad. Oper. Res. Soc.* **6**, 96–100.

Pearl, M. (1973). *Matrix Theory and Finite Mathematics.* McGraw-Hill, New York.

Pedler, P. J. (1971). Occupation times for two state Markov chains, *J. Appl. Probab.* **8**, 381–390.

Penrose, R. A. (1955). A generalised inverse for matrices, *Proc. Cambridge Philos. Soc.* **51**, 406–413.

Perron, O. (1907). Zur theorie der Matrizen, *Math. Ann.* **64**, 248–263.

Prabhu, N. U. (1958). Some exact results for the finite dam, *Ann. Math. Statist.* **29**, 1234–1243.

Prabhu, N. U. (1964). Time dependent results in storage theory, *J. Appl. Probab.* **1**, 1–46.

Prabhu, N. U. (1965). *Stochastic Processes, Basic Theory and Its Applications.* Macmillan, New York.

Pringle, R. M., and Rayner, A. M. (1971). *Generalised Inverse Matrices with Applications to Statistics.* Griffen, London.

Rao, C. R., and Mitra, S. K. (1971). *Generalised Inverse of Matrices and Its Applications.* Wiley, New York.

Robert, P. (1968). On the group inverse of a linear transformation, *J. Math. Anal. Appl.* **22**, 658–669.

Rohde, C. A. (1968). Special applications of the theory of generalised matrix inversion to statistics, *Proc. Symp. Theory Appl. Generalised Inverses Matrices, Lubbock, Texas,* pp. 239–266.

Rudin, W. (1953). *Principles of Mathematical Analysis.* McGraw-Hill, New York.

Searle, S. R. (1971). *Linear Models.* Wiley, New York.

Seneta, E. (1973). *Non-Negative Matrices—An Introduction to Theory and Applications.* Allen & Unwin, London.

Smith, W. L. (1958). Renewal theory and its ramifications, *J. Roy. Statist. Soc. Ser. B* **20**, 243–302.

Snell, J. C. (1975). *Introduction to Probability Theory with Computing.* Prentice-Hall, Englewood Cliffs, New Jersey.

Steffensen, J. F. (1930). Om sandsynligheden for at afkommet uddør, *Mat. Tidsskr. B* **1**, 19–23.

Titchmarsh, E. C. (1939). *The Theory of Functions,* 2nd ed. Oxford Univ. Press, London.

Uspensky, J. V. (1937). *Introduction to Mathematical Probability.* McGraw-Hill, New York.

Wyner A. D. (1974). On the probability of buffer overflow under an arbitrary bounded input–output distribution, *SIAM J. Appl. Math.* **27**, 544–570.

Index

283